Informatik-Fachberichte 306

Herausgeber: W. Brauer
im Auftrag der Gesellschaft für Informatik (GI)

D. Krönig   M. Lang (Hrsg.)

Physik und Informatik – Informatik und Physik

Arbeitsgespräch
München, 21. / 22. November 1991

Proceedings

ITG Φ

Springer-Verlag
Berlin Heidelberg New York London Paris
Tokyo Hong Kong Barcelona Budapest

Herausgeber

D. Krönig
Dornier GmbH
Postfach 1420, W-7990 Friedrichshafen 1

M. Lang
Technische Universität München
Lehrstuhl für Elektroakustik
Arcisstr. 21, W-8000 München 2

Veranstalter
Gesellschaft für Informatik e. V.
Deutsche Physikalische Gesellschaft e. V.
Informationstechnische Gesellschaft im VDE

ITG Φ

Tagungsleitung
Herr Dr. D. Krönig, Dornier GmbH, Friedrichshafen
Herr Prof. Dr. M. Lang, Technische Universität München

Programmausschuß
Frau Dr. I. Bausch-Gall, Bausch-Gall GmbH, München
Herr Prof. Dr. A. Bode, TUM München
Herr Prof. Dr. D. Heermann, Universität Heidelberg
Herr Prof. Dr. Hehl, Universität Köln
Herr Prof. Dr. Hilberg, TH Darmstadt
Herr Dipl.-Ing. R. Kober, Siemens AG, München
Herr Dr. D. Krönig, Dornier GmbH, Friedrichshafen
Herr Dipl.-Phys. U. Krüger, Universität Frankfurt / Main
Herr Prof. Dr. M. Lang, TUM München
Herr Prof. Dr. G. Mahler, Universität Stuttgart
Herr Prof. Dr. J. Mahrun, Universität Frankfurt / Main
Herr Prof. H. C. Mayr, Universität Klagenfurt / Österreich
Herr Dr. B. Schürmann, Siemens AG, München
Herr Dr. F. Schwarz, GMD, Birlinghoven
Herr Prof. Dr. K. Waldschmidt, Universität Frankfurt
Herr Prof. Dr. H. C. Zeidler, Universität Hamburg

Organisation
Herr H. Benesch, Siemens AG, München

CR Subject Classification (1991): I.2

ISBN-13:978-3-540-55298-7 e-ISBN-13:978-3-642-77382-2
DOI: 10.1007/978-3-642-77382-2

Satz: Reproduktionsfertige Vorlage vom Autor

33/3140-543210 – Gedruckt auf säurefreiem Papier

VORWORT

Die Forschungs- und Entwicklungsaufgaben der Informations- und Kommunikationstechnik werden immer vielschichtiger und komplexer. Technische Übertragung, Speicherung, Verarbeitung und Interpretation von Informationen sind Gegenstand von sowohl grundlagenorientierten als auch anwendungsbezogenen Projekten mit zumeist interdisziplinärem Charakter. Die Deutsche Physikalische Gesellschaft, die Informationstechnische Gesellschaft im VDE und die Gesellschaft für Informatik haben sich das Ziel gesetzt, ihre Zusammenarbeit auf eine breitere Basis zu stellen. Auftakt dazu war eine gemeinsam veranstaltete Tagung mit einem breiten Themenangebot, das von Neuen Rechnerprinzipien und Materialien über Algorithmen und Architekturen bis hin zum Wissenschaftlichen Rechnen (Computational Science) und zur Simulation von Modellen und Prozessen reichte.

Es wurden mehr hervorragend bewertete Vorträge vorgelegt, als bei der begrenzten Gesamtredezeit im Programm berücksichtigt werden konnten. Um bei der ersten Veranstaltung dieser Art möglichst viele Aspekte ansprechen zu können, bei denen interdisziplinäre Kontakte gewünscht sind, mußten Kürzungen, z.T. unter Verzicht auf Vertiefung, in Kauf genommen werden. Disziplinübergreifende Vorträge wurden bevorzugt, bei inhaltlich verwandten konnte nur einer ausgewählt werden. In den vorliegenden Band wurden jedoch einige Arbeiten zusätzlich aufgenommen, die nur der fehlenden Zeit wegen nicht vorgetragen werden konnten.

Die Tagung wurde eingeleitet vom Leiter der Zentralabteilung Forschung und Entwicklung und Vorstandsmitglied der Siemens AG, Herrn Prof. Dr. H. G. Danielmeyer. Er betonte u.a. die zunehmende wirtschaftliche und gesellschaftspolitische Bedeutung der Informations- und Kommunikationstechnik und forderte Wissenschaftler und Ingenieure zur kooperativen Zusammenarbeit auf. Sie können damit einen wesentlichen Beitrag zur Systemfähigkeit unserer Wirtschaft leisten.

Die Übersichtsvorträge der sieben eingeladenen Redner und die 22 Kurzvorträge waren, wie der vorliegende Band, in Themengruppen gegliedert. Die Themengruppen sind: **Modelle, Verfahren, Algorithmen und Architekturen** mit u.a. den Themen *Rechnender*

Raum (Übersichtsvortrag von K. Zuse), *Parallele genetische Algorithmen für Optimierungsprobleme in Technik und Wissenschaft* (Übersichtsvortrag von H. Mühlenbein), Flexibilisierung von Rechnerstrukturen, Der Zeitbegriff der Informatik, Informationsverarbeitung als algebraisches und physikalisches Phänomen; **Neue Rechnerprinzipien und Materialien** mit u.a. den Themen *Verteilte Repräsentation* (Übersichtsvortrag von G. Görz), *Physik lernender Netzwerke* (Übersichtsvortrag von W. Kinzel), Bildverarbeitung mittels Quantennetzwerken, optische Rechner, Neuronale Netze, Nanotechnologie; **Anwendungsbezogene Architekturen und Systeme** mit Vorträgen zu u.a. Signal- und Bildverarbeitung mit Methoden der nichtlinearen Dynamik und der statistischen Physik, Erkennen und Interpretieren durch adaptive Mustererkennung; **Simulation von Modellen und Prozessen** mit u.a. den Themen Simulation von Zellularautomaten (Übersichtsvortrag von D. Stauffer), Analyse komplexer physikalischer Systeme mit informationstechnischen Methoden, Quanten-Monte-Carlo-Simulationen für Hochtemperatur Supraleitung; **Computeralgebra** mit u.a. einer Übersicht durch F. W. Hehl; **Lösung rechenaufwendiger Probleme** mit u.a. dem Übersichtsvortrag von F. Hoßfeld über das Wissenschaftliche Rechnen, das Theorie und Experiment qualitativ und methodisch ergänzt.

Die Tagung und die lebhaften Diskussionen haben gezeigt, daß es eine Vielfalt bedeutsamer Themen gibt, in denen eine disziplinübergreifende Bearbeitung in Forschung und Entwicklung einen Mehrwert zu erbringen verspricht. In diesem Sinne wurde eine von der DPG, der GI, und der ITG gemeinsam getragene Fachgruppe "Physik, Informatik, Informationstechnik" gegründet, die das Begonnene weiterführen soll.

Wir danken abschließend allen, die bei der Vorbereitung und bei der Durchführung der Tagung mit großem Einsatz zum Gelingen beigetragen haben.

Friedrichshafen und München, November 1991 D. Krönig, M. K. Lang

Inhaltsverzeichnis

Modelle, Verfahren, Algorithmen und Architekturen

Arndt Bode
Technische Universität München

Dieter W. Heermann
Universität Heidelberg

Die Sitzung und anschließende Diskussion beleuchtete die Modelle, Verfahren, Algorithmen und Architekturen aus Physik und Informatik von den verschiedensten Blickwinkeln. Es wurde diskutiert, inwieweit die von Folberth geforderte Flexibilisierung der Hardware von Rechneranlagen zu einer neuen Schnittstellen-Problematik führt. Es wurde ferner die Frage aufgeworfen, inwieweit die Kooperation zwischen Physik und Informatik dazu führen kann, daß Europa eine Rolle beim Entwurf der Hardware von Systemen spielen kann.

Der Beitrag von W.A. Halang befaßt sich mit dem unterentwickelten Zeitbegriff der Informatik. In der Diskussion wurde darauf hingewiesen, daß neben der kurzfristigen Lösung des Zugriffs über Funk auf eine zentrale Uhr mit 100 Nanosekunden Auflösung auch eine theoretisch fundierte Auseinandersetzung mit dem Zeitbegriff notwendig ist.

Der Beitrag von K. Kornwachs: "Information, Informatik und der Begriff der Wirkung" führte zu einer Diskussion der unterschiedlichen Interpretation des Begriffs der Entropie in der Physik und in der Informationstheorie. Die Diskussion zum Beitrag von Thomas Beth: "Physik und Informatik - Synergie für Forschung, Lehre und Entwicklung" zeigt, daß eine intensive Kooperation von Physik und Informatik bezüglich des Curriculums in der universitären Ausbildung für das jeweils andere Fach notwendig ist.

Parallel Genetic Algorithms in Optimization

H. Mühlenbein
Gesellschaft für Mathematik und Datenverarbeitung
D-5205 Sankt Augustin 1

1 Introduction

Random search methods based on evolutionary principles have been proposed in the 60's. They did not have a major influence on mainstream optimization. We believe that this will change. The unique power of evolutionary algorithms shows up with parallel computers. Firstly, our parallel genetic algorithm PGA introduced in 1987 [MGSK87] runs especially efficient on parallel computers. Secondly, our research indicates that parallel searches with information exchange between the searches are often better than independent searches. Thus the PGA is a truly parallel algorithm which combines the hardware speed of parallel processors and the software speed of intelligent parallel searching.

We have successfully applied the PGA to a number of problems, including function optimization [MSB91] and combinatorial optimization [Müh91b] and game theory [Müh91a]. In this paper we summarize the results for two important benchmark problems, the traveling salesman problem (TSP) and the graph partitioning problem GPP. The quadratic assignment problem has been published elsewhere [Müh89].

The most difficult part of a random search method is to explain why and when it will work. This has been done for the TSP and binary functions in [Müh91b]. In this paper we will concentrate on the constructive part of the PGA.

Throughout the paper we will use biological terms. We believe, that this is a source of inspiration and helps to understand the PGA intuitively. We will use the following simple metaphor: The parallel search is done by N active individuals, x_i describes the position of individual i and $F(x_i)$ its current value, representing the height in an unknown landscape. The task is to find the global maximum or at least a good approximation of the maximum. We will investigate search algorithms which mimic evolutionary adaptation found in nature. Each individual is identified with an animal, which searches for food and produces offspring. In evolutionary algorithms, $F(x_i)$ is called the fitness of individual i, x_i^{t+1} is an offspring of x_i^t. A *selection schedule* selects the individuals which may create offspring.

The PGA is not based on a metaphor derived from physics like e.g. simulated annealing. Nevertheless, the analysis of the algorithm needs methods which are used in physics. We would like to mention the analysis of the diffusion of genes through a spatial population by a diffusion equation and the analysis of search spaces by a configuration space analysis. In this paper we will however concentrate on the problem how to apply the PGA to optimization problems. The question why the PGA gives such good results will be discussed rather informal.

2 Evolutionary algorithms and genetic algorithms

A survey of search strategies based on evolution has been done in [MGSK88]. We recall only the most important ones. A generic evolutionary algorithm can be described as follows

Evolutionary algorithm

STEP1: Create an initial population of size $M = N * O$

STEP2: Compute the Fitness $F(x_i) i = 1, \ldots, M$

STEP3: Select N individuals according to some selection schedule

STEP4: Create O offspring of each of the N individuals by small variation

STEP5: If not finished, return to STEP2

Variants of this algorithm have been invented by many researchers, see for example [Rec73], [Sch81]. An evolutionary algorithm is a random search which uses selection and variation. The linkage of the parallel searches is only implicit by the selection schedule. Searches with bad results so far are abandoned and new searches are started in the neighborhood of more promising searches.

In biological terms, evolutionary algorithms model natural evolution by asexual reproduction with mutation and selection. Search algorithms which model sexual reproduction are called genetic algorithms. They were invented by Holland [Hol75]. Recent surveys can be found in [Gol89] and [Sch89],[BB91].

Genetic Algorithm

STEP0: Define a genetic representation of the problem

STEP1: Create an initial population $P(0) = x_1^0, ..x_N^0$

STEP2: Compute the average fitness $\overline{F} = \sum_i^N F(x_i)/N$. Assign each individual the normalized fitness value $F(x_i^t)/\overline{F}$

STEP3: Assign each x_i a probability $p(x_j, t)$ proportional to its normalized fitness. Using this distribution, select N vectors from $P(t)$. This gives the set $S(t)$

STEP4: Pair all of the vectors in $S(t)$ at random forming $N/2$ pairs. Apply crossover with probability p_{cross} to each pair and other genetic operators such as mutation, forming a new population P_{t+1}

STEP5: Set $t = t + 1$, return to STEP2

In the simplest case the genetic representation is just a bitstring of length n, the *chromosome*. The positions of the strings are called *locus* of the chromosome. The variable at a locus is called *gene*, its value *allele*. The set of chromosomes is called the *genotype* which defines a *phenotype* (the individual) with a certain fitness. The crossover operator links two searches. Part of the chromosome of one individual (search) is inserted into the second chromosome giving a new individual (sample point). We will later show with examples why and when crossover guides the search.

A genetic algorithm is a parallel random search with centralized control. The centralized part is the selection schedule. For the selection the average fitness of the population is needed. The result is a highly synchronized algorithm, which is difficult to implement efficiently on parallel computers.

In our parallel genetic algorithm, we use a distributed selection scheme. This is achieved as follows. Each individual does the selection by itself. It looks for a partner in its neighborhood only. The set of neigborhoods defines a spatial population structure.

Our second major change can now easily be understood. Each individual is active and not acted on. It may improve its fitness during its lifetime by performing a local search.

A generic parallel genetic algorithm can be described as follows

Parallel genetic algorithm

STEP0: Define a genetic representation of the problem

STEP1: Create an initial population and its population structure

STEP2: Each individual does local hill-climbing

STEP3: Each individual selects a partner for mating in its neighborhood

STEP4: An offspring is created with genetic crossover of the parents

STEP5: The offspring does local hill-climbing. It replaces the parent, if it is better than some criterion (acceptance)

STEP6: If not finished, return to STEP3.

It has to be noted, that each individual may use a different local hill-climbing method. This feature will be important for problems, where the efficiency of a particular hill-climbing method depends on the problem instance.

The PGA can be described as a parallel search with a linkage of two searches. The linkage is done probabilistically, constrained by the neighborhood. The information exchange within the whole population is a diffusion process because the neighborhoods of the individuals overlap.

In a parallel genetic algorithm, all decisions are made by the individuals themselves. Therfore the PGA is a totally distributed algorithm without any central control.

There have been several other attempts to implement a parallel genetic algorithm. Most of the algorithms run k identical standard genetic algorithms in parallel, one run per processor. They differ in the linkage of the runs. Tanese [Tan89] introduces two *migration* parameters: the *migration interval*, the number of generations between each migration, and the *migrationrate*, the percentage of individuals selected for migration. The subpopulations are configured as a binary n-cube. A similar approach is done by Pettey et al. [PS89]. In the implementation of Cohoon et al. [CHMR87] it is assumed that each subpopulation is connected to each other. The algorithm from Manderick et al. [MS89] has been derived from our PGA. In this algorithm the individuals of the population are placed on a planar grid and selection and crossing-over are restricted to small neighborhoods on that grid.

All but Manderick's algorithm use subpopulations that are densely connected. We have shown in [Müh91b] why restricted connections like a ring are better for the parallel genetic algorithm. All the above parallel algorithms do not use hill-climbing, which is one of the most important parts of our PGA.

An extension of the PGA, where subpopulations are used instead of single individuals, has been described in [MSB91]. This algorithm outperforms the standard GA by far in the case of function optimization. It outperforms also many mathematical search methods on benchmark problems.

3 The search strategy of the PGA

The search strategy of the PGA is driven by three components - the spatial population structure, the crossover operator and the hill-climbing strategies. The importance of the spatial population structure and of a good hill-climbing strategy has been discussed in [Müh91b]. In this paper we concentrate on the problem dependent aspect, the crossover operator.

There have been attempts to "prove" that genetic algorithms make a nearly optimal allocation of trials. This result is called the "Fundamental Theorem of Genetic Algorithms" [Gol89]. We have shown in [Müh91b] that the above claim is only valid for simple optimization problems.

The search strategy of a genetic algorithm can be explained in simple terms. The crossover operator defines a *scatter search* [Glo77] where new points are drawn out of the area which is defined by the old or "parent" points. The more similar the old points are, the smaller will be the sampling area. Thus crossing-over implements an adaptive step-size control.

But crossing-over is also exploring the search space. Let us assume that the combinatorial problem has the *building block feature.* We speak of a building block feature if the substrings of the optimal solutions are contained in other good solutions. In this case it seems a good strategy to generate new solutions by patching together substrings of the old solutions. This is exactly what the crossover operator does.

This search strategy has been compared in [Müh91b] to simpler strategies like *multistart hill-climbing* and *iterated hill-climbing.* Iterated hill-climbing is a simple evolutionary algorithm with a population size of one. The result of the comparison is that crossing-over is a very robust search which does not depend critical on many parameters. The success of iterated hill-climbing depends on an optimal mutation rate for the given problem.

The major difficulty for applying the PGA to combinatorial problems is to define a crossover operator which creates valid solutions i.e. solutions which fulfil the constraints of the problem. We will explain this problem first with the TSP.

4 The traveling salesman problem

The famous travelling salesman problem (TSP) can be easily stated.

OPT 1 (TSP) *Given are n cities. The task of the salesman is to visit all cities once so that the overall tourlength is minimal.*

This problem has been investigated in [MGSK88] ,[GS89] and [GS91] with the PGA. The genetic representation is straightforward. The gene at locus i of the chromosome codes the edge (or link) which leaves city i. With this coding, the genes are not independent from each other. Each edge may appear on the chromosome only once, otherwise the chromosome would code an invalid tour. A simple crossing-over will also give an invalid tour. This is the reason why this simple genetic representation has not been used in genetic algorithms. The researchers tried to find a more tricky representation in order to apply a simple crossover operator.

We take the opposite approach. We use a simple representation, but an intelligent crossover operator. Our crossover operator for the TSP is straightforward. It inserts part of chromosome A into the corresponding location at chromosome B, so that the resulting chromosome is the most similar to A and B. A genetic repair operator then creates a valid tour.

We call our crossover operator MPX, the maximal preservative crossover operator. It preserves subtours contained in the two parents. The pseudocode can be found in [GS91].

In our first experiments we used a simple version of the famous 2-opt heuristic developed by Lin-Kernighan [Lin65] as local search. Two edges are exchanged randomly. If the resulting tour is shorter, the exchange is accepted. Then the next two edges are exchanged until no exchange of two edges gives a better tour. We have later found that a better local search method gives better results. This will be discussed in the next section.

5 Performance Evaluation for the TSP

A competition with newly developed neural network algorithms is shown in the table 1. For the competition, three random TSP problems of order 50,100 and 200 have been generated. HA is a hybrid approach, combining greedy search, simulated annealing and exhaustive search [KT85]. EN is the elastic net algorithm from Durbin [DSY89], MFA is the mean field approximation from Peterson [PS89], SA is a simple simulated annealing program [KGV83].

Alg	tsp-50	tsp-100	tsp-200
PGA	5.578	7.430	10.493
HA		7.480	10.530
EN	5.619	7.687	11.136
NN	6.610	8.584	12.660
SA	6.793	8.682	12.789

Table 1: Results of a TSP competition [Pet90]

Table 1 shows, that the other algorithms cannot compete with the PGA for large problems. For the tsp-200, the PGA got 10.52 after 100 seconds and 10.493 after 400 seconds on a 16 processor system with a population size of just 16. This shows how difficult it is to compute very good solutions of the TSP. The absolute differences are often misleading.

The above problems are far too small for a comparison. Therefore we turn to a popular benchmark problem, the ATT-532 problem solved to optimality in [PR87].

Ulder et al. [UPvL+91] investigated the question of the influence of the local search method on the quality of the solution. In their experimental setup, the genetic algorithm had to converge within a certain time limit. Convergence is achieved when either all tours in the current population have the same length or the length of the best tour did not improve within five successive generations. This condition forced the population sizes of the genetic algorithm to be very small.

Ulder et al. compared the standard 2-opt with the more complicated Lin-Kernigham (L-K) neighborhood [LK73]. In the latter case they used a pair of improvement operators, the dynamical k-swap and the additional 4-swap as described in [LK73]. In time t=8600 multiple runs of 2-opt (116) were 8.34% above optimal, multiple runs of L-K (77) were 0.37% above optimal, the genetic variants were 2.99% (954 local searches) and 0.17% (120 local searches) above optimal.

The results indicate that genetic search is more efficient than multiple runs. Furthermore, the efficiency of the genetic search increases with the quality of the local search. More precisely: Given a fixed amount of time on a sequential processor, a small number of individuals with a good local search method outperforms a larger number of individuals with a simpler local search method. This result has been independently confirmed for the PGA in [Müh91b].

The PGA with a population size of 64 and 2-opt as local search method got a tour length of 0.10% above optimal in ten runs of $t = 1080$ (1000 generations,15000 local searches) on a 64 processor system, the average final tour length was 0.19% above optimal [GS89]. This is a substantial improvement over the results of Ulder et al. for genetic 2-opt search.

We have not yet implemented the PGA with the L-K local search. In order to make the PGA still more efficient we will implement a number of changes. These will be outlined in the conclusion.

Inspired by genetic algorithms Johnson has developed a very fast and efficient heuristic, an *iterated Lin-Kernighan* search [Joh90]. In his implementation a new start configuration is obtained by an unbiased 4-opt move on the tour at hand. Then a new L-K search is started. If the search leads to a tour with a smaller tour length, the new tour is accepted.

Johnson reports the following results. In time $t = 2700s$ (500 L-K searches) the optimal tour (length 27686) was output 6 of 20 IterL-K runs, the average final tourlength was 0.05% above optimal. Multiple L-K runs gave much worse results. A single L-K run averages 0.98% above optimal in time $t = 120s$. 100 L-K runs gave a tour length of 0.25% above optimal. It needed 20000 L-K runs ($t = 530 hours$) to obtain a tour of length 27705.

Why is IterL-K more efficient than independent L-K runs? The success of IterL-K depends on the fact that good local L-K minima are clustered together and not randomly scattered. The probability to find a good tour is higher nearby a good tour than nearby a bad tour.

We have shown in [Müh91b] that 2-opt local minima are clustered. Furthermore we could show the following relation: The better the solutions are, the more similar they are. This relation is the reason for the success of Johnson's IterL-K. But the above relation also holds, if the problem has the *building block feature*, which is necessary for the success of the crossover operator of our genetic algorithm.

We have made a detailed comparison between iterated hill-climbing and the PGA on a simpler problem than the TSP, the optimization of a binary function of 60 binary variables [Mūh91b]. We have shown that an optimized iterated hill-climbing search needs less evaluations (configurations) to reach the optimum than the simple PGA. But it has to be mentioned that iterated hill-climbing is a sequential search method. The major problem of parallel searches like the PGA is that the same configurations are evaluated, only in parallel. Therefore it is no surprise that an optimized sequential search needs less evaluations.

The sequential efficiency of the PGA strongly depends on the population size. If the size of the population is too small, then it may be that the PGA does not find the optimum. If the size of the population is too large then a large number of identical configurations will be evaluated. If the optimal population size is used, the PGA needs less evaluations than an optimized iterated hill-climbing.

Furthermore the success of iterated hill-climbing depends on a fine tuned mutation rate. This mutation rate should lead with a high probability to the attractor region of a new local minimum. In the TSP case Johnson found that a simple 4-opt move is sufficient. In other combinatorial problems it is more difficult to find a good mutation rate and a good local heuristic like the Lin-Kernighan search for the TSP. Therefore we share the opinion of Johnson [Joh90] that the TSP is in practice much less formidable than its reputation would suggest.

We will now turn to another combinatorial problem, the graph partitioning problem GPP. Here the Lin-Kernighan heuristic is not as good as in the TSP case. We will show that the genetic search is very efficient for this problem. The major obstacle is to find a suitable crossover operator.

6 The graph partitioning problem

The m graph partitioning problem (m-GPP) is a fundamental problem which arises in many applications. The GPP is to divide a given graph into a number of partitions (m) in order to optimize some criterion e.g. to minimize the number of edges between partitions. More formally:

Let a graph $G = (V, E, w)$ be given. $V = \{v_1, v_2, ..., v_n\}$ is the set of nodes, $E \subseteq V \times V$ is the set of edges and $w : E \mapsto \mathbb{N}$ defines the weights of the edges.

The m-GPP is to divide the graph into m disjunct parts, such that some optimization criteria will be fulfilled. In this paper we will consider the following optimization criteria:

OPT 2 (m-GPP) *Let $\mathcal{P} = \{P_1, ..., P_m\}$ be a partition. Let $\mathcal{G} = (g_1 g_2 ... g_n)$ denote the partition to which the nodes belong $(1 \leq g_i \leq m)$. Then we look for*

$$\min_{\mathcal{P}} \sum_{\substack{1 \leq i < j \leq n \\ g_i \neq g_j}} w_{ij}$$

such that $\sigma(P)$ is minimal.

$\sigma(P)$ is defined as

$$\sigma^2(P) = \frac{1}{m} \sum_{i=1}^{m} |P_i|^2 - (\frac{1}{m} \sum_{i=1}^{m} |P_i|)^2$$

In order to solve the GPP, we have to define the genetic representation and the genetic operators. In the simplest representation, the value (allel) g_i on locus i on the chromosome gives the number of the partition to which node v_i belongs. But this representation is highly degenerate. The number of a partition does not have any meaning for the partitioning problem. An exchange of two partion numbers will still give the same partition. Alltogether m! chromosomes give the same fitness value.

$$F(\mathcal{G}) = \sum_{\substack{1 \leq i < j \leq n \\ g_i \neq g_j}} w_{ij}$$

All m! chromosomes code the same partitioning instance, the same "phenotype". The genetic representation does not capture the structure of the problem. We did not find a better genetic representation, so we decided that the crossover operator has to be "intelligent". Our crossover operator inserts complete partitions from one chromosome into the other, not individual nodes. It computes which partitions are the most similar and exchanges these partitions. Mathematically spoken, the crossover operator works on equivalence classes of chromosomes.

Figure 1 shows an example. The problem is to partition the 4×4 grid into four partitions.

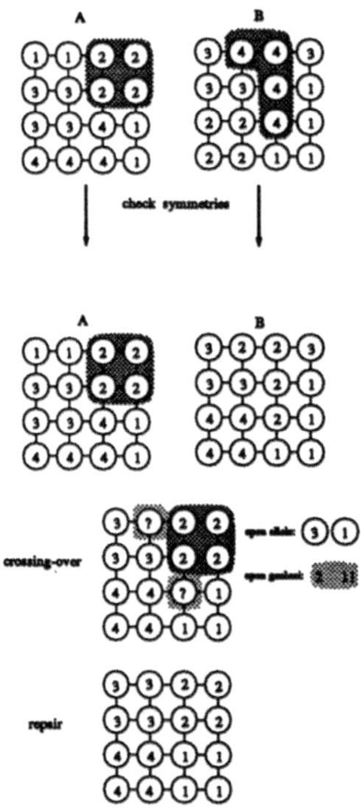

Figure 1: The crossover operator

The crossover operator works as follows. Partition 2 has to be inserted into $\mathcal{B}$. The crossover operator finds, that partition 4 of $\mathcal{B}$ is the most similar to partition 2 in $\mathcal{A}$. It identifies partition 2 of $\mathcal{A}$ with partition 4 of $\mathcal{B}$. Then it exchanges the alleles 2 and 4 in chromosome $\mathcal{B}$ to avoid the problems arising from symmetrical solutions. In the crossover step it implants partition 2 of chromosome $\mathcal{A}$ into $\mathcal{B}$.

After identifying all genloci and allels which lead to a nonvalid partition a repair operator is used to construct a new valid chromosome. Mutation is done after the crossover and depends on the outcome of the crossover. In the last step a local hill-climbing algorithm is applied to the valid chromosome.

For local hill-climbing we can use *any* popular sequential heuristic. It should be fast, so that the PGA can produce many generations. In order to solve very large problems, it should be of order

$O(n)$ where n is the problem size. Our hill-climbing algorithm is of order $O(n^2)$, but with a small constant. In order to achieve this small constant, a graph reduction is made. The general outline of our hill-climbing algorithm is as follows:

Local search for the GPP

1. Reduce the size of the graph by combining up to r nodes into one hypernode
2. Apply the 2-opt of Kernighan and Lin [KS70] to the reduced Graph. For the GPP it is defined as follows:
   (a) Select two hypernodes
   (b) Test if an exchange of this hypernodes gives a lower fitness
   (c) If this is the case, exchange the nodes
   (d) Terminate, if no exchange can be made over the set of nodes
3. Expand the resulting graph
4. Create a valid partition
5. Apply a further local hill-climbing algorithm to the valid partition

Step2 and step5 are of order $O(n^2)$. The constant is much smaller than doing 2-opt search on the original string.

The above local search is only done for the initial population. After the first generation we use a still faster search. We apply 2-opt only to the nodes which have connections to outside partitions.

7 Performance evaluation for the GPP

A detailed study of the graph bipartitioning problem can be found in [JAMS89]. In that paper random graphs and random geometric graphs up to 1000 nodes are used to compare different heuristics. We decided to make a performance analysis with real life graphs. Furthermore are we more interested in the general partitioning problem, not in the bipartitioning case. Detailed results can be found in [vL90].

We will give here the computational results for solving two of the largest GPP benchmarking problems. The problem are called *EVER918* and *EVER1005* [Eve79]. *EVER918* is a 3-D graph which consists of 918 nodes and 3233 edges. It has to be partitioned into 18 partitions. *EVER1005* has 1005 nodes and 3808 edges. It has to be partioned into 20 partitions.

The following figure shows the progress of the solution.

Figure 2 is typical for the progress of PGA's. The best progress is made at the beginning, then it decays exponentially if the combinatorial problem is NP. In the above example the best progress is made during the first generations, then it decays exponentially.

Table 2 gives a comparison to other solutions, which have been computed recently. It has to be mentioned, that the heuristic of Gilbert et al. [GZ87] and Moore [Moo88] do not use the constraint of equal partition size. This partitioning problem is simpler, furthermore the solutions should have a less cost. Nevertheless, the PGA found in one case the best solution (EVER918). In the second problem, the PGA found the best solution with minimal $\sigma(|P|)$.

Table 2 also shows that multiple 2-opt runs or multiple Lin-Kernighan runs give approximations which are far away from the optimum. The quality of the approximations seems to depend on the number of partitions. For the bipartitioning problem the quality of L-K solutions is much better [JAMS89].

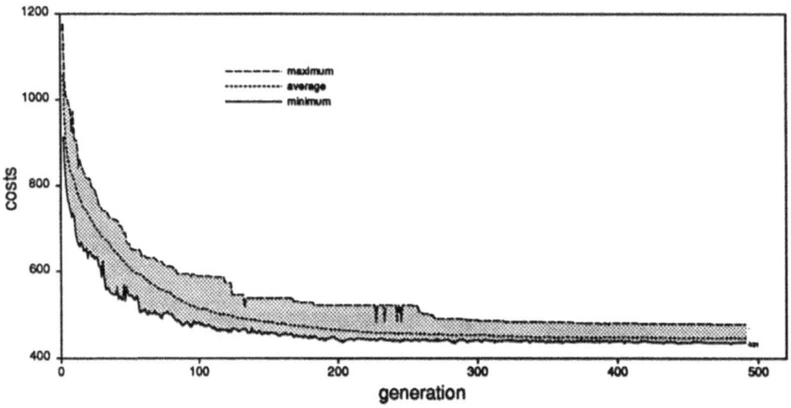

Figure 2: Problem EVER918, 64 individuals

problem	algorithm	minimal costs	σ ($\|P\|$)	run time
EVER918	MultOpt	1020	0.00	1010 s
	MultLK	625	0.00	26120 s
	GZ87	587	0.99	78 rounds
	Moore	453	0.99	78 rounds
	PGA	431	0.00	1750 s, 500 generations
EVER1005	MultOpt	1023	0.40	222 s
	MultLk	725	0.40	26280 s
	GZ87	696	1.29	91 rounds
	Moore	608	1.29	91 rounds
	PGA	631	0.40	1990 s 463 generations

Table 2: Comparison of GPP solutions for problems EVER918 and EVER1005

This indicates that it will be difficult to construct an efficient iterative Lin-Kernighan search for the m-GPP. There are two reasons for this conjecture. First, the quality of an average L-K solution is bad. Second, it is difficult to determine a good mutation rate which jumps out of the attractor region of the local minimum. This problem has been investigated in [vL91]. There the following relation is shown: the better the local minimum, the larger the attractor region. For the 2-GPP the quality of the L-K solutions seems to be better [JAMS89].

In summary: The m-GPP problem seems to be harder to solve than the TSP. The simple PGA got results which are better than other known heuristics.

8 Low autocorrelation binary sequences

The investigation of the properties of low autocorrelation binary sequences has a long history and is of great interest for technical as well as theoretical reasons. Autocorrelation sequences play an important role in several communication engineering applications [BCH85]. From the theoretical point of view it seems to be a very difficult optimization problem and the cost function seems to show a *golf course* like configuration space landscape with an irregular structure.

The autocorrelation function is defined on a Boolean hypercube of size n. The elements are binary squences $S = (s_1, ..., s_n)$, where s can have the value +1 or -1. To measure the quality of such a sequence, the following standard criterion, called *merit factor F* was introduced by Golay [Gol77].

$$F = N^2/(2 * \sum_{i=1}^{n-1} R_i^2) \tag{1}$$

$$R_i = \sum_{k=1}^{n-k} s_k * s_{k+i}; 1 \leq k \leq n-1 \tag{2}$$

OPT 3 (Auto) *The best autocorrelation sequence of size n is given by* $F^* = \max_S F(S)$

Skew-symmetric sequences of odd length n=2m-1 are defined by

$$s_{m+l} = (-1)^l s_{m-l}, 1 \leq l \leq n-1$$

These sequences seem to be good candidates for high merit factors. In the following we will only consider skew-symmetric sequences. The configuration space can then be reduced from 2^n elements to 2^m elements.

This problem has been investigated recently by Beenker [BCH85] , Golay [Gol82], Wang [Wan87] and de Groot [dGWH89]. Beenker used simulated annealing, Golay enumeration techniques, Wang and de Groot evolutionary algorithms.

The genetic representation of this problem seems to be straightforword. Just take the binary string of length m as chromosome. The phenotype is then the skew-symmetric chain of length $n = 2m - 1$. There are no constraints, each sequence is valid. So it seems that a plain PGA can solve this problem easily. But notice, that every configuration is 4-fold degenerated: inversion and sequence reversal give the same merit factor. The following four sequences give the same fitness value for the case $n = 5, m = 3$

s	1	-1	1	1	-1
s'	-1	1	-1	-1	1
s_{inv}	-1	1	1	-1	1
s_{inv}'	1	-1	-1	1	-1

With this representation, no structure of the sequences shows up. We have shown elsewhere [Müh89], that in such cases a simple crossover will not lead to a better maximum. For a genetic algorithm, it is a bad thing to make a crossing-over between e.g. s and s'. These sequences are opposite to each other located on a m-dimensional hypercube. Crossover tries to combine these searches which are phenotypically equal but genetically opposite to each other.

The above situation does only seldom occur with the PGA. In a PGA, crossing-over is only done between nearby individuals. Thus the individuals in a neighborhood get more and more similar. So it becomes unlikely that individuals in a neighborhood are genetically very different. The breeders call this effect inbreeding.

The above representation did not give good results, therefore we changed it slightly. The modification was motivated by an observation of Golay [GH89] . He showed that good skew-symmetric solutions of order n can be found by an interleaving of good symmetric and antisymmetric solutions of order $n/2$. We show for the case $n = 13, m = 7$ how this is done :

1	1	1	1	1	-1	-1	1	1	-1	1	-1	1
1		1		1		-1		1		1		1
		1		1		-1		1		-1		-1

n	Beenker	Golay	Wang	Groot	PGA
81	7.32	8.20	-	8.04	8.20
101	6.06	8.36	6.91	8.36	8.36
103	5.90	9.56	7.76	-	9.56
105	6.07	8.25	7.61	-	8.25
107	6.53	8.46	-	-	8.46
109	6.15	8.97	-	-	8.97
111	6.02	8.97	-	-	8.97
113	6.33	8.49	-	-	8.49
115	6.40	-	-	-	8.60
117	6.42	-	-	-	8.12
119	6.01	-	-	-	7.61
121	6.61	-	-	6.75	8.67
141	6.01	-	-	6.48	7.45
161	6.02	-	-	6.02	6.89
181	5.70	-	-	6.02	6.77
201	-	-	-	5.92	6.29

Table 3: Best solutions found

So we tested the idea, to use two chromosomes as genetic representation. In the above example we have the following two chromosomes $c_{sym} = (1,1,1,-1)$ and $c_{asym} = (1,1,-1)$ The phenotype defined by these chromosomes can be obtained by interleaving c_{sym} and c_{asym} and then expanding the string to a skew-symmetric string. This will give the string shown in the first line above.

Our local hill-climbing method is very simple. One bit is flipped on the skew-symmetric string. This change is accepted, if the fitness value increases. Then the next bit is flipped and so on.

In table 3 the computational results are given. The PGA found all solutions from Golay [GH89], who used an enumeration technique tailored to the autocorrelation problem. De Groot used an evolutionary algorithm without crossing-over. He kept the sample points far apart by accepting only points which had a minimum Hamming distance to all other points.

Why did the PGA perform so good? Our genetic representation does not lead to a building block feature. The two chromosomes encode only little information about the problem.

In the autocorrelation problem the secret of success seems to be our local hill-climbing. We will demonstrate this with a small example of size $n = 31$. For $n = 31$ there exist one optimal solution with $F = 6.08228$ and four suboptimal solutions with $F = 5.52299$. After only three generations the PGA always found the best solution and three of the suboptimal solutions. We will show how the best solution was found in a typical example. In the following table the parent chromosomes and the offspring chromosome are shown in various stages of the algorithm.

state	s_{sym}	s_{asym}	F
parent1	10000011	11000101	3.599
parent2	01011011	11011110	3.599
cross	11010011	11001111	2.235
mutat.	11000011	11011111	1.115
hill-cl.	00100111	11010111	6.082

In the table we use 0 instead of -1. Crossover is done between the underlined substrings. Crossing-over and a small mutation of 1 bit leads to a string with very low fitness value. Nevertheless the local search climbs from 1.115 up to the global optimum. This behavior is different to other combinatorial problems like the GPP, where crossing-over explored the building blocks of the solution.

We are now evaluating, if random insertion of a large string instead of crossingover will give the same or better results. If this is not the case, we have to analyze, what kind of structure crossing-over is exploring in the autocorrelation problem.

9 Conclusion

The parallel genetic algorithm has been very successful in benchmark combinatorial optimization problems. The algorithm uses a distributed selection schedule, it self-organizes itself. The crossover operators described in this paper are not the only ones possible. Glover has suggested *adaptive structured combinations* for the TSP and the bipartitioning problem [Glo91]. We have used a voting crossover operator for the quadratic assignment problem, which combines five solutions instead of two [Müh89]. A good problem dependent crossover operator will improve the performance of the PGA.

But the major problem of the *simple* PGA described in this paper is that in the last stage of the algorithm the same evaluations are done over and over again. The *simple* PGA does not have any memory. For a computational optimal PGA it seems to be very promising to implement some kind of memory, maybe controlled by *tabu search* [Glo91].

Other extensions of the *simple* PGA are also straightforward. We will implement this extensions in the course of applying the PGA to more and more challenging applications.

The success of the PGA suggest exploring other problem solving metaphors also. Why not use the market of economy as a parallel search method? A comparison of problem solving by a market framework and by biological evolution on the same set of artificial problems would also give further insight into economy and biology.

References

[BB91] R. K. Belew and L. Booker, editors. *Procedings of the Fourth International Conference on Genetic Algorithms.* Morgan Kaufmann, 1991.

[BCH85] G.F.M. Beenker, T.A.C.M. Claasen, and P.W.C Hermens. Binary sequences with a maximally flat amplitude spectrum. *Phill. J. of Research*, 40:289, 1985.

[CHMR87] J.P. Cohoon, S.U. Hedge, W.N. Martin, and D. Richards. Punctuated equilibria: A parallel genetic algorithm. In J.J. Grefenstette, editor, *Proceedings of the Second International Conference on Genetic Algorithms*, pages 148–154. Lawrence Erlbaum, 1987.

[dGWH89] C. de Groot, D. Würtz, and K.H. Hoffmann. Low autocorrelation binary sequences: exact enumeration and optimization by evolutionary strategies. Technical report, ETH Zürich IPS 89-09, 1989.

[DSY89] R. Durbin, R. Szelinski, and A.L. Yuille. An analysis of the elastic net approach to the travelling salesman problem. *Neural Comp.*, 1:348–358, 1989.

[Eve79] G.C. Everstine. A comparison of three resquencing algorithms for the reduction of matrix profile and wavefront. *Int. J. Numer. Meth. in Engin.*, 14:837–853, 1979.

[GH89] M.J.E. Golay and D. Harris. A new search for skewsymmetric binary sequences with optimal merit factors. Technical report, 1989.

[Glo77] F. Glover. Heuristics for integer programming using surrogate constraints. *Decision Sciences*, 8:156–166, 1977.

[Glo91] F. Glover. Tabu search for nonlinear and parametric optimization. Technical report, University of Boulder, 1991.

[Gol77] M.J.E. Golay. Sieves for low autocorrelation binary sequences. *IEEE Trans. on Inform Theory*, 23:43, 1977.

[Gol82] M.J.E. Golay. The merit factor of long low autocorrelation binary sequences. *IEEE Trans. on Inform. Syst.*, 28:543, 1982.

[Gol89] D.E. Goldberg. *Genetic Algorithms in Search, Optimization and Machine Learning.* Addison-Wesley, 1989.

[GS89] M. Gorges-Schleuter. Asparagos: An asynchronous parallel genetic optimization strategy. In H. Schaffer, editor, *3rd Int. Conf. on Genetic Algorithms*, pages 422–427. Morgan-Kaufmann, 1989.

[GS91] M. Gorges-Schleuter. *Genetic Algorithms and Population Structures - A Massively Parallel Algorithm.* PhD thesis, University of Dortmund, 1991.

[GZ87] J.R. Gilbert and E. Zmijewski. A parallel graph partitioning algorithm for message-passing multiprocessors. In *First Int. Conf. on Supercomputing*, 1987.

[Hol75] J.H. Holland. *Adaptation in Natural and Artificial Systems.* Univ. of Michigan Press, Ann Arbor, 1975.

[JAMS89] D.S. Johnson, C.R. Aragon, L.A. McGeoch, and C. Schevon. Optimization by Simulated Annealing: An Experimental Evaluation; Part I, Graph Partitioning. *Operations Research*, 37:865–892, 1989.

[Joh90] D.S. Johnson. Local Optimization and the Traveling Salesman Problem. In M.S. Paterson, editor, *Automata, Languages and Programming*, Lecture Notes in Computer Science 496, pages 446–461. Springer-Verlag, 1990.

[KGV83] S. Kirkpatrick, C.D. Gelatt, and M.P. Vecchi. Optimization by simulated annealing. *Science*, 220:671–680, 1983.

[KS70] B.W. Kernighan and S.Lin. An efficient heuristic procedure for partitioning graphs. *Bell Syst. Techn Journ.*, 2:291–307, 1970.

[KT85] S. Kirkpatrick and G. Toulouse. Configuration space analysis of travelling salesman problems. *J.Physique*, 46:1277–1292, 1985.

[Lin65] S. Lin. Computer solutions of the traveling salesman problem. *Bell. Syst. Techn. Journ.*, 44:2245–2269, 1965.

[LK73] S. Lin and B. W. Kernighan. An efficient heuristic for the traveling salesman problem. *Operations Research*, 21:298–516, 1973.

[MGSK87] H. Mühlenbein, M. Gorges-Schleuter, and O. Krämer. New solutions to the mapping problem of parallel systems - the evolution approach. *Parallel Computing*, 6:269–279, 1987.

[MGSK88] H. Mühlenbein, M. Gorges-Schleuter, and O. Krämer. Evolution algorithms in combinatorial optimization. *Parallel Computing*, 7:65–88, 1988.

[Moo88] D. Moore. A round-robin parallel partitioning algorithm. Technical Report 88-916, Cornell University, 1988.

[MS89] B. Manderick and P. Spiessens. Fine-grained parallel genetic algorithm. In H. Schaffer, editor, *3rd Int. Conf. on Genetic Algorithms*, pages 428–433. Morgan-Kaufmann, 1989.

[MSB91] H. Mühlenbein, M. Schomisch, and J. Born. The parallel genetic algorithm as function optimizer. *Parallel Computing*, 17:619–632, 1991.

[Müh89] H. Mühlenbein. Parallel genetic algorithm, population dynamics and combinatorial optimization. In H. Schaffer, editor, *3rd Int. Conf. on Genetic Algorithms*, pages 416–421. Morgan Kaufmann, 1989.

[Müh91a] H. Mühlenbein. Darwin's continent cycle theory and its simulation by the prisoner's dilemma. *Complex Systems*, 1991.

[Müh91b] H. Mühlenbein. Evolution in time and space - the parallel genetic algorithm. In G. Rawlins, editor, *Foundations of Genetic Algorithms*, pages 316–337, San Mateo, 1991. Morgan-Kaufman.

[Pet90] C. Peterson. Parallel Distributed Approaches to Combinatorial Optimization: Benchmark Studies on Traveling Salesman Problem. *Neural Computation*, 2:261–269, 1990.

[PR87] W. Padberg and G. Rinaldi. Optimization of a 532-city symmetric traveling saleman problem by branch and cut. *Op. Res. Let.*, 6:1–7, 1987.

[PS89] C. Peterson and B. Söderberg. A new method for mapping optimization problems onto neural networks. *Int. J. Neural Syst.*, 1:995–1019, 1989.

[Rec73] I. Rechenberg. *Evolutionsstrategie - Optimierung technischer Systeme nach Prinzipien der biologischen Information.* Fromman Verlag, Freiburg, 1973.

[Sch81] H.-P. Schwefel. *Numerical Optimization of Computer Models.* Wiley, Chichester, 1981.

[Sch89] H. Schaffer, editor. *Procedings of the Third International Conference on Genetic Algorithms.* Morgan Kaufmann, 1989.

[Tan89] R. Tanese. Distributed genetic algorithm. In H. Schaffer, editor, *3rd Int. Conf. on Genetic Algorithms*, pages 434–440. Morgan-Kaufmann, 1989.

[UPvL+91] N.L.J. Ulder, E. Pesch, P.J.M. van Laarhoven, H.-J. Bandelt, and E.H.L. Aarts. Genetic Local Search Algorithm for the Traveling Salesman Problem. In R. Maenner and H.-P. Schwefel, editors, *Parallel Problem Solving from Nature*, Lecture Notes in Computer Science 496, pages 109–116. Springer-Verlag, 1991.

[vL90] G. von Laszewski. Ein paralleler genetischer Algorithmus für das Graph Partitionierungsproblem. Master's thesis, Universität Bonn, 1990.

[vL91] G. von Laszewski. Intelligent Structural Operators for the k-way Graph Partitioning problem. In R.K. Belew and L. Booker, editors, *Procedings of the Fourth International Conference on Genetic Algorithms*, pages 45–52. Morgan Kaufman, 1991.

[Wan87] Q. Wang. Optimization by simulating molecular evolution. *Biol. Cybern.*, 57:95, 1987.

Vortrag anläßlich des Arbeitsgesprächs
"Physik und Informatik - Informatik und Physik"
am 22. November 1991

R e c h n e n d e r   R a u m

Schon vor Beginn der Entwicklung haben Computerpioniere sich aus ihrer Sicht Gedanken über die physikalischen Erscheinungen gemacht. So drang der berühmte Babbage, welcher als erster die Idee des Computers gedanklich gefaßt und auch bereits die ersten Versuchsgeräte gebaut hat, bis zu den letzten Konsequenzen der abendländischen Feld-Physik vor, aus welcher sich ergibt, daß einmal gesprochene Worte nie vollständig verlorengehen können. Diese Gedanken sind sicher auch an anderer Stelle verfolgt worden, ohne daß meiner Kenntnis nach brauchbare Lösungen für dieses Problem vorliegen. Auch die Quantentheorie kann wohl nur Ansätze für die Lösungen dieses Problems geben. Durch sie werden verschiedene Größen gequantelt, jedoch arbeitet man auch dort mit Differentialgleichungen, also mit kontinuierlichen Größen.

Sowie man aber einen wesentlichen Grundsatz der abendländischen Mathematik anzweifelt, nämlich, daß sich zwischen zwei beliebig nahe aneinanderliegende Werte stets noch beliebig viele weitere Zwischenwerte legen lassen, stößt man auf das Problem der Digitalisierung bzw. der Beschränkung auf diskrete Werte.

Alle unseren heutigen Computer arbeiten digital mit beschränkter Stellenzahl, so daß es z.B. kaum möglich ist, natürliche Werte exakt darzustellen.

Auch das Modell von Laplace geht von der Idee der Vorausberechenbarkeit des Weltgeschehens aufgrund der vollständigen Kenntnis der augenblicklichen Siuation aus und arbeitet mit einem Kontinuum von Werten. Die Chaos-Theorie zeigt uns heute allerdings, daß bereits sehr kleine Abweichungen in der Anfangssituation sehr große Auswirkungen haben können, womit die Vorausberechenbarkeit im Sinne von Laplace an die Feinheit des Aufbaus des Kosmos äußerste Ansprüche stellt.

Diesem "Analog-Modell" des Kosmos kann man im Sinne der heutigen Informatik ein digitales Modell gegenüberstellen. Der Gedanke, den gesamten Kosmos als eine große Rechenmaschine aufzufassen, wird heute schon ernsthaft diskutiert. Als einfachste Lösung erscheint es zunächst, den Begriff des Feldes durch den Zellularen Automaten (ZA) zu ersetzen. Ein zweidimensionales Feld kann dabei durch ein orthogonales Netz simuliert werden, in dessen Gitterpunkten je ein kleiner Computer sitzt, der mit seinen Nachbarcomputern Daten austauscht. Zur Berechnung von partiellen Differentialgleichungen wurden solche Netze schon lange benutzt, z.B. bei der Wetterrechnung. Das Bestreben geht dabei dahin, durch möglichst hohe Genauigkeit und hohe Dichte der Gitternetze, das Modell der Differentialgleichungen möglichst genau anzunähern.

Auch in diesem Fällen benutzt der Physiker also den Computer als Werkzeug, wobei er aber im Prinzip an dem traditionellen kontinuierlichen Modell festhält. Hingegen wird aber bei der digitalen Auffassung des Kosmos von dem bereits von Babbage angeregten Gedanken ausgehend, die Natur nicht unendlich fein angenommen. Es bietet sich der ZA als einfachste Form eines Rechnenden Raums an. Hinzu kommt die heute bereits weitentwickelte Automatentheorie. Sie arbeitet mit einer diskreten Anzahl von Zuständen eines Systems, wobei die Uebergangsregel den Algorithmus ergibt, gemäß dem aus dem gegebenen Zustand der folgende gebildet wird. Beim ZA kann man sich darauf beschränken, diese Uebergangsregel für einen Gitterpunkt und seine Nachbarschaft festzulegen, wobei diese Regel dann im gleichen Sinn für alle Gitterpunkte gilt. Es handelt sich im Sinne der Physik dabei also um ein ähnliches Gesetz, da Wirkungen nur zwischen unmittelbar benachbarten "Zellen" bestehen.

Die Theorie der Zellularen Automaten hat bereits großes Interesse gefunden, und es sind zahlreiche Varianten der Algorithmen durchgespielt worden. Am bekanntesten ist das Life-Spiel, bei dem gewissermaßen lebende Muster sich im Raum fortschalten und sich dabei verändern oder gegenseitig beeinflussen. Dabei ging das Bestreben zunächst mehr dahin, Prozesse lebender Mikroorganismen und ihr Verhalten untereinander zu simulieren. Zunächst habe ich mir demgegenüber mehr die Aufgabe gestellt, einfache physikalische Vorgänge, wenn auch nur annäherungsweise, zu simulieren, z.B. das Verhalten von Elementarteilchen in einem ZA zu studieren. Es stellt sich allerdings sehr schnell heraus, daß selbst die einfachsten Elementarteilchen, wie etwa das Photon, äußerst kompliziert in ihrem Verhalten sind, und wir ja auch nicht in der Lage sind, dafür genaue Erklärungen zu geben, sondern nur sehr pauschal die den Quanteneffekten unterliegenden Gesetze kennen. Der Gedanke, unmittelbar etwa die Fortschaltgesetze der physikalischen Elementarteilchen zu simulieren, erschien also zunächst viel zu kühn. Also entschloß ich mich, zunächst Muster zu entwickeln, die sich gewissermaßen als Digitalteilchen im Zellularen Raum fortpflanzen und miteinander reagieren.

BILD 1 zeigt das Beispiel für ein "Pfeil-Modell", bei dem sich die Pfeile entsprechend dem gegebenen Automatengesetz bewegen. Sie durchlaufen eine Folge von Zuständen, die sich periodisch wiederholen. Was passiert nun, wenn sich zwei solche Teilchen begegnen?

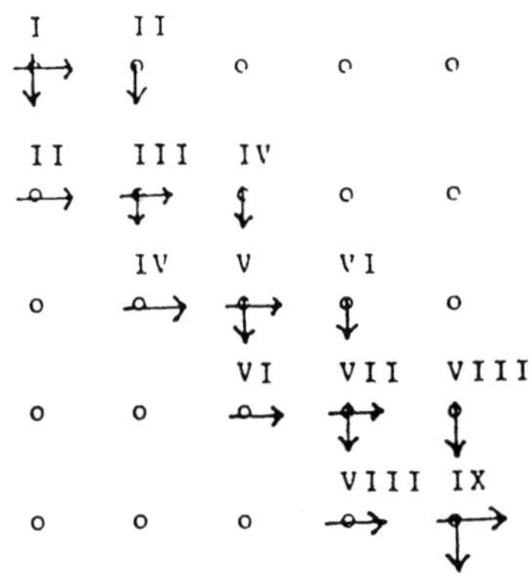

Bild 1

BILD 2 und BILD 3 zeigen zwei Beispiele. Es zeigt sich, daß je nach Phasenlage relativ zum Gitter des ZA die beiden Teilchen entweder durcheinander durchlaufen können, ohne sich zu beeinflussen (Bild 2) oder daß sie miteinander reagieren und als Ergebnis ein neues Teilchen entsteht (Bild 3). Ohne Kenntnis des

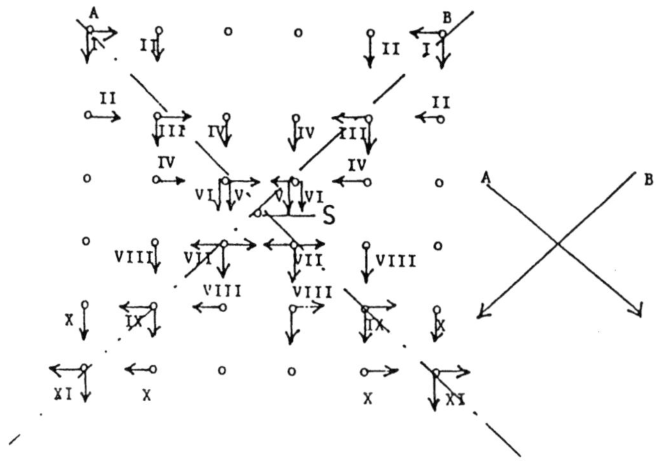

Bild 2

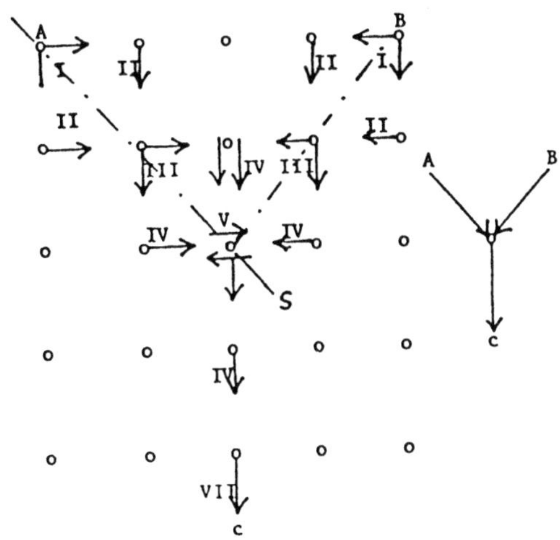

Bild 3

gitterförmigen Aufbaus des Automaten könnte man zu dem Schluß kommen, daß die Reaktion mit der Wahrscheinlichkeit 1:2 erfolgt. Damit haben wir einen an die Quantentheorie erinnernden Effekt. Der Unterschied ist nur, daß die Quantentheorie von echter Wahrscheinlichkeit spricht, und im ZA jedoch strenge Kausalität herrscht. Es ist also im Prinzip möglich, quantentheoretische Wahrscheinlichkeiten exakt determinierten Modellen gegenüberzustellen.

Das gezeigte Modell kann noch in verschiedener Weise verfeinert und ergänzt werden. Eine Reihe von solchen Versuchen sind durchgeführt worden, ohne jedoch in der Simulation physikalischer Erscheinungen durch ZA wesentlich weiterzukommen. Trotzdem hat die Analyse dieser Versuche zu einer Reihe von Erkenntnissen geführt, von denen ich Ihnen einige hier stichwortartig vortragen möchte:

Die Beschäftigung mit "Sich im Raum fortbewegenden digitalen und stabilen Mustern", läßt auch die "Lichtgeschwindigkeit" im anderen Licht erscheinen.

Um ein Muster größerer Ausdehnung stabil zu halten, bedarf es der Nachrichten zwischen den einzelnen Elementen. Diese müssen notwendigerweise schneller sein als die Fortbewegung des Musters im Raum. Das würde bedeuten, daß jenseits der oberen Grenzgeschwindigkeit der physikalischen Teilchen, also der Lichtgeschwindigkeit eine Ueberlichtgeschwindigkeit existieren muß, die den Zusammenhalt der Teilchen im ZA, also der Digitalteilchen, bewirkt. Als einfachstes Beispiel kann man eine Kolonne marschierender Soldaten nennen. Würde ich auf einem völlig freien Feld jeden Soldaten einzeln in Marsch setzen, so ist kaum zu erwarten, daß sie alle genau in derselben Richtung marschieren. Lasse ich sie aber alle in geschlossener Kolonne marschieren, so bleiben sie als geschlossene Formation zusammen. Das ist mögich, weil zwischen ihnen Informationen ausgetauscht werden (Ausrichten, Tuchfühlung), die schneller sind, als die Bewegung der Formation. Auf die Physik übertragen hieße dies, daß viele Erscheinungen, deren Grenze die Lichtgeschwindigkeit ist, eine Nachrichtenübertagung mit Ueberlichtgeschwindigkeit erfordern. Damit ist nicht gesagt, daß wir diese Verbindungen zur Nachrichtenübertragung ausnutzen könnten.

Es gibt ein Experiment in der Physik, das noch nicht zufriedenstellend gelöst ist, nämlich das Einstein-Podolski-Rosen-Paradox. Neben den zahlreichen Versuchen, es zu erklären, hat man auch die Möglichkeit der Ueberlichtgeschwindigkeit herangezogen. Das digitale Denken zeigt hier noch völlig neue Möglichkeiten.

Auch das Problem der Singularitäten erscheint im neuen Licht. Diese zeigen sich meistens bei mathematischen Ableitungen, die keinerlei Grenze in bezug auf die Größen aufweisen. Im ZA lassen sich diese Erscheinungen anders lösen, in einer Weise, die natürlichen Erscheinungen näherkommt.

Es gibt weitere interessante Unterschiede im Verhalten von Modellen der Differentialgleichungen und solche des ZA. Die spezielle Relativitätstheorie erlaubt es, grundsätzlich beliebig viele Inertialsysteme addativ aneinanderzusetzen, wobei die Gesetze (z.B. Lorentz-Transformationen) erhalten bleiben. Im ZA lassen sich auch diese Operationen simulieren. Aber aufgrund des Umstandes, daß die Maschenweite der Gitter nicht beliebig verkleinert werden kann, müssen wir bei hoher Energie zu einer Art "Verrechnen des Raumes" kommen. Der ZA zeigt auf natürliche Weise, daß bei sehr hoher Energie gewisse Phänomene anders verlaufen müssen. Wir können vielleicht erst heute erkennen, daß etwas Entsprechendes in der Natur bei Versuchen mit hoher Energie auftreten kann.

Ueber die Rolle des Begriffs "Information" in der Physik ist bereits viel diskutiert worden. Es ist nicht so, daß man sie als eine weitere physikalische Größe neben Masse, Energie, Impuls usw. setzen kann. Auch ein Gesetz von der Erhaltung der Information läßt sich nicht aufstellen. Die Beschreibung von Phänomenen mit unendlicher Variabilität (z.B. Länge) erfordert im Prinzip auch eine unendliche Information. Der Rechnende Raum bietet hier Möglichkeiten der Beschränkung auf endliche Informationen, z.B. zur Beschreibung des Zustands eines abgeschlossenen Systems. Das ist eine gute Voraussetzung für automatentheoretische Behandlung.

Im allgemeinen arbeiten rechnerische Modelle, also z.B. Computer streng deterministisch, und zwar in der positiven Zeitrichtung. Die Grundgesetze der Physik gelten im allgemeinen sowohl in der negativen als auch in der positiven Zeitrichtung. Obgleich schon gewisse Erscheinungen dafür sprechen, daß die volle Symmetrie in bezug auf die Zeitrichtung nicht immer gewahrt ist. Die uns gewohnten Vorgänge im täglichen Leben zeigen jedenfalls keine Symmetrie der zugrundeliegenden Gesetze in der Zeitrichtung. Wir brauchen nur an das Einschmelzen einer kompliziert gestalteten Figur oder an die Mischung verschiedener (zunächst getrennt gehaltener) Gase zu denken. Man spricht hier von Reversibilität und Irreversibilität. Die strenge (meiner Meinung nach nicht wirklich bewiesene) Reversibilität (Symmetrie des Verhaltens in beiden Zeitrichtungen) läßt uns durch Extrapolation der Abläufe in die Vergangenheit zur Theorie des Urknalls kommen. Dagegen erlaubt die Automatentheorie die Entwicklung einer Theorie der Keimzelle welche zu grundsätzlich anderen Perspektiven führen würde. Grundsätzlich kann man auch Automaten symmetrisch in beide Richtungen bauen. Hierfür haben Petri und Fredkin Beispiele gegeben.

Insbesondere in der Quantentheorie spielen Wahrscheinlichkeitsgesetze eine große Rolle. Auch in der Automatentheorie lassen sich Zufallselemente einbauen. Es wurde am Beispiel von Bild 2 und Bild 3 auch schon gezeigt, daß solche Wahrscheinlichkeitsphänomene sich auch streng determiniert erzeugen lassen. Für den Mathematiker ist die Annahme von Wahrscheinlichkeiten auf dem Papier verhältnismäßig einfach. Für den Ingenieur bietet die Konstruktion eines Wahrscheinlichkeitsmechanismus jedoch einige Schwierigkeiten. Ein Würfel muß sehr genau gebaut werden. Dasselbe gilt für Lotteriemechanismen, wie z.B. das Roulette.

Auch die heute viel genannte Chaos-Theorie läßt sich im Sinne des ZA determiniert behandeln. Die Informatik zeigt, daß bei sehr komplizierten Systemen und Netzen die Änderung eines einzelnen Bits den Ablauf völlig verändern kann. Trotzdem läuft alles determiniert ab.

Ein wichtiges Gegenargument gegen die Auffassung des Kosmos als Zellularer Automat ist die Isotropie. Alle unsere Naturgesetze und Erscheinungen kennen im Gegensatz zum ZA keine räumlichen Vorzugsrichtungen. Der regelmäßige ZA ist jedoch nur die einfachste Form einer Anordnung von Zellen. Es lassen sich jedoch Gesetze für unregelmäßige Strukturen entwickeln, die auch veränderliche und wachsende Automaten betreffen. Anstelle von gekrümmten Räumen können zerknitterte Räume treten.

Die Physiker haben auch schon die kleinste Länge und die kleinste Zeit diskutiert. Man hat jedoch nicht den Eindruck, daß diese Fragen schon geklärt sind. Im ZA ist aufgrund der regelmäßigen Struktur diese Frage einfach zu beantworten, da im allgemeinen durch den Gitterabstand in räumlicher und zeitlicher Hinsicht diese Werte gegeben sind.

Das Wirkungsquantum h kann im Zellularen Automaten die Bedeutung eines Schaltvorgangs annehmen. Energie ist dann Schaltvorgang pro Zeiteinheit, was eine sehr natürliche Erklärung ergibt.

Es hat wenig Sinn, den Rechnenden Raum dem seinerzeit von einigen Physikern konzipierten Äther gegenüberzustellen. Der Rechnende Raum betrifft rein logische Gesetze, bei denen die Informationsverarbeitung durch logische Schaltungen bewirkt wird. Der Äther-Begriff baut jedoch auf Eigenschaften von Stoffen und ihrem Verhalten auf.

Ich glaube, in dieser kurzen Abhandlung einige anregende Gedanken gegeben zu haben, die -wie ich hoffe- auf die zukünfige Entwicklung der physikalischen Theorie einigen Einfluß haben können. Ähnliche Ideen werden auch an anderer Stelle, insbesondere in den USA ernsthaft verfolgt, jedoch scheint noch viel Entwicklungsarbeit erforderlich zu sein, um zu brauchbaren Ergebnissen zu kommen.

Hünfeld, im November 1991

Flexibilisierung von Rechnerstrukturen

Eine Möglichkeit zur Verringerung der Hardware-Software-Lücke?

von O. G. Folberth, Böblingen

Kurzvortrag beim "Arbeitsgespräch" über
Physik und Informatik - Informatik und Physik
am 21. und 22. November 1991 in München

Im Laufe der Entwicklung der digitalen Daten- und Informationsverarbeitung wurden die Rechnerstrukturen zunehmend flexibler gestaltet. Historisch gesehen war dabei wohl der grösste und wichtigste Schritt die Einführung des gespeicherten Programms durch John von Neumann und die damit erzielte Beeinflussbarkeit des Programmablaufes durch Daten und Zwischenergebisse. Dieses Prinzip wurde später erweitert zur Flexibilisierung der Programmgestaltung durch gezielten Abruf und Anordnung von gespeicherten Mikroinstruktionen (Mikroprogrammierung). Wie schon 1973 erkannt wurde (1), kann die Flexibilisierung aber weiter getrieben werden, indem auch die Hardware diesem Prinzip unterworfen wird. Dabei handelt es sich um rekonfigurierbare Strukturen, wobei aus einer "grauen box" durch das Aufrufen von abgespeicherten Konfigurationen die für die gerade auszuführende Operation zweckmässige "schwarze box" - vorübergehend - gebildet wird.

Die folgende Übersicht über die zunehmende Flexibilisierung digitaler Systeme veranschaulicht diesen Trend:

Stufe	Bezeichnung	Art der Flexibilisierung
I.	starres Programm	keine; die Verarbeitung verläuft nach starrem, vorgegebenem Programm (Steckverbindungen)
II.	gespeichertes Programm ("von-Neumann-Maschine")	Daten <u>und</u> Programm im Speicher, Programmablauf durch Daten und Zwischenergebnisse beeinflussbar

III.	gespeicherte Funktion (mikroprogrammierbare Maschine)	Die informatorische Struktur des Systems kann durch einen Programmaufbau aus den Funktionen des Mikroprogrammes flexibel gestaltet werden
IV.	flexible Systeme (rekonfigurierbare Strukturen)	Die Hardware und/oder die Software kann - massgeblich eines "Superprogrammes" - gesteuert von Daten und Zwischenergebnissen flexibel rekonfiguriert und dem jeweils zu prozessierenden Algorithmus angepasst werden

Die Flexibilisierungsstufen II. und III. haben sich in der Praxis voll bewährt und gehören zum "Stand der Technik". Letztlich sind zur Stufe III. auch die neuerdings an Bedeutung gewinnenden RISC-Strukturen zu rechnen. (RISC = Reduced Instruction Set Computer). Diese, die Effizienz der Software steigerden Methoden wurden ermöglicht durch die zunehmende Verfügbarbeit von immer leistungsfähigerer umd preiswerterer Hardware. Für Stufe IV. findet man Ansätze bei Rechnernetzen für massive parallele Verarbeitung. Dabei werden ganze digitale Einheiten (Mikroprozessoren, Speicher, Hilfs- und Peripheriefunktionen) blockweise flexibel verdrahtet, beispielsweise durch Kreuzschienenverteiler oder baumartige Schaltwerke. Andererseits enthalten die bekannten und vielfach verwendbaren programmierbaren Gate-Arrays u. ä. Bausteine rekonfigurierbare Funktionen, die im Chip integriert sind.

Eine flexible, rekonfigurierbare Hardware ist dadurch gekennzeichnet, dass die digitalen Funktionen nicht fest verdrahtet sind, sondern erst durch Steuerimpulse gebildet - und natürlich auch wieder aufgelöst - werden können. Dabei kann man sich u.a. des bewährten LSSD-Prinzips bedienen (LSSD = Level Sensitive Scan Device). Hierbei werden funktionale Blöcke rekonfigurierbar verbunden, wobei dies Prinzip in letzter Konsequenz bis auf die Gatterebene ausgedehnt werden kann. Natürlich führt dieses zu einer grösseren Komplexität der Chips, auf die für diesen Zweck zusätzliche Leitungen und Steuerelemente aufgebracht werden müssen. In diesem Sinne lässt sich das Konzept einer "Super-von-Neumann-Maschine" etwa wie folgt skizzieren: Das System macht vollen Gebrauch von den aufgeführten Flexibilisierungsstufen. Ein "Super-Programm" steuert die Rekonfiguration der Hardware und der Software, ausgelöst durch Steuerimpulse des Objektprogrammes. Dieser zusätzliche Aufwand dürfte sich jedoch zunehmend lohnen, da man dadurch folgenden Vorteile einhandeln kann:

1. Während bisher immer die Software der vorgegebenen Hardware angepasst werden musste, kann nunmehr die Hardware an die Software angepasst werden. Die Hardware kann so rekonfiguriert werden, dass ein beliebiges Softwarepaket darauf laufen kann, ohne dass es hierfür gesondert emuliert oder compiliert werden muss.

2. Es können Softwareteile unterschiedlicher Provenienz und Struktur zu einem Paket zusammengeschnürt werden. Die Hardware wird jeweils dem gerade zu verarbeitenden Softwareteil entsprechned rekonfiguriert.

3. Im letzten Jahrzehnt hat das wissenschaftlich-technische Rechnen durch die Einführung der Vektorisierung und des Pipelinings eine beträchtliche Ausweitung und Kapazitätssteigerung gefunden. Folgerichtig führt das schliesslich zu Systemem, deren Prozessoren aus "Systolic Arrays" oder "Wavefront Arrays" bestehen (2). Dabei geht allerdings die Universalität verloren, da diese Einheiten jeweils auf spezielle Algorithmen optimal abgestimmt sind. Rekonfigurierbare Hardware könnte hier Abhilfe schaffen, indem dadurch die Universalität - wenigstens teilweise - wieder hergestellt werden kann (3). Hierfür gibt es experimentelle Ansätze (4).

Es ist daher anzunehmen, dass in Zukunft zunehmend mit der Entwicklung und Erprobung von rekonfigurierbaren Rechnerstrukturen zu rechnen ist. Dabei dürfte es zu tiefgreifenden Änderungen bei der Hardware und der Software kommen. Im Extremfall sollte dieser Trend zu einer "Erweichung" der Hardware ("Soft Hardware") und einer "Versteifung" der Software ("Hard Software") führen. Gekoppelt damit wird eine zunehmende Komplexität der Hardware zu beobachten sein, die aber - hoffentlich - zu weiteren

Hilfen für die Erarbeitung von Software führen sollte, um so die bekannte Hardware-Software-Lücke zu verringern, oder - mindestens - nicht noch weiter anwachsen zu lassen. Jedenfalls könnten so die weiterhin zu erwartenden Fortschritte in der Hardware der - in dieser Hinsicht nach wie vor - nachhinkenden Softwareentwicklung zugute kommen.

Diese Tendenz ist ja schon Tradition: Die dramatische Entwicklung von zunehmend leistungsfähigerer und preiswerterer Hardware - insbesondere im Zuge des rasanten Fortschritts in Dichte und Rechengeschwindigeit hochintegrierter Schaltungen - führte immer wieder zu neuen Hardwarehilfen für die Entwicklung und Anwendung von Software.

Da sich dieser Trend in etwa mit den aus der abgelaufenen Entwicklung bekannten Steigerungsraten mindestends in den nächsten ein bis zwei Jahrzehnten fortsetzen dürfte, ist diese Prognose mit einer hohen Wahrscheinlichkeit zu bewerten.

(1) K. Ganzhorn, Prinzipien in Rechnerstrukturen, Elektr. Rechenanl. 15 (1973), H. 6, S. 263 - 269

(2) siehe beispielsweise:
S.-Y. Kung, On Supercomputing with Systolic/Wavefront Array Processors, Proc. IEEE, 72 (1984), S. 867 - 884

(3) A. A. Sawchuk und T. C. Strand, Digital Optical Computing, Proc. IEEE, 72 (1984), S. 758 - 779

(4) A. Khurshid und P. D. Fisher, Design of a reconfigurable systolic array using LSSD techniques, Proc. ICCD '84: VLSI in Computers, New York, Oct. 8 -11, 1984, S. 171 -175

Adaptive Regelung eines linearen zeitinvarianten Systems dritter Ordnung mit Totzeit durch ein neuronales Netzwerk

Wolfram H. Schiffmann *

Zusammenfassung

Dieser Beitrag berichtet über den Einsatz neuronaler Netzwerke als adaptive Regler. An einem linearen System dritter Ordnung mit Totzeit wird demonstriert, daß ein neuronales Netzwerk in der Lage ist, eine Regelstrecke, über die nur minimales qualitatives Wissen vorliegt, selbständig zu identifizieren und zu regeln. Das Training wird mit dem weit verbreiteten Backpropagation–Algorithmus durchgeführt. Die Regelstrecke wird dabei als Teilnetz interpretiert, dessen Verbindungsgewichte nicht verändert werden können.

1 Einleitung

Konventionelle Reglerentwürfe setzen die Kenntnis eines linearen Modells der Regelstrecke voraus. Diese Voraussetzung ist in der Praxis fast nie gegeben, da die Regelstrecke meist nichtlineares Verhalten zeigt. Die Modellbildung ist also ungenau und erfordert erheblichen Rechenaufwand. Im folgenden werden existierende Ansätze zur Realisierung neuronaler Regler vorgestellt und miteinander verglichen. Zu nennen sind prädikative Modelle auf der Basis von assoziativer Speicherung bekannter Verläufe der Regelgrößen und dazu passender Stellsignale. Dieser AMS–Ansatz (Associative Memory Systems) wurde von Ersü et.al. entwickelt und benutzt die verteilte Speicherung in überlappenden Speicherzellen. Ein zweiter Ansatz stammt von Barto. Der Regler besteht aus einem ASE (Associative Search Element), das geeignete Stellsignale auswählt, und einer Bewertungskomponente ACE (Adaptive Critic Element), die das ASE steuert. Diese Art des Trainings wird als *Reinforcement learning* bezeichnet. Es handelt sich um unüberwachtes Lernen, d.h. die Stellsignale, die der Regler in einer bestimmten Situation ausgibt, werden nicht explizit vorgegeben.

Das von uns benutzte Verfahren basiert auf einer Idee von Psaltis et.al., nach der die Regelstrecke als zusätzliche, unveränderliche Schicht eines neuronalen Netzwerks betrachtet wird. Wir haben dieses Verfahren mit einer Lernratenadaption kombinert, um ein lineares zeitinvariantes System dritter Ordnung mit Totzeit zu regeln. Der neuronale Regler ist mit der Regelstrecke in Reihe geschaltet und bestimmt zu jedem Zeitpunkt die für einen bestimmten Sollwert notwendigen Stellgrößen. Zum Training wird ein modifizierter Backpropagation–Algorithmus benutzt, der einen Gradientenabstieg auf einer zeitveränderlichen Fehlerfunktion implementiert. Als Fehler wird die quadrierte Differenz zwischen der aktuellen Regelgröße und dem Sollwert benutzt (Regeldifferenz). Die Verbindungsgewichte werden so verstellt, daß der Fehler minimiert wird. Das Rückpropagieren der Regeldifferenz auf die Ausgänge des neuronalen Netzes erfordert lediglich qualitatives Wissen über die Regelstrecke. Für den allgemeinen Fall eines Mehrgrößensystems ist die benötigte Information gleich dem Vorzeichen der partiellen Ableitungen der Regelgrößen nach den Stellgrößen (Jakobimatrix).

Das oben skizzierte Verfahren zur neuronalen Regelung wurde mit zahlreichen Simulationen zur Regelung eines linearen, zeitinvarianten Systems dritter Ordnung mit unterschiedlichen Totzeiten und Zeitkonstanten getestet. Es zeigte sich, daß der adaptive Regler sofort funktionstüchtig ist, d.h. er regelt und adaptiert sich gleichzeitig an die Regelstrecke. Die Ergebnisse werden anhand von ausgewählten Signalverläufen der Soll- und Istwerte vorgestellt.

---

*Universität Koblenz–Landau, Institut für Physik, Rheinau 3-4, D-5400 Koblenz, e-mail: schiff@infko, Tel.: 0261–9119–358

2 Neuronale Netze als Regler

In diesem Abschnitt wird ein Überblick über existierende Ansätze zur Regelung mit Hilfe neuronaler Netze gegeben. Dieser Überblick soll dem Leser helfen, das von uns benutzte Verfahren richtig einzuordnen und dessen Vorteile zu erkennen. Wir können drei Arten von neuronalen Reglern unterscheiden:

1. Associative Memory Systems,
2. Reinforcement Learning Controller,
3. Backpropagation–Regler.

Die beiden erstgenannten Verfahren können im Rahmen dieses Textes nur skizziert werden. Da wir einen Backpropagation–Regler realisiert haben, werden wir diesen Ansatz ausführlicher behandeln.

2.1 Associative Memory Systems (AMS)

Der AMS–Ansatz wurde von Ersü et.al. entwickelt (vgl. [Ersü, 1982–1988]). Er benutzt einen *lokal verallgemeinernden* Assoziativ–Speicher, um durch Beobachtung der Regelstrecke ein nichtlineares Streckenmodell aufzubauen. Dieses prädikative Modell wird dann zusammen mit einem Gütekriterium dazu verwendet, einen nichtlinearen Regler zu erzeugen, der ebenfalls durch einen Assoziativspeicher realisiert wird. Für den momentanen Zustand der Regelstrecke, der durch eine Folge von Meßwerten der Regelgrößen (oder/und Störgrößen) vor dem Beobachtungszeitpunkt bestimmt ist, wird ein optimierter (diskreter) Stellgrößenverlauf berechnet und zusammen mit dem Meßgrößenverlauf im assoziativen Regelspeicher abgelegt. Der numerische Optimierungsprozeß benutzt dazu den prädikativen Streckenspeicher. Er muß nur dann ausgeführt werden, wenn für einen bestimmten Meßgrößenverlauf kein entsprechender Stellgrößenverlauf gespeichert ist. Im Idealfall wird das prädikative Streckenmodell überflüssig und es werden zu jedem Zeitpunkt vorberechnete Stellgrößenverläufe aus dem Regelspeicher abgerufen.

Nachteilig am AMS–Verfahren ist der hohe Speicheraufwand sowie der hohe Zeitbedarf zur assoziativen Suche und zur Berechnung neuer Stellgrößenverläufe. Dies ist vor allem problematisch, wenn industrielle Prozesse in Echtzeit geregelt werden müssen. Interessant ist, daß Speicher– bzw. Suchaufwand mit dem Rechenaufwand austauschbar sind. Beim Entwurf eines AMS–Reglers muß die Speicherkapazität von Strecken– und Regelspeicher richtig gewählt werden. Der Regelspeicher soll viele bewährte Stellgrößenverläufe aufnehmen können. Gleichzeitig muß jedoch erreicht werden, daß „ähnliche“ Meßgrößenverläufe nicht in verschiedenen Speicherzellen abgelegt werden. Das gleiche gilt für den Streckenspeicher. Einem bestimmten Regelgrößenverlauf (bzw. Meßgrößenverlauf) müssen mehrere ähnliche Stellgrößenverläufe zugeordnet werden. Dieses lokale Verallgemeinern (Generalisierung) wird durch komplizierte Codierungsschemata für die Assoziativspeicher erreicht. Im Laufe der Zeit wurden zwei von Varianten von Assoziativspeichern entwickelt: SOC–AMS und MIAS [Tolle, 1989].

2.2 Reinforcement Learning Controller

Dieses Verfahren geht auf Barto zurück und arbeitet mit zwei neuronenähnlichen Elementen, die auch als *adaptive* Elemente bezeichnet werden. Es ist kein a–priori Wissen über die zu regelnde Strecke erforderlich [Barto, 1983–1987].

Unter Reinforcement Learning versteht man die Bewertung der Auswirkungen einer Stellgrößenänderung auf die Regelgrößen. Die Stellgrößenverläufe werden im Gegensatz zum AMS–Verfahren *nicht* optimiert, sondern erfolgen zufällig. Es handelt sich um *unüberwachtes* Lernen, da lediglich die Regelgüte mit den zufällig gefundenen Stellgrößenverläufen von einem „Kritiker" bewertet wird. Die Stellgrößenverläufe werden den neuronenähnlichen Elementen nicht explizit vorgegeben, d.h. der Regler erhält keinerlei Informationen darüber, wie die Stellgrößen aussehen müßten, um eine hohe Regelgüte zu erreichen. Die Rückmeldung zur Adaption der Verbindungsgewichte besteht lediglich aus einem skalaren (Reinforcement)-Signal, das mit den ACE (Adaptive Critic Element) aus erlaubten Zuständen gebildet wird. In erlaubten Zuständen wird das ASE (Adaptive Search Element) „belohnt", sonst wird es „bestraft".

Der Zustandsraum der Regelstrecke wird mit einem Decoder in sogenannte Boxes zerlegt. Das ASE speichert letztendlich für jeden diskreten Zustand ein optimales Stellsignal, das in den Verbindungsgewichten vom Decoder zum ASE codiert wird. Die Gewichte bestimmen nicht direkt die Stellgröße, sondern die Wahrscheinlichkeit für die Ausgabe einer Stellgröße. Das ASE bildet ähnlich wie beim AMS Verfahren einen Regelspeicher (mit jeweils nur einem Paar von Regelgröße und Stellgröße), der mit Hilfe des Reinforcement-Signals des ACE und einem *Eligibility*-Signal eingestellt wird. Dieses „Eignungs"-Signal hängt von der Zeit ab, während der sich ein zu einer Box gehörendes Gewicht bewährt hat. Jedes Verbindungsgewicht hat einen eigenen Eligibility-Trace. Das ACE kann auch mit dem AMS-Verfahren in Verbindung gebracht werden. Es stellt ein prädikatives Modell für das Reinforcement-Signal dar.

Nachteilig beim Reinforcement Learning ist die willkürliche Unterteilung des Zustandsraums (in boxes) und die komplizierte Berechnung des Eligibility-Signals. Ein bekanntes Beispiel für Reinforcement Learning Controller ist das Balancieren eines Stabes durch Hin- und Herbewegen eines Wagens (oder Seehunds). Es wird in [Barto, 1983] zusammen mit der Theorie vorgestellt.

3 Backpropagation-Regler

Die oben beschriebenen neuronalen Regler haben den Nachteil, daß der Lernprozeß bzw. Suchprozeß im Assoziativspeicher sehr lange dauert. Sie sind daher für Anwendungen mit hoher Geschwindigkeitsanforderung (Echtzeit) schlecht geeignet. Ein adaptiver Regler soll sofort funktionstüchtig sein, d.h. er soll regeln und sich gleichzeitig an die Regelstrecke adaptieren. Der im folgenden beschriebene neuronale Regler erhält diese Eigenschaften, indem ein trainierbares feed-forward Netz in geeigneter Weise in den Regelkreis integriert wird. Zum Lernen wird der sogenannte Backpropagation-Algorithmus (BP) benutzt [Rumelhart, 1986]. Man kann zeigen, daß BP-Netze mit nur einer verdeckten Schicht (hidden layer) in der Lage sind, *jede* beliebige Abbildung von Eingabevektoren auf Ausgabevektoren zu lernen [Hornik, 1989].

In einem Regelkreis sind die Regelgrößen meßbar und es werden entsprechende Stellgrößen gesucht, die im zeitlichen Mittel die Differenz zwischen Ist- und Sollwerten minimieren. Der Regler hat die Aufgabe, aus den Regeldifferenzen geeignete Stellgrößensignale abzuleiten.

Es gibt drei Möglichkeiten, ein BP-Netz in einem Regelkreis als Regler zu betreiben: *indirektes, generalisiertes* und *spezialisiertes* Lernen [Psaltis, 1987]. Die beiden erstgenannten Verfahren identifizieren die Regelstrecke, indem sie Ein- und Ausgänge der Strecke während eines bestimmten Zeitintervalls beobachten (Learning by Examples). Das BP-Netz dient dann als inverses Streckenmodell, das die Sollwerte und Regeldifferenzen in Stellgrößen umformt. Wir beschränken uns hier auf die letztgenannte Regelkreis-Struktur, da sie einen direkt funktionsfähigen Adaptiv-Regler liefert (Learning by Doing).

Beim spezialisierten Lernen (Abb. 1) erhält das BP-Netz Sollwerte $\hat{y}$ und Istwerte (Regelgrößen y). Mit jedem neuen Abtastschritt werden die Regeldifferenzen e bestimmt und als Fehler zunächst durch die Strecke und dann durch das Netz „zurückpropagiert". Die Strecke wird dabei als eine zusätzliche Schicht des BP-Netzes interpretiert, deren Verbindungsgewichte *fixiert* sind. Nur die Verbindungsgewichte des neuronalen BP-Reglers werden mit jedem Abtastschritt verändert. Der

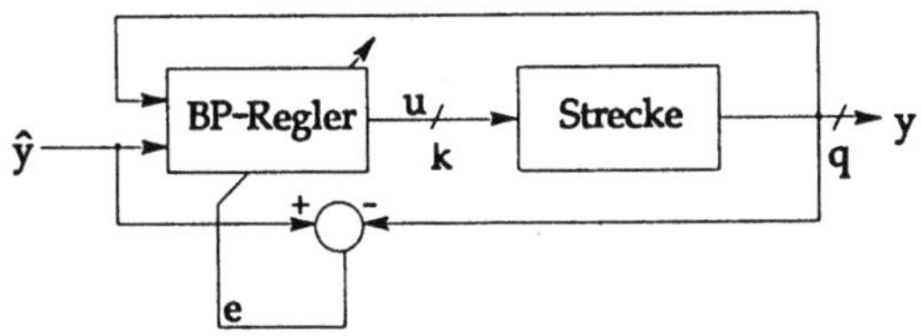

Abbildung 1: Spezialisiertes Lernen eines neuronalen Reglers (BP–Netz) im Mehrgrößen-Regelkreis: Regelgrößen y; Sollwerte ŷ; Regeldifferenzen e; Stellgrößen u

gesamte Regelkreis kann als ein großes BP–Netz aufgefaßt werden, das Ist- und Sollwerte an seinem Eingang auf den Nullvektor der Regeldifferenzen abbilden soll. Nur der Reglerteil kann mit Hilfe des BP–Algorithmus modifiziert werden. Dadurch wird gleichzeitig die Strecke identifiziert und geregelt.

Bei herkömmlichen Digital–Reglern geht die Abtastzeit in die Reglerparameter ein. Da es bei BP–Reglern keine vergleichbaren Parameter gibt, kann die Abtastung und Verarbeitung der Regeldifferenzen in beliebigen Zeitabständen erfolgen. Die verfügbare Rechenleistung wird somit jederzeit optimal genutzt, da keine Notwendigkeit zur Echtzeitverarbeitung besteht. Die Adaption des BP–Reglers an den momentanen Arbeitspunkt der Strecke erfolgt natürlich umso schneller, je mehr Abtastwerte pro Zeiteinheit zu Gewichtsveränderungen verarbeitet werden können. Das bedeutet, daß die Adaptionsgeschwindigkeit an wechselnde Streckeigenschaften direkt zur verfügbaren Rechnerleistung proportional ist.

Der BP–Algorithmus ermöglicht das Lernen von geschichteten feed–forward Netzen, die verdeckte Schichten (hidden layer) enthalten dürfen. Er minimiert schrittweise den quadratischen Fehler der Abbildung von Beispielen zur gewünschten Ein–/Ausgabe des Netzes (überwachtes Lernen). Für jedes Beispielpaar eines Eingabe- und Ausgabevektors verändert man die Verbindungsgewichte in Richtung der negativen partiellen Ableitung des quadratischen Fehlers E bezogen auf das betreffende Verbindungsgewicht. Hierdurch wird ein Gradientenabstieg im Fehlergebirge über den Verbindungsgewichten realisiert. Die Schrittweite, mit der die Gewichtsänderungen vorgenommen werden, bezeichnet man als Lernrate η. Für den Regelkreis gilt:

$$E = \frac{1}{2} \sum_q (y_q - \hat{y}_q)^2$$

und $$w_{ij}(t+1) = w_{ij}(t) - \eta \frac{\partial E}{\partial w_{ij}(t)}$$

BP arbeitet in zwei Phasen. Im *forward-pass* wird der anliegende Eingangsvektor durch das Netz zu einem Ausgangsvektor transformiert. In unserem Fall ist die Regelstrecke die letzte Schicht des Netzes (Abb. 1). Im Arbeitspunkt wird jede einzelne Regelgröße als Ausgang einer linearen Unit modelliert, deren Verbindungsgewichte sich aus der partiellen Ableitung $\partial y_q / \partial u_k$ der Regelgrößen zu Stellgrößen ergeben. Es handelt sich also um eine lineare Approximation der Strecke im Arbeitspunkt.

Im *backward-pass* erfolgt das eigentliche Lernen, indem der am Ausgang angetroffene Fehlervektor (hier Regeldifferenzen) zurückpropagiert wird. BP benutzt hierzu die verallgemeinerte Delta–Regel. Die Fehler in der Ausgangsschicht des Regel*kreises* werden wie folgt berechnet (vgl. [Rumelhart, 1986]).

$$\delta_q = (\hat{y}_q - y_q) \cdot \frac{\partial\, o_q}{\partial\, net_q}$$

Hierbei bezeichnet net_q die Netzeingabe der einzelnen Ausgabe–Units (Regelgrößen):

$$net_q = \sum_k \frac{\partial\, y_q}{\partial\, u_k} \cdot u_k$$

Da die Aktivierungsfunktion mit der Netzeingabe übereinstimmt, ergibt sich in der Ausgabeschicht:

$$\delta_q = (\hat{y}_q - y_q)$$

Dieser Fehler wird nun schichtweise zurückpropagiert. Für die Ausgangsschicht des BP–Reglers gilt:

$$\delta_k = f'(net_k) \cdot \sum_{x \in S(k)} \delta_x \cdot w_{jx} \tag{1}$$

Hierbei bezeichnet $S(k)$ die Menge der Nachfolger der k Stellgrößen–Units und $f'(net_k)$ die Ableitung der Aktivierungsfunktion nach der Netzeingabe. Meist wird die sigmoide Aktivierungsfunktion $o = 1/(1 + e^{-net})$ benutzt. Für sie gilt:

$$f' = \frac{\partial\, o}{\partial\, net} = o(1 - o)$$

Damit ergibt sich für die Gleichung 1:

$$\delta_k = o(net_k) \cdot (1 - o(net_k)) \sum_q (\hat{y}_q - y_q) \cdot \frac{\partial\, y_q}{\partial\, u_k} \tag{2}$$

Die Änderung der Verbindungsgewichte wird nach der verallgemeinerten Delta–Regel wie folgt bestimmt:

$$\Delta w_{ij} = -\eta \frac{\partial\, E}{\partial\, w_{ij}} = \eta \cdot \delta_j \cdot o_i \tag{3}$$

Die angegebenen Gleichungen (1) und (3) für die k Stellgrößen–Units (Ausgangsschicht) gelten sinngemäß für alle vorangehenden verdeckten Schichten des BP–Reglers.

Die partiellen Ableitungen $\partial\, y_q / \partial\, u_k$ bilden die Elemente der ***Jakobi–Matrix***. Sie sind im allgemeinen nicht vollständig bekannt, d.h. man kennt nicht ihre genauen Werte. In den meisten Fällen sind jedoch die Vorzeichen der partiellen Ableitungen bekannt und über den gesamten Arbeitsbereich konstant. Das bedeutet, daß der Zusammenhang zwischen einer Regelgröße und einer Stellgröße monoton ist, und daß man weiß, in welcher Richtung eine Regelgröße durch das Vergrößern einer bestimmten Stellgröße verändert wird. Ersetzt man die partiellen Ableitungen $\partial\, y_q / \partial\, u_k$ in (2) durch ihre Vorzeichen, so kann mit minimalem a–priori Wissen über die Regelstrecke ein BP–Netz als neuronaler Adaptiv–Regler eingesetzt werden [Saerens, 1989]. Die verbleibende Unsicherheit über die exakten Werte der Jakobi–Matrix wird durch den Regelkreis ausgeglichen.

Wichtig für die praktische Anwendung des oben beschriebenen BP–Reglers ist eine schnelle Konvergenz des Gradientenabstiegs. Die Geschwindigkeit von Backpropagation kann erhöht werden, wenn die Lernrate η an die lokalen Eigenschaften des Fehlergebirges angepaßt wird. Ein erster derartiger Ansatz ist die Einführung eines Trägheitsanteils α (momentum term) in Δw_{ij}, um Oszillationen in engen Schluchten zu vermeiden und die Konvergenz in flachen Plateaus des Fehlergebirges zu erhöhen:

$$\Delta w_{ij}(t) = \eta \cdot \delta_j \cdot o_i + \alpha \cdot \Delta w_{ij}(t-1)$$

Üblicherweise wird $\alpha = 0,7 \ldots 0,9$ gewählt [Rumelhart, 1986]. Ein anderes Verfahren zur Adaption der *globalen* Lernrate η wird in [Chan, 1987] vorgeschlagen.

Wenn die Zahl der Verbindungsgewichte nicht zu groß ist, kann eine *lokale* Lernratenadaption erfolgen. Dieses in [Jacobs, 1988] und [Silva, 1990] beschriebene Verfahren sieht für jedes Verbindungsgewicht eine eigene Lernrate vor. Wenn das Verbindungsgewicht in zwei aufeinanderfolgenden Lernschritten sich nicht in die gleiche Richtung verändert wird, verringert man die Lernrate dieses Verbindungsgewichts um einen gewissen Prozentsatz. Damit sollen Oszillationen der Verbindungsgewichte, die in tiefen schmalen Tälern auftreten können, vermieden werden. Wird ein Verbindungsgewicht in zwei aufeinanderfolgenden Lernschritten in die gleiche Richtung verändert, so erhält man die entsprechende Lernrate von einem bestimmten Prozentsatz. Die Simulationsergebnisse in [Silva, 1990] zeigen, daß ein down–Faktor von 0,7 und ein up–Faktor von 1,05 zu empfehlen ist, und daß zusätzlich ein Momentum benutzt werden sollte.

4 Simulationsergebnisse

Um den BP–Regler zu testen und den Einfluß der Lernratenadaption zu untersuchen, wurde eine lineare Regelstrecke 3. Ordnung mit Totzeit simuliert (PT_3 T_t). Das Streckenmodell arbeitet mit einer Abtastrate von 1 s. und ist durch folgende Parameter bestimmt: $T_1 = 3s$, $T_2 = 6s$, $T_3 = 9s$ und $K_p = 1$. Der Stellgrößenbereich ist von 0 bis 100 begrenzt.

Es wurden verschiedene Netzwerktopologien untersucht, um ihren Einfluß auf die Stabilität und Adaptionsgeschwindigkeit zu ermitteln. Generell kann festgestellt werden, daß der BP–Regler sehr empfindlich ist gegenüber den Anfangswerten der Gewichte. Abhängig von der Anfangsinitialisierung neigt der Regler bei einer großen Zahl von Schichten zu Schwingungen. Als Netztopologie für das hier vorgestellte Beispiel wurde ein 1-3-3-3-3-1 feed forward Netzwerk gewählt, dessen aneinandergrenzenden Schichten *fully-interconnected* sind.

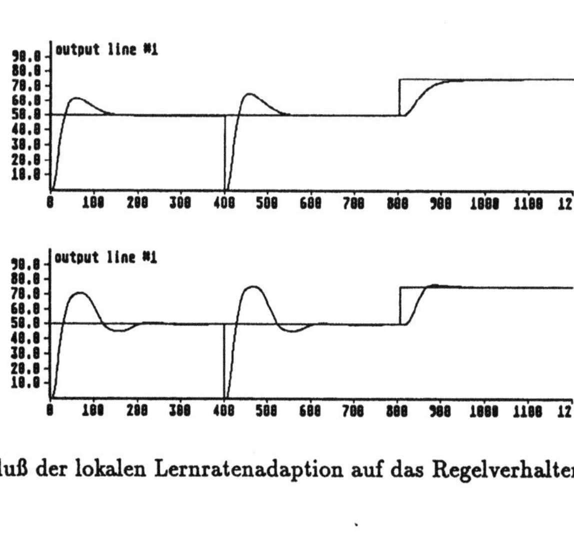

Abbildung 2: Einfluß der lokalen Lernratenadaption auf das Regelverhalten eines BP–Reglers

In Abb. 2 wurde der Einfluß einer Lernratenadaption untersucht. Das obere Signal zeigt den Verlauf der Regelgröße bei fester Lernrate und bei einem Sollwertsprung. Im unteren Teil wurde eine lokal adaptive Schrittweitensteuerung mit den oben angegebenen up/down–Parametern benutzt. Es zeigt sich, daß bei „kleinen" Sollwertänderungen der lokal adaptive BP–Regler schneller ausregelt. Dabei ist ein leichtes Überschwingen erkennbar.

Danksagung

Ich danke Herrn Dipl.–Inform. W. Geffers, der im Rahmen seiner Diplomarbeit das oben beschriebene Simulationsprogramm implementiert hat.

Literatur

Barto, 1983 Barto A. G., Sutton R. S., Anderson C. W.: *Neuronlike Adaptive Elements that can Solve Difficult Learning Control Problems*, IEEE Transactions on Systems, Man and Cybernetics, Vol. SMC–13, No. 5, pp. 834–846

Barto, 1987 Barto A. G.: *An Approach to Learning Control Surfaces by Connectionist Systems* in: Arbib M. A.: *Vision, Brain, and Cooperative Computation*, MIT Press, Cambridge, Massachusetts, London, pp. 665–701

Chan, 1987 Chan L. W., Fallside F.: *An adaptive training algorithm for back propagation networks*, Computer Speech and Language, Cambridge, pp. 205–218, 1987

Ersü 1982 Ersü E., Militzer J.: *Software Implementation of a Neuron–Like Associative Memory System for Control Applications*, Proceedings of the second IASTED Conf. on Mini– and Micro–Computer Applications, Davos

Ersü 1983 Ersü E., Mao X.: *Control of pH by Use of a Self–Organizing Concept with Associative Memories* ACI '83, Copenhagen

Ersü 1988 Ersü E., Tolle H.: *Learning Control Structures with Neuron–Like Associative Memory Systems* in: Seelen, W. et . al. : *Organization of Neural Networks, Structures and Models*, VCH–Verlagsgesellschaft Weinheim

Hornik, 1989 Hornik K.: *Multilayer Feedforward Networks are Universal Approximators*, Neural Networks, Vol. 2, pp. 359-366

Jacobs, 1988 Jacobs R.A.: *Increased Rates of Convergence Through Learning Rate Adaptation*, Neural Networks, Vol.I, Pergamon Press, pp. 295–307, 1988

Psaltis, 1987 Psaltis d., Sideris A., Yamamura A.: *Neural Controllers*, Proceedings of the first ICNN, Vol. 4, San Diego, pp. 551–558

Rumelhart, 1986 Rumelhart D.E. , Hinton G.E., Williams R.J.: *Learning internal representations by error propagation*, Parallel Distributed Processing: Explorations in the Microstructures of Cognition, Vol.I, MIT Press, pp. 318–362, 1986

Saerens, 1989 Saerens M., Soquet A.:*A Neural Controller Based on Backpropagation Algorithm*, Proceedings of the first IEEE International Conference on Artificial Neural Networks, London

Silva, 1990 Silva F.M., Almeida L.B.: *Speeding up Backpropagation*, Advanced Neural Computers, North–Holland, 1990

Tolle, 1989 Tolle H., Militzer J., Ersü E.: *Zur Leistunsgfähigkeit lokal verallgemeinernder assoziativer Speicher und ihrem Einsatzmöglichkeiten in lernenden Regelungen*, msr. Berlin 32 (1989) 3

Zum unterentwickelten Zeitbegriff der Informatik

Wolfgang A. Halang
Reichsuniversität zu Groningen
Fachgruppe Informatik
Postfach 800
NL-9700 AV Groningen

Zusammenfassung

Durch Gegenüberstellung mit der täglichen Praxis wird deutlich, daß die Informatik nur einen sehr rudimentären Zeitbegriff besitzt. Es wird die Notwendigkeit herausgearbeitet, den Begriff Zeit explizit einzuführen. Darauf aufbauend wird ein Konzept zur Synchronisation paralleler Prozesse auf der Basis der Angabe von Zeitfenstern vorgeschlagen. Zur Organisation des zeitlichen Ablaufes ganzer Anwendungssysteme sowie zur Planung der Kooperation und Konkurrenz paralleler Prozesse wird ein graphisches Verfahren vorgestellt. Durch Zusammenstellung und Diskussion ihrer Struktureigenschaften wird gezeigt, daß die Antwortzeitsteuerung das vorteilhafteste und effizienteste Prozessorzuteilungsverfahren ist. Es wird vorgeschlagen, die Zeitgebung in Rechnern von offiziellen, externen Referenzen abzuleiten, wodurch das Problem der Uhrensynchronisation in verteilten Systemen entfällt. Die Zusammenarbeit mit der Physik wird angeregt, um deren hochentwickelten Zeitbegriff in die Informatik zu übertragen.

1. Einleitung

In der Physik kommt der Zeit als unabhängiger Variable vieler (der meisten?) Gesetzmäßigkeiten und als Dimension, in der sich die Änderungen in der Natur vollziehen, fundamentale Bedeutung zu. Demgegenüber ist der Zeitbegriff der Informatik kaum ausgeprägt und eigentlich gar nicht existent — man könnte sogar sagen, daß die Zeit bewußt verdrängt wird. Deshalb erscheint eine ganze Reihe von Konzepten der Informatik sehr primitiv im Vergleich zu Methoden des Alltagslebens zu sein, die gleichen Zwecken dienen. Dieser Beitrag soll in erster Linie dazu dienen, über bisher kritiklos angewandte Methoden zu reflektieren und Anregungen zur Entwicklung neuer — an der Denkweise von Ingenieuren und Physikern orientierter — Verfahren zu geben. Insbesondere wird die Notwendigkeit herausgearbeitet, den Begriff Zeit explizit einzuführen und zu unterstützen und darauf aufbauende Konzepte zur Synchronisation und Organisation paralleler Prozesse zu entwickeln.

Zuerst wird der Zeitbegriff konkretisiert. Dabei wird die Zeit als Dimension unseres Erfahrungsraumes und als absoluter Maßstab erkannt, der es erlaubt, zukünftige Abläufe zu planen. Dann wird obige Aussage mit einigen, wenigen Beobachtungen untermauert. Es ist bezeichnend, daß die Definition des für die Informatik zentralen Algorithmenbegriffes sowie die der Korrektheit der von einem Algorithmus gelieferten Resultate völlig unabhängig von der Zeit sind, weshalb es eigentlich keine informatische Disziplin der Echtzeitsysteme geben könnte. Die zeitliche Ordnung von Prozessen wird auf die Vorher-Nachher-Relation reduziert. Selbst in der Theorie (physisch) paralleler und verteilter Systeme wird die Zeitdimension wegabstrahiert. Spezifikationsmethoden lassen keine absoluten Zeitangaben zu. In funktionellen und logischen Programmiersprachen bestehen noch geringere Möglichkeiten zur Kontrolle der Verarbeitungsgeschwindigkeit als in den klassischen imperativen. Rechnerinterne Uhren sind häufig erschreckend primitiv und ungenau. Zwar wird einem mit der Dimension Zeit zusammenhängenden Schnelligkeitsbegriff größte Bedeutung beigemessen und zum Fetisch von Vertriebsbeauftragten und Anwendern erhoben. Da er jedoch auf nicht vergleichbaren MIPS-Zahlen basiert, verliert er sich im Irrationalen. Als Ergebnis dieser Ignoranz des Zeitbegriffes in der Informatik, die in dem Titel "Time Considered Irrelevant for Real-Time Systems" einer Veröffentlichung [12] von W. M. Turski gipfelt, läßt sich bei frisch diplomierten Informatikern im ersten Jahr ihrer Industrietätigkeit beobachten, daß ihnen im Gegensatz zu Elektrotechnikern und Physikern die Denkkategorie Zeit fehlt.

In Echtzeitumgebungen manifestiert sich die Dimension Zeit in den beiden Hauptforderungen *Rechtzeitigkeit* und *Vorhersehbarkeit*, die auch unter Worst-Case-Bedingungen gelten müssen. Neben vielen anderen Ursachen führen jedoch vor allem Synchronisationsoperationen zu Systemreaktionen mit unvorhersagbaren Anfangszeiten und Dauern. Während einige Synchronisationsprimitive über Wartezeitüberwachungsmechanismen verfügen, erlaubt kein Synchronisationskonzept die Angabe — und damit verbunden die Forderung nach Einhaltung — von Zeitpunkten zur Durchführung bestimmter Operationen. Die Synchronisationskonzepte der Informatik schneiden im Vergleich mit solchen aus anderen Bereichen und insbesondere auch aus dem Alltagsleben, wie z.B. dem Zusammenspiel in der Musik durch Angabe eines strengen Taktes, der Regelung des Strassenverkehrs durch Ampeln, der Verabredung von Besprechungs- und Sitzungsterminen, oder mit modernen Produktionsverfahren nach dem Kanban-System ("Just-in-Time") auf Grund ihrer Primitivität äußerst schlecht ab. Deshalb kann die Betrachtung der Alltagsmethoden hilfreich bei der Definition neuer Konzepte zur Prozeßsynchronisation sein.

So führen Analogiebetrachtungen zum Vorschlag der gegenseitigen Synchronisation paralleler Prozesse auf der Grundlage der Angabe von Zeitfenstern. Die grundlegende Idee dieser Synchronisationsmethode ist, daß die parallelen Prozesse Zeitfenster für zukünftige Betriebsmittelzugriffe anfordern. Das Betriebssystem prüft die Realisierbarkeit der Anforderungen gemäß der Betriebsmittelcharakteristika und verwaltet Belegungspläne. Eventuell auftretende Konflikte werden auf der Basis relativer Dringlichkeit aufgelöst und zu nicht zugeteilten Zeitfenstern werden Alternativen angeboten. Falls notwendig, lassen sich Zeitfenster schließlich noch umverteilen. Auf diese Art und Weise können genaue Zeitinformationen in den Anwenderprogrammen zur Verbesserung der Ergebnisse, wie z.B. von Regelungen, eingesetzt werden und es wird die volle Vorhersehbarkeit ihres Zeitverhaltens erzielt. Geeignete Prozeduren zur Implementation des vorgeschlagenen Synchronisationsverfahrens werden in Form des sichtbaren Teils eines Ada-Paketes vorgestellt. Die Synchronisation des Ablaufs paralleler Prozesse auf der Basis von Zeitfenstern hat eine Reihe bedeutender Vorteile. So ist die Methode problemorientiert und entspricht der Denkweise des Benutzers, was zur erhöhten Zuverlässigkeit der Programme beiträgt, Verzögerungen unbestimmter Länge sind ausgeschlossen, die Zuverlässigkeit und Vorhersehbarkeit der Programmausführung wird verbessert und weiterhin wird die aktive Teilnahme der parallelen Prozesse an der Auflösung von Konflikten ermöglicht.

Basierend auf dem oben eingeführten Synchronisationsverfahren mit Hilfe von Zeitfenstern und bei Kenntnis der maximalen Ausführungszeiten der Prozesse und Teilprozesse in einem Anwendungssystem läßt sich eine ganz einfache und naheliegende Methode zur graphischen Ablauforganisation ("Programming in the Large") angeben. Entsprechend ihrer Aktivierungszeiten werden die einzelnen Prozesse als Balken in ein Zeitdiagramm eingezeichnet. Vorgänger-Nachfolger-Relationen lassen sich auf Aktivierungszeiten abbilden, da obere Schranken für die Laufzeiten der einzelnen Prozesse bekannt sind. Die Betrachtung der gegenseitigen Abhängigkeiten und der Zeitschranken für die Beendigung der einzelnen Prozesse erlaubt die Bestimmung der minimal nötigen und maximal möglichen Anzahl physischer Prozessoren zur Ausführung eines Prozeßsystems. Dieses graphische

Verfahren erlaubt die Behandlung von Zuteilungsfragen in anschaulicher Art und Weise unter zwangloser Berücksichtigung der sich als völlig unkompliziert darstellenden Planung der Kooperation und Konkurrenz paralleler Prozesse. Da diese Planung relativ zum Maßstab Zeit erfolgt, ergibt sich die Vorhersehbarkeit des Programmverhaltens von selbst. Für die zeitliche Planung kann auf bewährte Verfahren der Wirtschaftsmathematik und des Operations Research zurückgegriffen werden.

Weiterhin soll eine Lanze für den praktischen Einsatz zeit- und damit anwendungsorientierter Zuteilungsverfahren gebrochen werden, um so zur lange überfälligen Ablösung von Verfahren ("Prioritäten") beizutragen, die sich nur durch ihre leichte Implementierbarkeit auszeichnen. Durch Zusammenstellung und Diskussion ihrer Struktureigenschaften wird gezeigt, daß die Antwortzeitsteuerung das vorteilhafteste und effizienteste Prozessorzuteilungsverfahren ist. Insbesondere läßt es zu, das Systemverhalten vorhersehbar zu gestalten und anwendungsorientierte Überlasthandhabungsverfahren zu definieren. Diese erlauben lastabhängige, adaptive und dynamische Zuteilung auf der Grundlage eines Kriteriums zur zeitgerechten Verarbeitbarkeit, das zukünftige Überlastfälle so frühzeitig wie möglich zu erkennen gestattet.

Schließlich werden wir noch anregen, die von offiziellen Eichinstituten über Rundfunk angebotenen Zeitinformationen zur Synchronisation rechnerinterner Zeitgeber zu verwenden. Dadurch lassen sich auf einfache Weise die wegen Software-Steuerung recht ungenauen Rechneruhren ablösen. Das Problem der Uhrensynchronisation in verteilten Systemen verschwindet sogar völlig.

2. Der Zeitbegriff in der Informatik

Die Primitivität der Begriffes von Zeit in der Informatik ist umso verwunderlicher, als Zeit ein sehr klares Konzept ist, mit dem wir im täglichen Leben mit großer Selbstverständlichkeit umgehen. Sie ist mehr als nur eine physikalische Dimension, in welcher sich die Abfolge von Ereignissen darstellen läßt. So wie im täglichen Leben gebraucht, ist Zeit ein in hohem Maße allgemeines and abstraktes Konzept, das zur physikalischen Umwelt einen vereinheitlichenden Rahmen von Ordnung und Regelmaß hinzufügt, der für die Organisation der Gesellschaft eine lebenswichtige Funktion hat, indem er Mittel bereitstellt für:

Orientierung Da hinreichend abstrakt, bietet der Zeitbegriff ein Referenzsystem zur Beschreibung der historischen Aspekte der Welt, in gleicher Weise wie deren geographische Aspekte mit Raumkoordinaten angegeben werden können. Zeit ist auch die Dimension, in der sich zukünftige Entwicklungen annehmen, planen oder vorhersagen lassen, womit sie in aktiver Weise mit der Gegenwart in Verbindung gebracht werden.

Einordnung Schon früh lernt jeder Einzelne, sein Verhalten den Zeitbedingungen anzupassen, die die Gesellschaft setzt. Diese Zeitdisziplin ist eine Voraussetzung für die erfolgreiche Integration in unsere Umgebung.

Koordinierung Die von der Uhr gegebene objektive Ordnung ist zur sozialen Institution gewachsen und lebenswichtig zur Organisation der Gesellschaft als Ganzes geworden. Sie stellt ein mächtiges Hilfsmittel zur Koordinierung des Verhaltens vieler Menschen dar.

Im Gegensatz zu ihrer gesellschaftlichen Bedeutung ist der Zeitbegriff in der Informatik nur gering ausgeprägt und in vielen Teilbereichen sogar nicht existent. Bis in die Gegenwart dominiert das der von-Neumann-Architektur zugrundeliegende Denkmodell die Informatik. Dieses Modell beruht auf einer sequentiellen Ablaufsteuerung. Instruktionen auf elementaren Daten werden eine nach der anderen abgearbeitet. Als einzige Relation ist die Abfolge, d.h. die Vorher-Nachher-Beziehung, auf der Maschinenebene vorhanden. Sie reicht aus, um Programme zu spezifizieren. Deshalb war es bei Einprozessorsystemen nicht erforderlich, den Zeitparameter explizit einzuführen, weshalb Programmausführungszeiten i.a. a priori unbekannt sind und a posteriori nur durch sehr grobe Meßmethoden bestimmt werden können. In systemunabhängigen Sprachen sind absolute Zeitangaben und die zeitliche Überwachung von Aktivitäten bisher noch nicht möglich. Im Rahmen des Sprachentwurfs wurden andere Modelle auf der Basis mathematischer Funktionen und logischer Zusicherungen entwickelt. Daraus entstanden funktionale und logische Programmiersprachen, deren Ziel es ist, von den prozeduralen Aspekten der Programmausführung zu abstrahieren. Deshalb erlauben diese hohen Programmiersprachen noch geringere Kontrollmöglichkeiten der Verarbeitungszeiten als imperative Sprachen. Sogar in parallelen Rechensystemen wird die Zeitdimension durch Abstraktion eliminiert. Der Programmierer hat keine Kontrolle über die Ausführungsgeschwindigkeiten seiner parallelen Prozesse. Insbesondere ist er nicht für die Organisation des Zustandsübergangs von Tasks von 'bereit' nach 'ablaufend' verantwortlich. In der kürzlich erschienenen Übersicht von Arbeiten zur Parallelverarbeitung [6] wird dem Zeitbegriff keine Aufmerksamkeit geschenkt. Als zeitliche Aspekte treten nur Sequentialität und Nichtsequentialität sowie Verarbeitungsgeschwindigkeit in Erscheinung. Da wundert es dann wenig, daß Spezifikationsmethoden keine Möglichkeiten bieten, Handlungen mit absoluten Zeitangaben zu versehen, daß sich die Ignoranz des Zeitbegriffes in Titeln von Zeitschriftenartikeln wie dem oben zitierten von Turski äußert und daß auch die rechnerinternen Uhren häufig erschreckend primitiv und ungenau sind. Damit wird dann die Uhrensynchronisation in verteilten Systemen zum Problem, das mit verschiedenen und meistens recht komplizierten Methoden angegangen wird. Wir werden später zeigen, daß das letztgenannte Problem eigentlich nicht existiert, da hochgenaue Funkuhren zur Verfügung stehen.

Allein der *Echtzeitbetrieb* unterscheidet sich von der allgemeinen Datenverarbeitung durch das explizite Hinzutreten der Dimension *Zeit*. Dieser wird in der Norm DIN 44300 "Informationsverarbeitung" vom März 1972 unter der Nummer 161 (bzw. als "Realzeitverarbeitung" unter der Nummer 9.2.11 in einem neueren Entwurf der Norm vom Oktober 1985) folgendermaßen definiert:

> *Ein Betrieb eines Rechensystems, bei dem Programme zur Verarbeitung anfallender Daten ständig betriebsbereit sind, derart, daß die Verarbeitungsergebnisse innerhalb einer vorgegebenen Zeitspanne verfügbar sind. Die Daten können je nach Anwendungsfall nach einer zeitlich zufälligen Verteilung oder zu vorherbestimmten Zeitpunkten anfallen.*

Die in dieser Betriebsart arbeitenden Digitalrechner haben mithin die Aufgabe, Programme auszuführen, die mit externen technischen Prozessen assoziiert sind. Die Verarbeitung der Programme muß zeitlich mit den in den externen Prozessen auftretenden Ereignissen synchronisiert werden und muß schritthaltend mit diesen Prozessen erfolgen, was sich in den folgenden beiden Hauptanforderungen des Benutzers an Echtzeitsysteme ausdrückt, die auch unter extremen Lastbedingungen erfüllt sein müssen:

- **Rechtzeitigkeit** und
- **Vorhersehbarkeit.**

Auf Anforderung durch den externen Prozeß müssen die Erfassung und Auswertung von Prozeßdaten sowie geeignete Reaktionen pünktlich ausgeführt werden. Dabei steht nicht die Schnelligkeit der Bearbeitung im Vordergrund, sondern die Rechtzeitigkeit der Reaktionen innerhalb vorgegebener und *vorhersehbarer* Zeitschranken. Echtzeitsysteme sind mithin dadurch charakterisiert, daß die funktionale Korrektheit eines Systems nicht nur vom Resultat einer Berechnung bzw. einer Verarbeitung, sondern auch von der Zeit abhängt, wann dieses Resultat produziert wird. Korrekte Zeitpunkte werden von der Umwelt der Echtzeitsysteme vorgegeben, d.h. diese Umwelt kann nicht wie von Batch- und Time-Sharing-Systemen dazu gezwungen werden, sich der Verarbeitungsgeschwindigkeit von Rechnern unterzuordnen.

Die Vorhersehbarkeit des Systemverhaltens ist von zentraler Bedeutung für den Echtzeitbetrieb. Sie ergänzt die Rechtzeitigkeitsforderung, da letztere nur dann garantiert werden kann, wenn das Systemverhalten exakt vorhersehbar ist, und zwar sowohl in der Zeit als auch bzgl. der Reaktionen auf externe Ereignisse. Da das zeitliche Verhalten heute verfügbarer Rechensysteme höchstens in Ausnahmefällen vorhersehbar ist, fehlt noch die technologische Basis für eine systematische a-priori-Bewertung von Anwendungssystemen. Damit ist auch eine wichtige Voraussetzung für die sicherheitstechnische Abnahme von rechnergestützten Steuer- und Regelsystemen noch nicht erfüllt.

3. Synchronisation mit Hilfe von Zeitfenstern

Die in der Informatik gebräuchlichen Synchronisationshilfsmittel sind den von Menschen angewandten Praktiken gegenüber deutlich unterlegen. Durch den Vergleich der Semaphor- und Rendezvous-Konzepte mit ihren Pendants aus dem täglichen Leben werden wir nun zeigen, daß dies auf dem Fehlen der Zeitdimension beruht. Führt ein Prozeß eine P-Operation auf einem Semaphor aus, so muß er so lange warten, bis ein anderer Prozeß eine V-Operation bzgl. desselben Semaphors beendet hat. Die Wartezeit ist durch keine Schranke begrenzt und ist i.a. unbestimmt. Ein Semaphor ist mit einer Verkehrsampel vergleichbar (im Spanischen heißt eine Ampel auch semaforo). Bei letzterer ist die maximale Wartezeit jedoch beschränkt, um den Verkehrsfluß sinnvoll zu regeln. Ein Rendezvous im täglichen Leben ist ein Arrangement, das durch die betroffenen Personen, einen vereinbarten Ort und eine abgesprochene Zeit bestimmt ist. Dagegen fehlen im gleichnamigen Synchronisationskonzept der Sprache Ada zwei dieser Parameter. Somit wird das Ada-Rendezvous zum "blind date", das zu unbestimmter Zeit stattfindet.

Im Gegensatz zu diesen primitiven Methoden wird die Zeit im täglichen Leben sehr häufig in effizienter Weise zur Koordinierung von Vorgängen aller Art benutzt. An dieser Stelle soll nur an Fahr- und Flugpläne, Platzreservierungen und das Kanban-Produktionssteuerungssystem erinnert werden. Durch diese Methoden werden die Abläufe planbar und vorhersehbar und Wartezeiten lassen sich zumindest reduzieren.

Basierend auf der obigen Analyse wollen wir nun ein Konzept zur Synchronisation paralleler Prozesse in Echtzeitsystemen vorstellen, das expliziten Gebrauch von der Zeit macht. Wir betrachten eine Menge von Tasks, die auf gemeinsame Betriebsmittel zugreifen. Zur Vermeidung von Konflikten stelle das Betriebssystem einen Verwaltungsdienst zur Verfügung, dem die Synchronisation der Zugriffsoperationen obliege. Dazu unterhält dieser für jedes Betriebsmittel eine Liste von Zeitfenstern, was den Zeilen für Stunden oder Tage in einem Kalender entspricht. Solche Fenster sind Zeitintervalle, für die das entsprechende Betriebsmittel reserviert werden kann. Ihre Längen sind gerätespezifisch. Eine Task kann nun zu beliebiger Zeit vor dem eigentlichen Gebrauch ein Betriebsmittel unter Angabe des oder der Zeitfenster, in denen der Zugriff erfolgen soll, anfordern. Durch Auswertung der entsprechenden Liste im Betriebssystem wird entschieden, ob die Zugriffe zugelassen werden können. Dabei können natürlich Konflikte auftreten, die durch Betrachtung der relativen Dringlichkeiten der Tasks aufgelöst werden:

- Ist die Dringlichkeit einer anfordernden Task geringer als die der Task, der bereits ein bestimmtes Zeitfenster zugewiesen wurde, so wird die Anforderung abgewiesen und es werden andere geeignete Zeitfenster vorgeschlagen.
- Hat aber die anfordernde Task eine höhere Dringlichkeit als der bisherige Besitzer eines Zeitfensters, so wird letzteres entzogen und der dringlicheren Task zugewiesen. Der ursprüngliche Besitzer wird über ein Ausnahmesignal informiert.

In jeder Task muß für den letztgenannten Fall eine Ausnahmefallbehandlungsroutine vorhanden sein, die unter den vom Betriebssystem angebotenen Zeitfenstern ein neues aussucht und reserviert. Falls nötig können auch Ausgabewerte der Task unter Verwendung der geänderten Zeitangaben neu berechnet werden. Falls keine neuen Zeitfenster zugewiesen werden können, haben Tasks geringer Dringlichkeit zumindest die Möglichkeit, frühzeitig Ausweichreaktionen einzuleiten, die ohne die belegten Betriebsmittel auskommen. Die sichtbaren Teile der zur Implementierung der oben beschriebenen Funktionalität benötigten Prozeduren werden im folgenden in Form eines Ada-Pakets angegeben.

```
package TIMED_SYNC is
  - - Deklarationen verschiedener device-types
  - - Deklarationen von Prozeduren für diese Typen:
    - - Anforderung einer periodischen Menge von Zeitfenstern:
    procedure SEIZE(DEVICE: device...type; BEGIN_OF_SCHEDULE: clock;
      INCREMENT: duration; END_OF_SCHEDULE: clock);
    - - Freigabe einer periodischen Menge von Zeitfenstern:
    procedure FREE(DEVICE: device...type; BEGIN_OF_SCHEDULE: clock;
      INCREMENT: duration; END_OF_SCHEDULE: clock);
    - - Angabe des letzten belegten Zeitfensters:
    function FREE_AFTER (DEVICE: device...type) return clock;
    - - Angabe des COUNT-ten freien Zeitfensters:
    function NEXT_SLOT_AVAILABLE (DEVICE: device...type;
      COUNT: integer) return clock;
  - - Ausnahmesignale:
    RESERVATION_IMPOSSIBLE: exception;
    RESERVATION_CONFLICT: exception;
    RESERVATION_WITHDRAWN: exception;
end TIMED_SYNC;
```

Die dargestellte Methode illustriert, wie das abstrakte Zeitkonzept die aktive Gestaltung der Zukunft ermöglicht. Gegenüber dem konventionellen Vorgehen hat sie eine Anzahl von Vorzügen:

- größere Problemorientierung, da die Probleme mit Zeitangaben gestellt werden,
- entspricht der Denkweise des Benutzers,
- verbesserte Vorhersehbarkeit,
- keine Wartezeiten unbestimmter Länge,

- erhöhte Verläßlichkeit durch Möglichkeiten zur Prüfung und frühzeitigen Konfliktauflösung,
- Synchronisation und Zuteilung werden auf der gleichen konzeptionellen Grundlage durchgeführt.

4. Zeitliche Planung von Task-Abläufen

Das bisher übliche Paradigma der Ablaufkoordination kooperierender sequentieller Prozesse ist schlichtes und eventuell unbegrenztes Warten aufeinander. Dies ist äußerst primitiv und nur durch den mangelnden Zeitbegriff der Informatik zu erklären. Die Analyse der Planungsmethoden, die wir alle im täglichen Leben, bei der Planung von Produktion und Verkehr sowie von zu koordinierenden Handlungen jeglicher Art völlig selbstverständlich anwenden, lehrt uns, die Zukunft durch die Zugrundelegung des Maßstabs Zeit in exakter Weise aktiv zu gestalten. Was liegt daher näher, als für das "Programming in the Large", d.h. für die Organisation des Ablaufes von Task-Systemen, einfache graphische Balkendiagramme zu verwenden, in denen die Abfolgen der einzelnen Aktivitäten relativ zur Zeitachse eingezeichnet werden? Die Abbildung zeigt ein typisches Beispiel. Für jeden Prozeß und jeden Teilprozeß eines Anwendungssystems ist dafür nur die Kenntnis der entsprechenden Aktivierungszeiten und der maximalen Ausführungszeiten notwendig. Alles andere ergibt sich dann von selbst. Parallele Prozesse werden durch parallele Balken dargestellt. Die Betrachtung ihrer gegenseitigen Abhängigkeiten und der Zeitschranken für die Beendigung der einzelnen Prozesse erlaubt die Bestimmung der minimal nötigen und der maximal möglichen Anzahl physischer Prozessoren zur Ausführung eines Prozeßsystems. Vorgänger-Nachfolger-Relationen werden auf Aktivierungszeiten abgebildet, da obere Schranken für die Laufzeiten der einzelnen Prozesse bekannt sind. Die Synchronsationszeitpunkte und die notwendigen Wartezeiten lassen sich schließlich den Diagrammen entnehmen. Sie stellen wichtige Parameter für Fehlererkennungsmaßnahmen dar. Die graphische Planung der Kooperation und Konkurrenz paralleler Prozesse relativ zur Dimension Zeit ist äußerst anschaulich und darum anwenderorientiert. Im Gegensatz zu den in der Informatik gebräuchlichen Methoden ist sie auch unkompliziert und allgemeinverständlich. Da das Verfahren konstruktiv ist, ergibt sich die für die Echtzeitverarbeitung grundlegende Vorhersehbarkeit des Programmverhaltens von selbst. Alle diese wichtigen Vorteile werden sehr preiswert erkauft, und zwar durch nicht optimale Auslastung der Prozessoren.

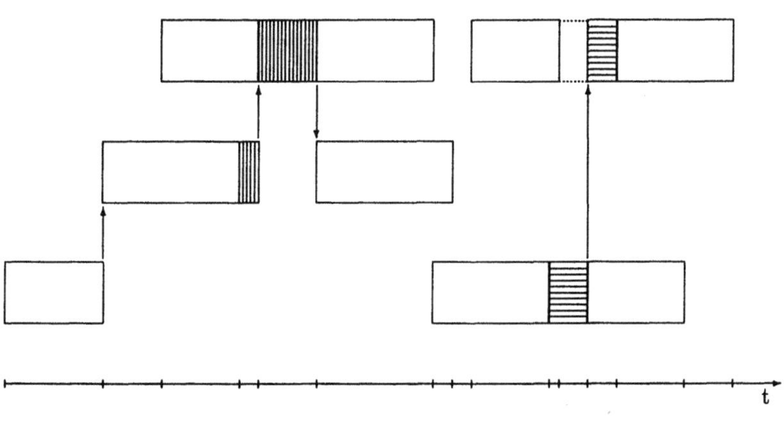

Abbildung. Graphische Planung des zeitlichen Task-Ablaufs

5. Terminbezogene Prozessorzuteilung

Parallele Rechenprozesse (Tasks) innerhalb vorgegebener Zeitschranken auszuführen ist — sofern überhaupt möglich — die grundlegende Anforderung an in harten industriellen Echtzeitumgebungen eingesetzte Prozeßrechensysteme. Algorithmen, die geeignete Prozessorbelegungspläne (Schedules) für alle unter Beachtung ihrer Zeitbedingungen verarbeitbaren Task-Mengen erzeugen, werden *zeitgerecht* genannt. Die allgemeine Situation ist dadurch charakterisiert, daß zu jedem beliebigen Zeitpunkt eine Anzahl ablauffähiger Tasks um die Zuteilung des Prozessors konkurriert, unabhängig davon, ob es sich dabei um sporadische Tasks oder um die Inkarnationen zyklisch aktivierter Tasks handelt. In diesen Zustand gelangen Tasks durch explizite Aktivierung, Fortsetzung oder durch Freigabe von Synchronisationselementen. Zur Zuteilung solcher "freier" Task-Mengen kennen wir für Einprozessorsysteme den Antwortzeitalgorithmus (Earliest Deadline First), den darauf aufbauenden Vorhaltealgorithmus für symmetrische Mehrprozessorsysteme sowie die Zuteilungsmethode gemäß minimalen Spielraums (Minimum Laxity) für beide Typen von Rechensystemen. In [3, 4, 11] wurde gezeigt, daß diese Verfahren zeitgerecht sind. Leider ist die derzeitige Praxis dadurch gekennzeichnet, daß die beiden am häufigsten — bzw. bisher ausschließlich — eingesetzten Zuteilungsverfahren, nämlich First-Come-First-Served und statische Prioritäten, die von allen höheren Echtzeitprogrammiersprachen unterstützt werden und in allen kommerziell verfügbaren Echtzeitbetriebssystemen implementiert sind, *nicht zeitgerecht* sind. Zwar läßt sich zeitgerechte Zuteilung jederzeit mit dynamischen Veränderungen der Task-Prioritäten durch den Benutzer erreichen. Dieses Vorgehen ist jedoch nicht angebracht, weil so die ganze Verantwortung für die Prozessorzuteilung dem Anwendungsprogrammierer aufgebürdet wird. Er könnte sonst seine Software-Module nicht unabhängig von anderen schreiben und benötigte Informationen über alle im System befindlichen Tasks, die normalerweise nicht verfügbar ist. Obwohl er zeitgerecht ist, hat der Spielraumalgorithmus keine praktische Bedeutung, da er präemptiv ist und die gemeinsame Benutzung der Prozessoren durch alle solche Tasks verlangt, die den gleichen, d.h. dann minimalen, Spielraum besitzen. Beide genannten Eigenschaften führen zu häufigen Kontextumschaltungen, die die Systemleistung durch unproduktiven Verwaltungsaufwand deutlich herabsetzen. Schließlich ergeben sich durch die (quasi-) parallele Task-Verarbeitung in vielen Fällen Synchronisationskonflikte in Bezug auf die durch die einzelnen Tasks angeforderten Betriebsmittel. Demgegenüber ist der Antwortzeitalgorithmus nicht präemptiv, wenigstens so lange bis keine weitere Task in den Zustand ablaufbereit übergeht. Das Verfahren erweist sich sogar als optimal [2], wenn man die Anzahl der durch eine Zuteilungsdisziplin erzwungenen Präemptionen als Auswahlkriterium betrachtet. Es bewahrt seine Eigenschaften und erzeugt dann optimale präemptive Belegungen, wenn eine weitere Task während der Bearbeitung einer freien Task-Menge dieser auf Grund des Übergangs in den ablaufbereiten Zustand hinzugefügt wird [7, 8]. Die Übertragung des Antwortzeitalgorithmus auf homogene Mehrprozessorsysteme ist jedoch nicht mehr zeitgerecht. Diese Eigenschaft wird erst durch eine modifizierte Version, und zwar den Vorhaltealgorithmus [3, 4, 5], erzielt, indem auf die Nichtpräemptivität verzichtet und eine deutlich höhere Komplexität in Kauf genommen werden.

Antwortzeitgesteuerte Task-Zuteilung in einem Einprozessorsystem hat eine Reihe von bedeutenden Vorzügen, die wir in der folgenden Übersicht zusammengestellt haben:

- das Antwortzeit- bzw. Fertigstellungsterminkonzept ist problemorientiert,
- das Antwortzeitkonzept erlaubt die Formulierung von Tasks sowie die Erweiterung und Modifikation existierender Software ohne die Kenntnis des gesamten Task-Systems,

- sporadische, periodische und durch Präzedenzrelationen verbundene Tasks können mit einem einheitlichen Zuteilungsverfahren behandelt werden,
- die Zuteilungsdisziplin ist zeitgerecht,
- lineare Komplexität, beschränkte Ausführungszeit und einfache Implementation,
- Antwortzeiten lassen sich dynamisch garantieren wegen der Möglichkeit zur einfachen Überprüfung der zeitgerechten Verarbeitbarkeit einer Task-Menge mit geringer Komplexität, d.h. das Betriebssystem kann die Einhaltung der fundamentalen Rechtzeitigkeitsforderung kontinuierlich überwachen,
- erleichtert frühzeitige Überlasterkennung und -behandlung durch dynamische Lastanpassung und unterstützt so die Forderung nach "graceful degradation" der Systemleistung,
- Minimierung der Anzahl der zur Erzeugung einer zeitgerechten Prozessorbelegung erforderlichen Verdrängungen,
- Maximierung der Prozessorauslastung unter Beachtung der zeitgerechten Verarbeitbarkeit einer Task-Menge,
- ist prinzipiell nicht präemptiv, d.h. Task-Verdrängungen können nur durch Änderung der Konkurrenzsituation hervorgerufen werden, also wenn ruhende Tasks aktiviert oder suspendierte Tasks fortgesetzt werden,
- die Reihenfolge der Task-Ausführungen wird zu den Zeitpunkten der Task-(Re-)Aktivierungen bestimmt und bleibt danach unverändert, d.h. wenn eine neue Task zu einer freien Task-Menge hinzutritt, bleibt die Reihenfolge der anderen untereinander konstant,
- die Ordnung der Task-Bearbeitung ist grundsätzlich sequentiell,
- Betriebsmittelzugriffskonflikte und Verklemmungen werden inhärent vermieden,
- unproduktiver Aufwand wird inhärent minimiert,
- das "Priority Inversion Problem", dem in letzter Zeit in Theoretikerkreisen sehr viel Beachtung geschenkt worden ist, tritt überhaupt nicht auf und
- verdrängbare und (teilweise) nicht verdrängbare Tasks können auf eine gemeinsame Weise zugeteilt werden.

Trotz bester Systemplanung besteht immer die Möglichkeit temporärer Überlastzustände. Diese lassen sich mit Hilfe der folgenden Bedingungen für zeitgerechte Ausführbarkeit von Task-Mengen frühestmöglich erkennen, wodurch der Antwortzeitalgorithmus in geeigneter Weise um eine Zuteilbarkeitsprüfung ergänzt wird. Für jeden Zeitpunkt $t, 0 \leq t < \infty$, und jede Task T mit der Zeitbedingung $t_z > t$, sei

$a(t) = t_z - t$ ihre Antwortzeit,
$l(t)$ ihre (restliche) bis zur Fertigstellung benötigte Ausführungszeit und
$s(t) = a(t) - l(t)$ ihr Spielraum.

Gemäß [3, 4] gelten dann für die zeitgerechte Verarbeitbarkeit einer nach aufsteigenden Antwortzeiten ihrer n Elemente indizierten Task-Menge auf Einprozessorsystemen die notwendigen und hinreichenden Bedingungen:

$$a_k \geq \sum_{i=1}^{k} l_i, k = 1, ..., n. \qquad (1)$$

Erkennt das Betriebssystem beim Hinzufügen einer neuen ablauffähigen Task zur freien Task-Menge, daß mindestens eine der Ungleichungen des Systems (1) nicht erfüllt ist, so ist mit einer Überlastung des Prozessors zu rechnen. Durch die frühzeitige Erkennung ist gewährleistet, daß Überlastsituationen in geordneter und vorhersehbarer Weise unter Beachtung der Forderung nach allmählicher Reduzierung der Systemleistung ("Graceful Degradation") behandelt werden können. Viele Methoden sind denkbar, die für diesen Zweck geeignet sind. Dazu gehören Lastabwurf verbunden mit der Aktivierung von Notfall-Tasks oder die spezifische Leistungseinschränkung innerhalb einzelner Tasks.

6. Externe Zeitreferenzen

Die Uhrensynchronisation in verteilten Systemen erscheint vielen Forschern als ein großes Problem, das mit verschiedenen und meistens recht komplizierten Methoden angegangen wird. Das Gegenteil ist der Fall, da in verteilten Systemen eine gemeinsame, zuverlässige Zeitbasis höchster Genauigkeit dadurch realisiert werden kann, daß jeder Knoten mit einem Empfänger für die offizielle Atomzeit ausgestattet wird, die von den Satelliten des General Positioning Systems (GPS) [9, 10] oder von der Physikalisch-Technische Bundesanstalt in Braunschweig über den Langwellensender DCF 77 [1, 13] kontinuierlich ausgestrahlt werden. Das GPS verbreitet die offizielle Atomzeit des amerikanischen National Bureau of Standards mit Hilfe von 14 Satelliten (18 im Endausbau) über die ganze Erde. Mit Hilfe des Systems, das sich bereits unter schwierigen Bedingungen (Golfkrieg) hervorragend bewährt hat, lassen sich lokale Uhren bis zu 100 nsec genau mit dem internationalen Zeitstandard synchronisieren. Der Langwellendienst der PTB steht schon seit vielen Jahren zur Verfügung und wird von der Industrie, öffentlicher Verwaltung, Bundesbahn usw. genutzt. In letzter Zeit sind sogar Armbanduhren auf den Markt gekommen, die diese Funkuhr als Zeitreferenz verwenden. Damit ist gezeigt, daß die international koordinierte, und damit einzige globale Zeitbasis durch Integration von Empfängerelektronik, lokaler quarzgesteuerter Uhr und entsprechender Synchronisationslogik in einem Schaltkreis preisgünstig in elektronischen Systemen zur Verfügung gestellt werden kann. Es erhebt sich mithin die Frage, warum man sich bei Computern im Gegensatz zu Armbanduhren immer noch mit höchst ungenauen internen Uhren begnügt, die allein auf einem Intervall-Timer beruhen und alle übrigen Funktionen per Software erledigen.

7. Lernen von der Physik

Wir meinen, im wenig ausgeprägten Zeitbegriff der Informatik einen wesentlichen Mangel der wissenschaftlichen Grundlagen der Informatik identifiziert zu haben, und sehen die dringende Notwendigkeit, neue, auf der expliziten Anwendung der Dimension Zeit basierende, Konzepte in der Informatik zu entwickeln. Solche neuen Konzepte müssen in die Programmiersprachen aufgenommen und sowohl von der Hardware als auch von den Betriebssystemen unterstützt werden. Ein klares Verständnis dafür, weshalb und in welcher Weise die Zeit für den Systementwurf von Bedeutung ist, wird als Voraussetzung für eine ausgereifte Methodik der Informatik gebraucht. Es wurde eine kleine Auswahl von Problembereichen angegeben, in denen deutliche Verbesserungen durch einsichtigere und anwendungsorientiertere Konzepte basierend auf einem expliziten Zeitbegriff erzielt werden könnten. Die Auswahl erhebt keinerlei Anspruch auf Vollständigkeit und soll nur als Anregung und als Erläuterung der hier gemachten Aussage dienen.

Zur Erreichung des umrissenen Forschungsziels erscheint eine enge Zusammenarbeit mit der Physik besonders vielversprechend zu sein. Durch Einbringung ihrer Methodik, ihres Erfahrungsschatzes und insbesondere ihrer Sachkompetenz bzgl. der Größe Zeit und deren Darstellung könnte die Physik so einen bedeutenden Beitrag zu den Grundlagen der Informatik leisten. Es wird der Hoffnung Ausdruck gegeben, daß diese Anregung zur Zusammenarbeit Früchte tragen wird und daß die obigen Ideen von der wissenschaftlichen Gemeinschaft aufgegriffen und zusammen mit anderen weiter ausgearbeitet werden.

Literatur

[1] G. Becker, "Die Sekunde", *PTB-Mitteilungen*, 85, 14 – 28, 1975.

[2] R. Henn, "Deterministische Modelle für die Prozessorzuteilung in einer harten Realzeit-Umgebung", Dissertation, Technische Universität München, 1975.

[3] R. Henn, "Zeitgerechte Prozessorzuteilung in einer harten Realzeit-Umgebung", GI — 6. Jahrestagung, Informatik-Fachberichte 5, 343 – 359, Berlin-Heidelberg: Springer-Verlag 1976.

[4] R. Henn, "Feasible Processor Allocation in a Hard-Real-Time Environment", *Real-Time Systems* 1, 77 – 93, 1989.

[5] H. H. Johnson und M. Maddison, "Deadline Scheduling for a Real-Time Multiprocessor", Eurocomp. Conf. Proceedings, 139 – 153, 1974.

[6] E. V. Krishnamurthy, "Parallel processing: principles and practice", Addison-Wesley Publishers Ltd. 1989.

[7] J. Labetoulle, "Ordonnancement des Processus Temps Reel sur une ressource pre-emptive", These de 3me cycle, Université Paris VI, 1974.

[8] J. Labetoulle, "Real Time Scheduling in a Multiprocessor Environment", IRIA Laboria, Rocquencourt, 1976.

[9] W. R. Lange, "Synchronisation verteilter Rechner mit Hilfe von standardisierten Zeitcodes", LANGE-ELECTRONIC GmbH, Olching.

[10] D. C. Robinson, "Using Bus Level IRIG Time Code Translators to Time Tag Data and Synchronize Multiple Processing Stations", Bancomm, Division of Datum Inc., San Jose, CA, U.S.A.

[11] G. Schrott, "Ein Zuteilungsmodell für Multiprozessor-Echtzeitsysteme", Dissertation, Technische Universität München, 1986.

[12] W. M. Turski, "Time Considered Irrelevant for Real-Time Systems", *BIT* 28, 473 – 486, 1988.

[13] "Zur Zeit", Physikalisch-Technische Bundesanstalt, Braunschweig und Berlin, 1989.

Information und der Begriff der Wirkung

Klaus Kornwachs
FHG-IAO, Stuttgart;
Universität Ulm

Abstract

Die Ansätze, über die bisherige Kritik an der Shannonschen Entropie als Informationsbegriff hinauszugelangen bis hin zum Begriff der Pragmatischen Information, werden kurz dargestellt. Es wird erläutert, inwiefern die Pragmatische Information die Dimension einer Wirkung haben könnte. Der pragmatische Informationsbegriff ist nur sinnvoll im Kontext einer dazugehörenden Systemtheorie zu definieren. In speziellen Fällen sind dies physikalische Systeme. In anderen Fällen können dies auch symbolverarbeitende Systeme sein. In beiden Fällen treten komplementäre Begriffspaare auf, die die gleiche formale Grundstruktur aufweisen. Die Theorie der Pragmatischen Information ist bei weitem noch nicht abgeschlossen, jedoch verspricht sie, zu einer interessanten Brücke zwischen Physik, Informatik und einem revidierten Informationsbegriff zu werden.

1 Einleitung

Der folgende Beitrag bezieht sich auf das Grenzgebiet zwischen Physik und Informatik. Um eine Abgrenzung vorzunehmen, ist es wichtig, den Begriff der "Computational Physics" von dem zu unterscheiden, was hier gemeint ist. Im Bereich der "Computational Physics" wird untersucht, welche physikalischen Probleme sich mit Hilfe moderner Rechenmethoden und Computer durch Steuerung, Auswertung und Simulation von Experimenten und theoretisch-numerischen Berechnungen behandeln lassen. Der hier angesprochene Bereich der physikalischen Betrachtung der Informationsverarbeitung läßt sich grob in drei Teile gliedern:

1. Die Untersuchung der physikalischen Bedingungen der Signal - und Informationsübertragung, die Klärung der dabei auftretenden Prozesse, die Bedingungen für Speicherung, Weiterleitung und Verarbeitung (Berechnung) von Information einschließlich thermodynamischer und synergetischer Betrachtungen (***HAKEN 1988; ROTHSTEIN 1982; ATMANSBACHER 1989; ATMANSBACHER, SCHEINGRABER 1987***).

2. Die von der mathematisch-statistischen Informationstheorie und der mathematischen Theorie der Kommunikation (***SHANNON, WAEVER 1949***) ausgehenden Versuche, die Bedeutung und die Wirkung von Information zu verstehen und Modelle für die Bestimmung hinreichender Kriterien für Wirkung ermitteln zu können (vgl. ***KLIR 1985; THIELE 1972***).

3. Die Versuche, ausgehend von der Allgemeinen Systemtheorie, einen Begriff der Pragmatischen Informationsbegriff zu entwickeln, die im Rahmen von definierten Systemen bzw. Systemklassen die Entstehungsbedingungen, die Verstehensbedingungen und die Wirkung von Information quantitativ und qualitativ zu beschreiben versuchen (***WEIZSÄCKER, E.U. v. 1974; WEIZSÄCKER, C.F. v. 1985; KORNWACHS 1987, 1988, 1990; LUCADOU 1989***).

Obwohl alle drei Ansätze sehr eng zusammenhängen und sich vom formalen Apparat her ergänzen, geht der dritte Ansatz von der Annahme aus, daß Systeme, die in der Physik beschrieben werden und Systeme, in denen Symbole vorkommen oder verarbeitet werden, zunächst voneinander verschieden sind und eine Untersuchung der Bedingungen, unter denen in realen (also physikalischen) Systemen eine Wirkung aufgrund von Symbolen oder deren Verarbeitung festgestellt werden kann, diese Unterschiedlichkeit herausarbeiten muß.

Insbesondere geht der dritte Weg, der hier eingeschlagen wird, davon aus, daß der Entropiebegriff bei *SHANNON* und *WAEVER* (1949) nicht ohne weiteres mit dem thermodynamischen Entropiebegriff gleichgesetzt werden darf (vgl. hierzu auch *ZUCKER 1974*).

1.1 Die Kritik am statistischen Informationsbegriff

SHANNON und *WAEVER* (1949) haben bei ihrer mathematischen Analyse des Informationsbegriffs radikal von der möglichen Bedeutung der Information abgesehen und lediglich eine Quantifizierung der Informations**menge** I_{Sh} vorgeschlagen, wobei sich der mittlere Informationsgehalt eines Zeichens x_i aus einem diskreten Repertoire von n Zeichen aus der Auftretenswahrscheinlichkeit $p(x_i)$ des jeweiligen Zeichens sich nach

$$I_{Sh} = -\sum_{i=1}^{n} p(x_i) \cdot lnp(x_i) \tag{1}$$

bestimmt.[1] Dieses Maß hat vieldiskutierte Eigenschaften[2], und es läßt sich auch für Zeichen, die mit Hilfe einer kontinuierlichen Variablen realisiert werden können, anwenden: Man hat dann statt Häufigkeiten resp. Wahrscheinlichkeiten $p(x_i)$ Wahrscheinlichkeitsdichten $\rho(x)$, die die Wahrscheinlichkeit angeben, daß ein Zeichen aus einem infinitesimal kleinen Intervall $x + dx$ stammt, so daß der mittlere Informationsgehalt für Zeichen aus dem Intervall $[x_0, x_1]$ zu

$$H(x) = -\int_{x_0}^{x_1} \rho(\xi) ln\rho(\xi) d\xi \qquad x_0 \leq \xi \leq x_1 \tag{3}$$

wird.

Die hauptsächliche Anwendung dieses Begriffes findet sich in der Codierungstheorie, der Berechnung von Kanalkapazitäten zur Informationsübertragung bei Kanälen, deren Störungen statistisch beschreibbar sind.

1.2 Modifikationen

Es hat viele Versuche gegeben, den nachrichtentechnischen Informationsbegriff zu modifizieren, um dadurch die Bedeutung einer Nachricht quantifizieren zu können. So haben *KULLBACK* und *LIEBLER (1951)* einen verallgemeinerten Ausdruck für eine "Entropie vom Grade β" eingeführt, die nicht mehr additiv ist, aber für $lim_{\beta \to 1}$ den Shannonschen Ausdruck bis auf eine Konstante enthält.[3]

Eine andere Möglichkeit, zu einem modifizierten Informationsbegriff zu gelangen, besteht darin, die Frage nach der Quantifizierung der Unsicherheit in einer bestimmten Situation nicht mehr mit

[1] *HARTLEY (1928)* ging bei der Begründung für das logarithmische Maß von folgender Überlegung aus: Wenn man die Information dadurch mißt, daß man die Variantenanzahl von Symbolen, die in einer Signalfolge vorkommen, zählt, dann müßte man diese Maße für verschiedene Signalfolgen A und B multiplizieren, um das Gesamtmaß zu erhalten. Bei einer aufeinanderfolgenden Übertragung von zwei Informationen müssten diese Maße aber addiert werden. Da nur für den Logarithmus gilt, daß

$$log(A + B) = logA \cdot logB \tag{2}$$

liegt es nahe, für die Informationsmenge das logarithmische Maß $I_{Hartley} = -logA$ anzusetzen.

[2] Wie z.B. Funktionalität, Additivität, Normierung, Symmetrie und Monotonie

[3] Eine Reihe weiterer modifizierter Maße wird in *KORNWACHS (1987)* diskutiert.

Hilfe des Wahrscheinlichkeitsmaßes, sondern mit Hilfe von anderen Maßen zu beantworten. Hierzu gehören die Fuzzy-Maße (vgl. *KAUFMANN 1975)*, das Plausibilitätsmaß (*SHAFER 1976)* und das Possibilitätsmaß (*HIGASHI, KLIR 1983; KLIR 1985, 1987*).[4]

Die letzte Klasse der Verallgemeinerungen des Shannonschen Informationsbegriffs läßt sich durch die Einführung von Faktoren charakterisieren, die den Quantifizierungen der Erwartung verschiedener Ereignisse ein verschiedenes Gewicht geben sollen. Diese Faktoren beziehen sich auf Nutzen, Kosten und dergleichen und verlangen daher eine weitere semantische Beschreibung der fraglichen Situation. Man kann z.B. zu einem vorliegenden Verteilungsgesetz $\Pi(x_i)$ über Ereignisse x_i, die eine Situation charakterisieren und das probabilsitisch, fuzzy, plausibilistisch oder possibilistisch realisiert sein kann, ein weiteres, subjektives Verteilungsgesetz $\sigma(x_i)$ hinzufügen (vgl. *THIELE 1972*). Dies würde dann den Informationsgehalt relativ zu einer subjektiven Kenntnis ausdrücken.

Es sind hier viele Möglichkeiten und Kombinationen der genannten Ansätze denkbar; die angestrengte Erweiterung des Begriffes in Richtung auf "Bedeutung" wird jedoch nicht geleistet, da

- alle verwendeten Größen und Terme auf der syntaktischen Ebene verbleiben,
- die Einführung faktorieller Bewertungen zur Wichtung von subjektiven oder objektiven Verteilungsfunktionen aus der Situation selbst nicht ableitbar ist - diese Faktoren definieren lediglich das Vorwissen oder Interesse des Empfängers,
- alle Begriffe davon ausgehen, daß die Gleichung

$$Mass(Unsicherheit) = -Mass(Information) \tag{4}$$

  generell gilt und dies lediglich zu Metaaussagen über die Bewertung einer Situation hinsichtlich ihrer Unsicherheit führt,
- von einer zeitlichen Konstanz der betrachteten Verteilungen ausgegangen wird und
- die (formalen) Grundeigenschaften des Shannonschen Ausdrucks erhalten bleiben.

Diese Umstände haben schon früh zu einer radikalen Kritik an der Shannonschen Entropie geführt, wonach es unmöglich sei, mit diesem Maß strukturelle Fragen nach Sender und Empfänger sinnvoll stellen und beantworten zu können.[5]

1.3 Physikalische(r) Informationsbegriff(e)

Im Rahmen der Physik hat die formale Ähnlichkeit des Begriffes der Entropie in der Thermodynamik mit dem Ausdruck in Gleichung (1) bzw. (3) zu einer Debatte geführt, ob im Sinne der Gleichsetzung (4) die Entropie, als einem Maß für die "Unordnung" in einem System, durch "Zuführen" von Information erniedrigt werden kann. *C.F.v. WEIZSÄCKER (1974,1985)* und *EBERT (1974)* haben an Beispielen[6] zeigen können, daß eine Zunahme der Ordnung sehr wohl mit Entropiezunahme verbunden sein kann und damit die intuitive Gleichsetzung von Entropie und "Unordnung" nicht korrekt ist.[7]

---

[4]Eine ausführliche Darstellung der Eigenschaften findet sich ebenda.

[5]"Shannons Werk ist gut für den Übertragungskanal - angesichts des Addierwerks ist die Entropie sinnlos"(*ZEMANEK 1972*).

[6]Vorgang der Kristallbildung und Entstehung eines Planetensystems.

[7]Auf den notwendigen philosophischen Hintergrund dieser Debatte kann hier nicht eingegangen werden (vgl. insbesondere *ZUCKER 1974*).

Physikalisch orientierte Ansätze fragen naturgemäß nach der Wirkung von Information. Für Systeme, deren zeitliche Entwicklung sich in einen deterministischen (N) und eine fluktuativen Term (F) aufspalten läßt, so daß für für die zeitliche Entwicklung des Zustandsvektors $\mathbf{q}$ mit dem Parameter $\tilde{\alpha}$ gelten soll

$$\frac{d\mathbf{q}}{dt} = N(\mathbf{q}, \tilde{\alpha}) + F(t), \tag{5}$$

hat *HAKEN (1988)* vorgeschlagen, daß die Wirkung einer Information auf ein solches System darin besteht, daß sie das System aus einem "neutralen Zustand" $\mathbf{q}_o$ in einen anderen Attraktor $\mathbf{q}_k$ überführt. Je nachdem, wie "wichtig" dieser Zustand im Hinblick auf den "Zweck" des Systems ist, kann die Wichtigkeit der Information bestimmt werden. *HAKEN (1988)* definiert im Rahmen eines solchen Modells die synergetischen Systeme danach, daß sie in der Lage sind, durch das Entstehen von geordneten Strukturen selbst "Bedeutung" im Sinne von Wichtungen zu erzeugen. Im Rahmen dieses Modells wird auch die Vernichtung, Erhaltung und Erzeugung von Information definiert. Bei der Enstehung von Strukuren laufen die Prozesse unter dem Regime des Prinzips der maximalen Entropie ab.[8]

Beim Prozeß der Beobachtung hat *ATMANSBACHER (1989)* aufgezeigt, daß Information beim Transfer ΔI eine quantisierte Größe darstellt, die mit einem endlichen Energiebetrag ΔE in einem endlichen Zeitintervall Δt übertragen wird.[9] Die Bedingungen werden durch das Wirkungsquantum $\hbar$ begrenzt und festgelegt:

$$\Delta I \propto \frac{\Delta E \Delta t}{\hbar} \tag{6}$$

Das legt nahe, daß Information selbst, da sie an eine physikalische Wirkung gebunden ist, die Dimension der Wirkung hat.[10]

Die Trennung in verschiedene Modalitäten von Prozessen hat *PATTEE (1990)* dadurch eingeführt, daß er die informations- und symbolverarbeitenden Systeme und die physikalischen Systeme nach folgenden Merkmalen unterscheidet:

	"Rechnende Systeme"	Physikalische Systeme
Gesetze	Algorithmus als Ablaufgesetz	dynamische Gesetze
Prozesse	ratenunabhäng	ratenabhängig Differentialgleichungen
Art	deterministisch auf der Makroebene	nicht-deterministisch in QM und ThD deterministisch in klass. Physik
Prognose	durch "laufen" lassen	Rechnung, Simulation
Retrodiktion	i.a. nicht möglich	nur klassisch möglich
Zeitumkehr	irreversibel	klassisch: reversibel

Die notwendigen Bedingungen für die Informationsübertragung aus physikalischer Sicht sind formuliert in *BREMERMANN (1982), LEVITIN (1982), LANDAUER (1982), HAKEN (1978)*; sie geben aber keine Bedingungen an, die ein Kriterium abschätzen lassen, unter denen eine **Wirkung** von Information zu erwarten ist.

2 Der Begriff der Pragmatischen Information

Der Begriff der Pragmatischen Information geht davon aus, daß eine Quantifizierung und damit auch eine Bewertung von Information nur über die Wirkung, die sie auf ein System ausübt, geleistet

[8] Vgl. *HAKEN (1988), JAYNES (1978), CHRISTENSEN (1985)*.

[9] Die von *ATMANSBACHER (1989)* behandelten Übertragungsbedingungen für sog. Subquantenprozesse ändern an dieser Festellung nichts.

[10] Darauf hat m.W. *E.U. v. WEISZÄCKER (1974)* erstmals hingewiesen.

werden kann. Deshalb ist der Empfänger das entscheidende System. *E.U.v. WEIZSÄCKER (1974)* hat hierfür den Begriff der Pragmatischen Information vorgeschlagen.[11]

Um die Darstellung knapp zu halten, wird eine eher axiomatische Darstellung gewählt.

Pragmatische Information ist zu definieren anhand eines Systems, das als Quelle der Information und/oder als Empfänger von Information fungiert.

2.1 Systemdefinition

Es gibt verschiedene (unterschiedlich tiefe) Systemdefinitionen, die sich aber zueinander in Verbindung bringen und sich als eine fortlaufende Spezifizierung interpretieren lassen (*KLIR 1969*). Die weitaus größte Klasse von Systemen läßt sich auf folgende verallgemeinerte Systemdefinition zurückführen:

Ein System S sei ein Quintupel

$$S = < \alpha, \phi, \sigma, \pi, T > \tag{7}$$

wobei die Variablen des Systems gegeben sind durch

$$\alpha = \{\alpha_x, \alpha_y, \alpha_z, t\} \tag{8}$$

mit

$$\alpha_x \in x \qquad \text{als der Inputvariablen,} \tag{9}$$

$$\alpha_y \in y \qquad \text{als der Ausgangsvariablen} \tag{10}$$

und

$$\alpha_z \in z \qquad \text{als der Zustandsvariablen.} \tag{11}$$

Das Verhalten des Systems bestimmt sich dann aus der Zustandsüberführungsfunktion δ und der Ausgangsfunktion λ durch

$$\phi = \{\delta, \lambda\} \tag{12}$$

mit

$$\delta = \alpha_x \bigotimes \alpha_z \longrightarrow \alpha_z \tag{13}$$

und

$$\lambda = \alpha_x \bigotimes \alpha_z \longrightarrow \alpha_y \tag{14}$$

Ist ein System zerlegbar in Subsysteme, dann sind diese Subsysteme durch eine Struktur π miteinander verbunden, die sich durch eine Verbindungsmatrix K beschreiben läßt als

$$\pi = \{\mathbf{K_{ij}}\} = \begin{pmatrix} 1 & \kappa_{12} & \dots \\ \kappa_{21} & 1 & \dots \\ \vdots & \vdots & 1 \end{pmatrix} \tag{15}$$

mit der Diagonale aus mit sich selbst verbundenen Subsystemen (als Konvention) und entsprechend einer 1 oder einer 0, je nachdem, welches Subsysysteme mit welchen verbunden sind. Diese Verbindungen können zeitliche, kausale oder andere Verknüpfungen repräsentieren.

[11] "Am saubersten wird die Begrifflichkeit, wenn wir den *Begriff Kanal fallenlassen* und jeden Informationsvermittler, der also kein völliges Informationsgrab sein darf, als Empfänger betrachten. Zugleich konzedieren wir damit, daß *jeder Empfänger* alsbald (oder nach einer zeitlichen Verzögerung) wieder Sender wird. Dies gilt dann auch theoretisch für jedes differentiell kleine Stück Kabel oder für jedes Stück Raum, durch das Signale gehen. Diese Aussage erinnert stark an das Huygen'sche Prinzip der Wellenausbreitung. Die zweifellos anschließende elektrodynamische und quantentheoretische Seite dieser Aussage verstehen wir noch nicht."(*WEIZSÄCKER, E.U. 1974, S.92f*)

Die Subsysteme selbst können wiederum weiter (Sub)subsysteme enthalten, so daß sie analog definiert werden können mit

$$\sigma = \{S_s'\} \tag{16}$$

mit

$$S_s' = < \alpha_s', \phi_s', \sigma_s', \pi_s', T > \tag{17}$$

und für die zweite Hierachieebene darunter

$$\sigma_s' = \{S_s''\} \tag{18}$$

mit

$$S_s'' = < \alpha_s'', \phi_s'', \sigma_s'', \pi_s'', T > \tag{19}$$

und so fort (nach *MESAROVIC 1972*).

2.2 Systemoperatoren

Ein System kann eine Quelle von Informationen (im Sinne von Beobachtung) in mehrfacher Weise sein, in dem folgende Größen beobachtet werden können (und/oder):

$$\{\alpha_x, \alpha_y, t\} \quad \text{Black-box-System, t als Zeitvariable} \tag{20}$$

$$\{\alpha_y\} \quad \text{Datenquelle} \tag{21}$$

$$\{\sigma, \pi\} \quad \text{Struktursystem} \tag{22}$$

Um Information über das System (oder vom System) zu erhalten, benutzt man folgende Operationen (inferenzprozeduren), um aus den Beobachtungen Schlüsse ziehen zu können:

$$\mathbf{I} : \{\alpha_x, \alpha_y, t\} \longrightarrow \bar{\phi} \quad \text{Systemidentifikation} \tag{23}$$

$$\mathbf{A} : \{\sigma, \pi\} \longrightarrow \bar{\phi} \quad \text{Systemsynthese} \tag{24}$$

$$\mathbf{E} : \phi \longrightarrow \{\alpha_x, \alpha_y, t\} \quad \text{Systemerklärung} \tag{25}$$

wobei ϕ wie oben (Gleichung (12)) definiert ist, und $\bar{\phi} : \alpha_x(t) \rightarrow \alpha_y(t)$ das nur beobachtete Verhalten darstellt. Es ist einfach von ϕ auf $\bar{\phi}$ durch das Anlegen einer Maske zu schließen, die lediglich die Paare α_x, α_y auch zeitlich anordnet.

Wenn eine vollständige Systembeschreibung S vorliegt, dann kann man auf diese Beschreibung Operationen anwenden, um zu Aussagen über Verhalten und Struktur zu gelangen. Solche Operationen liefern dann

$$V(S) = \{\alpha_x, \alpha_y, t\} \quad \text{Verhaltensoperator} \tag{26}$$

$$C(S) = \{\sigma, \pi\} \quad \text{Strukturoperator} \tag{27}$$

d.h., daß die Anwendung dieser Operatoren beobachtbare Größen liefern soll. Mit den Inferenzprozeduren **I** und **E** kann man nun substituieren:

$$\mathbf{I} : V(S) \rightarrow \bar{\phi} \tag{28}$$

$$\mathbf{E} : \phi \rightarrow V(S) \tag{29}$$

Es ist einleuchtend, daß $\mathbf{I} = \mathbf{E}^{-1}$ und vice versa, wenn $\phi = \bar{\phi}$, d.h. wenn sowohl die Systemidentifikation wie die Systemerklärung vollständig sind. In der Praxis ist das selten der Fall.

Auf der Ebene der Information über die Struktur eines Systems ist

$$C(S) = \{\sigma, \pi\} = \{\kappa_{i,j}, S_i\} \qquad \text{für die erste Ebene} \tag{30}$$

während eine Systemanalyse erfordern würde

$$\mathbf{A} : \{\sigma, \pi, \phi_{S_i}\} \rightarrow \phi, \tag{31}$$

wobei ϕ eine Lösung der fundamentalen Systemgleichung

$$\mathbf{O}_i \alpha_x = \alpha_y - \mathbf{O}_i\{\mathbf{K}_{ij}\}\alpha_y \tag{32}$$

ist, die gestattet, aus der Kenntnis der Struktur $\{\mathbf{K}_{ij}\}$ und dem Verhalten der Elemente das Gesamtverhalten zu bestimmen. $\mathbf{O}_i$ symbolisiert die Operatoren, die die Eigendynamik der i-ten Elemente beschreiben (vgl. *KORNWACHS 1988; ANSCHÜTZ 1970.*

Will man die gesamte Information über das System erhalten, muß man die Operatoren $V(S)$ und $C(S)$ hintereinander auf das System anwenden. Man kann nun zeigen (*KORNWACHS 1988*), daß die Reihenfolge der Anwendung nicht gleichgültig ist, d.h. man bekommt eine andere Information über das System, wenn man zuerst das Verhalten und dann die Struktur beobachtet und Schlüsse daraus zieht, als wenn man dies in der anderen Reihenfolge tut. Betrachtet man die Struktur, muß man die Elementarfunktionen kennen, um zum Verhalten zu gelangen, betrachtet man das Verhalten, so ist dieses Verhalten durch eine Reihe von möglichen Strukturen realisierbar.

Dies verweist darauf, daß die genannten Systemoperatoren nicht vertauschbar sind und es ist die Vermutung geäu—ssert worden, daß Struktur und Verhalten eines Systems im Rahmen einer solchen Systembeschreibung komplementäre Größen darstellen.[12]

2.3 Erstmaligkeit und Bestätigung

Will man die Wirkung von Information objektivieren, bietet sich herkömmlicherweise das Kanalmodell von *SHANNON* an, allerdings nun mit der Modifikation, daß der Empfänger auch Sender werden muß, um die Wirkung der Information beobachten zu können. Da der Sender aber durch seine von ihm erhaltbare Information repräsentiert wird, und der Empfänger ebenfalls durch die Information dargestellt werden kann, die er aufgrund der Wirkung von Information abgibt, scheinen die Beschreibungen von System und Information zwei Modalitäten einer fundamentaleren Größe zu sein. Dies würde bedeuten, daß System und Information ineinander überführbare Konzepte sind.

Anstatt dies näher zu vertiefen, seien zwei Komponenten eingeführt, die eine Bedingung für die Wirkung der Pragmatischen Information darstellen: Erstmaligkeit und Bestätigung. Diese beiden Komponenten sind ausführlich diskutiert und beschrieben in *E.U v. WEIZSÄCKER (1974), KORNWACHS (1987,1988); KORNWACHS, LUCADOU 1982, 1984.* Erstmaligkeit bewertet die Komponente der Information, die den überraschenden, neuartigen Teil enthält, Bestätigung den Teil, der auf dem Vorwissen des Empfängers aufbaut. Information, die lediglich Bestätigung enthält, löst keine Wirkung aus, ebensowenig eine Information, die nur Erstmaligkeit enthält - dies wäre vergleichbar mit einem weißen Rauschen in der Signaltheorie.

Im Rahmen der oben gegebenen systemtheoretischen Definitionen kann man nun Erstmaligkeit definieren als ein Maß für die größte auftretende Differenz zwischen dem beobachteten Verhalten eines Systems $\bar{\phi}$ und dem aus dem Vorwissen des Empfängers resultierenden erwarteten Verhalten

[12]Komplementarität im formalen Sinne wie in der Physik äußert sich u.a. darin, daß die zugehörigen Operatoren nicht mehr vertauscht werden können und daß das Produkt aus zwei solchen Operatoren die Dimension einer Wirkung hat (vgl. *KORNWACHS 1988, 1989*; weitere Literatur siehe dort).

ϕ_r. Dabei steht der Index r für die möglichen distinkten Erwartungen des Empfängers. Vereinfacht geschrieben:

$$E(S) = \max_r(\phi_r - \bar{\phi}) \quad (33)$$

Stehen nur beobachtete Daten zur Verfügung, ist $\bar{\phi}$ sinngemäß durch α zu ersetzen.

Entsprechend ist die Bestätigung zu definieren. Hätte man die volle Information (symbolisch geschrieben als $\boxed{1}$) über das System, dann wäre ein Maß für die Bestätigung die minimale Differenz zwischen erwarteter und beobachtbarer Struktur. Wiederum vereinfacht geschrieben:

$$B(S) = \boxed{1} - \min_r(\{\sigma_r, \pi_r\} - \{\sigma, \pi\}) \quad (34)$$

Für bestimmte Systeme und Informations"arten" kann man diese Ausdrücke operationalisieren. Wenn z.B. die Strukturen (erwartete wie beobachtete) in Form von Verteilungen gegeben sind, kann man Distanzmaße benutzen, um die Differenzen auszudrücken.[13] Dies ist bei der Analyse von Simulatonsexperimenten sehr nützlich (vgl. *KORNWACHS 1990*).

2.4 Dimensionsbetrachtung: Information als Wirkung

Bisher wurde die Pragmatische Information als empfangene Information zu quantifizieren versucht. Die Wirkung auf ein System, die sie ausübt, ist, neben den Ansätzen von *HAKEN (1988*, s.o.), weitaus schwerer zu erfassen. Phänomenologisch kann man sich folgende Effekte vorstellen:

- P.I. (Pragmatische Information) verändert die eigene Quantifizierungsbedingung, d.h. sie läßt die Erwartungswahrscheinlichkeiten im Empfänger nicht unverändert - eine Voraussetzung, die der Shannonsche Informationsbegriff nicht macht.
- P.I. verändert die Struktur und/oder das Verhalten des Systems, auf das sie wirkt. Dadurch erzeugt sie neue Information, sofern diese Veränderung beobachtbar ist.[14]
- P.I. erweitert oder reduziert Strukturen, die für den Transfer von P.I. notwendig sind, d.h. verändert Kanalkapazitäten.

Pragmatische Information ist als "Produkt" aus Erstmaligkeit und Bestätigung zu definieren versucht worden. Auch hier ist es möglich zu zeigen, daß die Reihenfolge der Operatoren, die zur Ermittlung der Maße für Erstmaligkeit und Bestätigung führen sollen, nicht gleichgültig ist.

Der vermutete fundamentale Zusammenhang zwischen System und Pragmatischer Information ist durch einen funktionellen Zusammenhang zwischen den Komponenten der Information und der Systembeschreibung vorstellbar, so daß durch

$$E(S) \bigotimes B(S) = \mathcal{F}[V(S) \bigotimes C(S)] \quad (36)$$

bei Kenntnis von $\mathcal{F}$ eine Bestimmung der Wirkung für einfache Beispiele möglich sein müßte.

Folgende Überlegung soll zeigen, daß sowohl die Pragmatische Information als auch das "Produkt" aus Verhalten und Struktur, also all dessen, was man von einem System wissen kann, die Dimension einer Wirkung, also $Energie \cdot Zeit$ haben könnten:

---

[13]So haben *HAI, KLIR (1985)* das Maß für die Differenz zwischen zwei Verteilungen α und β über $\{A\}$ mit

$$d(\alpha, \beta) = \frac{1}{log[card(A)]} \sum_{\xi} \alpha(\xi) \frac{\alpha(\xi)}{\beta(\xi)} \quad (35)$$

mit $\xi \in \{A\}$ angegeben.

[14]*C.F. v. WEIZSÄCKER (1970)*: "Information ist, was Information erzeugt."

1. Wegen der Nichtkommutativität von E und B resp. V und C liegt die Analogie zu den quantenmechanischen Operatoren nahe, für die gilt, daß $[A_1A_2 - A_2A_1] \geq \hbar$, mit $Dim[...] = Dim[\hbar]$.

2. Die Dimension $Dim(A \cdot B) = F(\frac{m \cdot l^2}{t})$, entspricht der Dimension der Wirkung, nämlich *Energie·Zeit*.

Für die Begriffe der Struktur C und und des Verhaltens V kann man sich überlegen, ob beide Bedingungen zutreffen. Jede Eingangs-, Ausgangs- oder Zustandsüberführungsfunktion, die in V beschrieben ist, läßt sich ausdrücken als eine Funktion der Zeit oder von Schrittweiten. In beiden Fällen enthalten sie Information über Änderungen in der Zeit, wie Raten, Zählraten, Geschwingkeit etc. Die Dimension dieser Raten ist t^{-1}. Wenn Struktur begrifflich gesehen keine Zeit beinhaltet, was nicht heißt, daß es keine zeitabhängigen Strukturen geben könnte, dann benötigen Strukturen räumliche Ausdehnung. Die Verbindungsmatrix, die die Verbindung zwischen den Elementen des Systems beschreibt, gibt an, welche Elemente nicht zugleich am selben Ort sein können, da sie entweder getrennt oder durch irgendeine Relation miteinander verbunden sind. Dies impliziert die Dimension einer Distanz, gewissermaßen einer generalisierten Länge. Die Verbindungsmatrix definiert nicht nur die Natur der Verbindungen, sondern auch die Gesamtheit der Elemente und könnte deshalb in ihrer Dimension wie eine verallgemeinerte Masse behandelt werden. Elemente in ihrer Gesamtheit als verallgemeinerte Masse und die Verbindungen (da bilateral) als Quadrat der generalisierten Länge genommen, bedeutet, für die Existenz von raumfüllenden Strukturen in der Physik ein Trägheitsmoment als Voraussetzung auch in der Dimension anzusehen.[15] Deshalb wird angenommen, daß die Struktur eines Systems die Dimension eines Trägheitsmoments hat. Damit wird die Dimension von $V(S) \otimes C(S)$ entsprechend der obigen zweiten Bedingung ebenfalls als die der Wirkung angenommen (vgl. auch *LUCADOU 1989*).

Entspechend den Definitionen für Erstmaligkeit (Gleichung (33) und Bestätigung (Gleichung (34)) kann man weiter analogisieren, daß die Erstmaligkeit die Dimension einer Rate hat, also $Dim(E) = t^{-1}$, während die Bestätigung als die "zeitlosere" Komponente die Dimension des verallgemeinerten Trägheitsmoments $Dim(B) = m \cdot l^2$ haben könnte, da sie sich auch in ihrem semantischen Gehalt eher auf die Struktur bezieht.

3 Ausblick

Die Theorie der Pragmatischen Information bezieht eine Reihe ihrer Einsichten aus Analogien mit der Quantentheorie und der Theorie der nicht-klassischen Systeme (also solche Systeme, die nicht vollständig beschreibbar sind; vgl. *KORNWACHS, LUCADOU 1984*). Es zeigt sich, daß auch eine Reihe anderer Begriffe aus der Systemtheorie sich komplementär in dem geschilderten Sinne zueinander zu verhalten scheinen. Die nachfolgende Tabelle gibt einen groben Überblick (vgl. *KORNWACHS 1989*)[16]:

Struktur	Verhalten
Zuverlässigkeit	Autonomie
Gesetz	Randbedingung
Bestätigung	Erstmaligkeit
Effizienz	Fehlerfreundlichkeit
Unabhängigkeit	Separabilität
Prozessor	Symbol

[15] Dabei wird unterstellt, daß Verbindungen zwischen den Elementen mindestens in einer Fläche realisiert sein müssen, um alle potentiellen Verbindungen aufbauen zu können.

[16] Zum Begriffspaar Effizienz/Tüchtigkeit vs. Fehlerfreundlichkeit vgl. *WEISZÄCKER E.U. und CH. 1984*.

In diesem Sinne kann das Auftreten von solchen Komplimentaritäten als Hinweis gedeutet werden, daß sich in der systemtheoretischen Beschreibung nicht-klassische Strukturen finden lassen, cor allem dann, wenn es sich um Systeme handelt, in denen der Begriff der Information eine Rolle spielt.

Die Begriffe der Pragmatischen Information sind möglicherweise geeignet, einen umfassenderen Informationsbegriff zu gewinnen, der eine Verbindung zwischen einer Theorie der Informatik und der Physik schaffen könnte. Es sollte deshalb versucht werden, weiter interdiszipinär an dieser Theorie zu abeiten, die vermutlich auch tiefere Einsichten in die Art und Weise freigibt, wie wir Systeme definieren und ihre "Oberfläche" festlegen. Denn die Oberfläche eines Systems ist der Ort, an dem, physikalisch wie begrifflich gesprochen, die Information in ihrer Wirkung observabel werden muß (*HINDERER, KORNWACHS 1991*).

Literatur

Anschütz, H.: Kybernetik, kurz und bündig. Vogel-Verlag, Würzburg 1970

Atmansbacher, H.: The Aspect of Information Production in the Process of Information. *Foundation of Physics* 19(1989), p. 553-577

Atmansbacher,H.; Scheingraber, H.: A Fundamental Link between System Theory and Statistical Mechanics. *Foundation of Physics* 17(1987), p. 939-963

Bremermann, M.J.: Minimum Energy Requirement of Information Transfer and Computing. *Int. J. Theoretical Physics* **21** (1982 3/4), p. 203-217

Christensen, R.: Entropy Minimax Multivariate Statistical Modelling. I. - Theories. Int. J. Gen. Systems 11(1985) pp.231-277

Ebert, R.: Enropie und Struktur kosmischer Systeme. In: Weizsäcker, E.U. v. (ed.): Offene Systeme I. Klett, Stuttgart 1974, pp.222-228

Hai, A.; Klir,G.J.: An empirical investigation of reconstructability analysis: Probabilistic Systems. Int. J. Man Machine Studies 22 (1985), pp.163-192

Haken, H.: Synergetics. An Introduction. Springer, Berlin Heidelberg u.a. 1978

Haken, H.: Information and Self-Organization. Springer Berlin u.a. 1988

Hartley, R.V.L.: Transmission of Information. *Bell Syst. Techn. Journal* **7** (1928/3), p.535-563

Higashi, M.; Klir, G.J.: On the notion of distance representing information closeness: possibility and probability distributions. Int.J.Gen.Systems 9(1983) pp.103-115

Hinderer, W.; Kornwachs, K.: Physik der Symbolverarbeitung. *Newsletter der Deutschen Gesellschaft für Systemforschung* **2** (1991), Heft 1, S. 25

Jaynes, E.T.: Maximum Entropy Principle. In: Levine, R.D.; Tribus, M.(eds.): The Maximum Entropy Formalism. MIT Press, Cambridge, Mass. 1978

Kaufmann, A.: Introduction into Theory of Fuzzy Sets. Academic Press, New York San Francisco 1975

Klir, G.J.: An Approach to General System Theory. Van Nostrand, New York 1969

Klir, G.J.: Architecture of Problem Solving. Plenum Press, New York, London 1985

Klir, G.J.: Where do we stand on measures of uncertainty, ambiguity, fuzziness and the like? Fuzzy Sets and Systems 24 (1987) pp.141-160

Kornwachs, K.: Offene Systeme und die Frage nach der Information. Habilitationsschrift Universität Stuttgart 1987

Kornwachs, K.: Cognition and Complementartity. In: M. Carvallo (ed.): Nature, Cognition and System I. Kluwer Ac. Publ., Amsterdam 1988, p. 95-127

Kornwachs, K.: Complementarity Relations within Cognitive System Theory. *Cognitive Systems* **2-3** (1989), p.251-260

Kornwachs,K.: Reconstructability Analysis and its Re-Interpretation in Terms of Pragmatic Information. In: Pichler, F.; Moreno-Dias, R. (ed.): Computer Aided System Theory: Eurocast '89. Lecture Notes in Computer Science 140, Springer, Berlin u.a. 1990, p. 170-181

Kornwachs, K.; Lucadou, W. v.: Pragmatic Information and Non-Classical Systems. In: Trappl, R. (ed.): Cybernetic and Systems Research. North Holland, Amsterdam 1982, pp.191-197

Kornwachs, K.; Lucadou, W. v.: Some Notes on Information and Interaction. In: Trappl, R. (ed.): Cybernetic and Systems Research 2. North Holland, Amsterdam 1984, pp.9-14

Kornwachs, K.; Lucadou, W.v.: Komplexe Systeme. In: Kornwachs, K. (Hrsg.): Offenheit - Zeitlichkeit - Komplexität. Campus, Frankfurt 1984, p.110-165

Kullback, S.; Liebler, R.: On Information and Sufficiency. *Annales for Mathematical Statistics* **22** (1951), p.79-86

Landauer, R.: Uncertainty Principle and Minimal Energy Dissipation in the Computer. *Int. J. Theoretical Physics* **21** (1982 3/4), p. 283-297

Levitin, L.B.: Physical Limitations of Rate, Depth and Minimum Energy in Information Processing. *Int. J. Theoretical Physics* **21** (1982 3/4), p. 299-309

Lucadou, W. v.: Non-Locality in Complex Systems. In: Rosseel, E.; Heylighen, F.; De Meyere, F. (eds.): Selfsteering and Cognition in Complex Systems. Studies in Cybernetics Vol II. Gordon and Breach, N.Y., London 1989

Mesarovic,M.D.: A Mathematical Theory of General Systems. In: Klir, G.J. (ed.): Trend in General System Theory. Wiley, New York u. a. 1972, p. 251-269

Pattee, H.H.: The Measurement Problem in Physics, Computation and Brain Theories. In: Carvallo, M.E. (ed.): Nature, Cognition and Systems II. Kluwer, Amsterdam 1990. Preprint

Rothstein, J.: Physics of Selective Systems: Computation and Biology. *Int. J. Theoretical Physics* **21** (1982 3/4), p. 327- 350

Shafer, G.: A Mathematical Theory of Evidence. Princeton University Press, Princeton, N.J. 1976

Shannon, C.E.; Weaver,W.: The Mathematical Theory of Communication. Urbana, Chicago, London 1949.

Thiele,H.: Einige Bemerkungen zur Weiterentwicklung der Informationstheorie. In: Scharf, H. (Hrsg.): Informatik. *Nova Acta Leopoldina* **37** (1972/1), Nr. 206, p.473-502

Weizsäcker, C.F. v. : Aufbau der Physik. Hanser, München 1985

Weizsäcker, C.F. v.: Evolution und Entropiewachstum. In: Weizsäcker, E.U. v. (ed.): Offene Systeme I. Klett, Stuttgart 1974, pp.200-221

Weizsäcker, E.U. v.: Pragmatische Information. In: Weizsäcker ,E.U. v. (ed.): Offene Systeme I. Klett, Stuttgart 1974, pp.82-113

Weizsäcker, E.U. v.; Weizsäcker, Ch. v.: Fehlerfreundlichkeit. In: Kornwachs, K.(Ed.): Offenheit - Zeitlichkeit - Komplexität. Zur Theorie der offenen Systeme. Campus, Frankfurt-New York 1984, pp. 166-201

Zemanek, H.: Informale und formale Beschreibung (1) und (2). *IBM Nachrichten*, Heft 211 und 212 (1972), p.175-179 und 279-283

Zucker, F.J.: Information, Entropie, Komplementarität und Zeit. In: Weizsäcker,E.U. v. (ed.): Offene Systeme I. Klett, Stuttgart 1974, pp.35-81

Hans-Joachim Dubrau

Bit, Entropie und Information

1. Einleitung

Über 50 Jahre nach der Veröffentlichung Shannons zur Kommunikations-Theorie [Sha 48][Sha 49], die bis etwa 1980 weltweit als allgemeine Informationstheorie angesehen wurde, sind heute die Standpunkte zur Frage, was ist Information und ist sie meßbar bzw. welche ihrer Eigenschaften sind meßbar, unterschiedlicher, als je zuvor. Bereits 1983 stellte Völz [Völ 83] fest, daß man drei Gebiete unterscheiden sollte: Informations-Übertragung (Kommunikation), Informations-Speicherung und Informations-Verarbeitung. Dabei sei durch Shannon die Theorie lediglich für das erste Gebiet, die Informations-Übertragung, geschaffen wurde, was seinen Audruck auch in der Benennung "Kommunikations-Theorie" durch Shannon fand. Für die Informations-Speicherung konstatierte Völz zu dieser Zeit einen Stand, der der Situation auf dem Gebiet der Kommunikation kurz vor Shannons Arbeit entsprach, d.h. er erwartete eine theoretische Fundierung dieses Gebiet in nächster Zeit (die jedoch bis heute nicht erfolgte), während er auf dem Gebiet der Informations-Verarbeitung reinen Pragmatismus sah.

1989 stellten A. und T. Schileyko in ihrem Buch "Gespräche über die Informatik" [Sch3 89] resignierend fest:" Wir müssen nun nicht nur konstatieren, daß wir nicht wissen, in welchen Einheiten wird die Information messen sollen, sondern daß wir überhaupt nicht wissen, was Information ist."

Im "Lexikon der Informatik und Kommunikationstechnik" ist Hans-Jochen Schneider 1990 [Sch1 90] der Meinung "Der Begriff Information wird heute zu abstrakt und allgemein verwendet, so daß die Definitionen des Begründers der Informationstheorie Claude E. Shannon für die Informatik zu speziell sind".

Schließlich zitiert R. Sietmann 1991 [Sie 91] die Feststellung von Jeffrey Wicken, USA,: "Als Resultat unabhängiger Entwicklungslinien in der Thermodynamik und in der Informationstheorie gibt es heute zwei 'Entropien' in der Wissenschaft, das ist eine zuviel."

Diese Zitate sollen nur stellvertretend für eine ganze Reihe ähnlicher Bemerkungen stehen, die immer häufiger in der Literatur auftauchen und die derzeitige Krise beim Verständnis des Phänomens Information einschließlich der damit verbundenen Begriffe Bit und Entropie verdeutlichen.

2. Das Bit

Hans-Jürgen Schneider [Sch2 90] gibt für das Bit zwei Bedeutungen an: "Ein Bit ist ein Zeichen in einer binären Zahlendarstellung, d.h. eine 0 oder 1. ... In der Informationstheorie wird bit (klein geschrieben) als Maßeinheit für den Informationsgehalt einer Nachricht verwendet." Allgemein verbreitet ist die Auffassung, daß das Bit eine Maßeinheit ist ,wie z.B.das Dezibel oder Meter, und bit die entsprechende Schreibweise in Gleichungen analog zu db oder m. So findet man im"Fachlexikon Meßtechnik" [Aut1 84] S.50: "Bit,...: kleinste Einheit einer Information, ...Einheitenzeichen: bit." Allerdings wird weder in diesem Lexikon noch an anderer Stelle eine eindeutige Zuordnung des Bit entsprechend dem Internationalen Einheitensystem SI vorgenommen. Eine Betrachtung der "gültigen Einheiten für Verhältnisgrößen (Tabelle 4)" des SI zeigt, daß das Bit eindeutig diesen Größen zugeordnet werden kann:

Größe	Einheit	Einheitenzeichen	Definition
Verhältnisgröße	Eins	1	-
	Prozent	%	1% $= 1*10^{-2}$
	Promille	‰	1‰ $= 1*10^{-3}$
	Millionstel	ppm	1 ppm $= 1*10^{-6}$
	Milliardstel	ppb	1 ppb $= 1*10^{-9}$
Logarithmierte Verhältnisgröße	Bel	B	1 B = lg 10
	Dezibel	dB	1 dB = 0,1 B = 0,1 lg 10
	Neper	N	1 N = ln e

Tabelle 1: Gültige Einheiten für Verhältnisgrößen des SI[Autl 84]

Eine Betrachtung dieser Tabelle ergibt drei Erweiterungsmöglichkeiten:
- Die Einschränkung der "Verhältnisgrößen" auf dezimale Bruchteile von 1 erscheint ungerechtfertigt . Ebenso, wie 1 oder % usw. können Tausend, Million, Milliarde, Billion und Billiarde aufgeführt werden. Es gibt jedoch keine zwingende Bedingung, nur Verhältnisse als Zehnerpotenzen zuzulassen, auch Paar (2),Dutzend (12), Mandel (15), Schock (60) oder Gross (144) lassen sich zwanglos ergänzen.
- Unter den "Logarithmierten Verhältnisgrößen" läßt sich offensichtlich auch das Bit einorden mit 1 bit = ld 2.
- Die Bezeichnung Verhältnisgrößen wird weder der Tatsache gerecht, daß es sich um die Tabelle 4 des Internationalen Einheitensystems handel, die der Gruppe "SI-fremde Einheiten" zugeordnet wurde (alle anderen Gruppen enthalten Maße bzw. Maßeinheiten) noch ist der Ausdruck "Verhältnis" glücklich gewählt.Wir schlagen daher vor, diese Gruppe als Zählmaße zu bezeichnen.

Alle angegebenen Einheiten sind Zählschritte, die entweder linear oder logarithmisch vorgenommen werden.Somit sollte die erste Gruppe in Tabelle 4 als Lineare und die zweite Gruppe als Logarithmische Zählmaße bezeichnet werden. Charakteristisch für die erste Gruppe ist die additive Summenbildung: 1 + 1 = 2 oder 2 ppm + 3 ppm = 5 ppm.

Der Vorteil Logarithmischen Zählmaßen besteht in der Zurückführung multiplikater Zusammenhänge auf die Addition von Logarithmen.So ergibt sich die Gesamtverstärkung von zwei in Reihe geschalteten Verstärkern mit den Einzelverstärkungen

$V = 31{,}6$ (= 30 dB) zu $V_{ges} = 31{,}6 * 31{,}6 = 1000$ bzw. $V_{ges} = 30$ dB + 30 dB = 60 dB und die Verdopplung einer Wortbreite von 8 bit (= 256 unterscheidbare Worte) auf 16 bit

als 256 * 256 = 65 536 unterscheidbare Worte bzw. 8 bit + 8 bit = 16 bit

Die jeweils rechts stehenden Gleichungen sind mathematisch einfacher auszuführen und leichter zu merken, was auch der Grund für die Nutzung logaithmischer Zählmaße ist.

Betrachtet man aus dieser Sicht das Bit, so kann man feststellen, daß es eine reine Zahlengröße ist und somit ein Maß für die Mächtigkeit von Mengen, wie auch jede andere Zahl. Da man zwischen Zählsystemen eindeutige Umrechnungen ausführen kann, ergibt sich, daß 8 bit identisch sind mit der Dezimalzahl 256, oder, anders gesagt, ein 8-Bit-Bus kann auch als 256-Zustandsbus mit binärer Kodierung bezeichnet werden ohne daß sich an der Aussage irgendetwas ändert.
Somit ist festzustellen, daß die Behauptung, das Bit sei ein Maß für die Information, zu der Schlußfolgerung führt, daß die Zahlen ein Maß für die Information sind.
Die von Rényi [Ren 82,S.17] aufgestellte Vermutung, "daß der Mathematischen Informationstheorie die Erkenntnis zu verdanken sei, daß sich Informationsmengen numerisch erfassen lassen, ähnlich, wie mann Entfernungen, Zeit, Masse oder Temperatur in Zahlen ausdrücken kann." ist nur zur Hälfte wahr, denn alle angeführten Größen haben als Meßwerte ein Produkt aus Zahlenwert und Einheit. Da das Bit jedoch ein Zahlenwert ("Zählmaß") ist, fehlt offensichtlich die Einheit der Information.

3. Die Entropie

Der Begriff Entropie wurde von Rudolf Emanuel Clausius (1822-1888) geprägt für das Verhältnis von Wärmemenge zu Temperatur:

$$H = Q/T \qquad (1)$$

(H = Entropie, Q = Wärmemenge, T = abs. Temperatur). Da Q in Joule (J) und T in Kelvin (K) gemessen werden, hat S die Dimension J/K.
Ludwig Boltzmann nutzte den Begriff Entropie für die Beschreibung des Zustandes eines idealen Gases mit der Gleichung

$$H = k \ \ln w , \qquad (2)$$

wobei $k= 1{,}38*10^{-23}$ J/K ein konstanter, ebenfalls dimensionsbehafteter Faktor ist, der bewirkt, daß auch die Entropie H nach Gl.(2) diese Dimension hat. Mathematisch interessant ist hierbei, daß als 2. Faktor auf der rechten Seite der Gleichung für die thermodynamische Entropie ein logarithmisches Zählmaß ln w auftritt.
Shannon stellte nun fest [Sha 48], daß die mittlere Wahrscheinlichkeit $H_o(n)$ für das Auftreten eines Symbols bei dessen zufälliger Auswahl aus einer Menge von n Symbolen

$$H_o(n) = \mathrm{ld}\ n \qquad (3)$$

beträgt, wenn die Auswahl binär erfolgt. Diese Wahrscheinlichkeit nannte er Entropie.
Die Anwendung des Begriffes Entropie auf die Information durch Shannon hat nach seinen eigenen Aussagen keinen anderen Grund, als das dabei zur Anwendung kommende logarithmische Zählmaß damit zu charakterisieren . Sietmann [Sie 91] schreibt dazu: "Shannon hat einmal bemerkt, daß der Physiker John v. Neumann ihn zur Namensgebung 'Entropie' angeregt hatte - weil ohnehin niemand verstünde, was das sei, wäre das von Vorteil in Diskussionen seiner Theorie." Diese leichtfertige Übertragung des in der Physik eindeutig definierten Begriffs Entropie in die Informatik hat zu einer Unzahl spekulativer Arbeiten über den Zusammenhang von Energie und Information geführt, die natürlich zu keinem

wissenschaftlich tragbaren Ergebnis kommen konnten, da lediglich der formale Zusammenhang besteht. daß in der Boltzmannschen Entropie-Gleichung (2) der Thermodynamik ein Term in Form eines logarithmischen Zählmaßes für Wahrscheinlichkeiten auftaucht und auch Shannon einen derartigen Term in Form des logarithmischen Zählmaßes in Gl.(3) bei der Formulierung der Kommunikationstheorie fand. Der entscheidende Unterschied besteht jedoch darin, daß der Term ln w bei Boltzmann zusammen mit dem Zahlenwert des Faktors k der Größe $1{,}38 * 10^{-23}$ den Zahlenwert des Meßwertes für die Entropie bildet. die Maßeinheit jedoch als Dimension Joule/Kelvin des Faktors k ebenfalls vorhanden ist.
Die informationelle Entropie ist dagegen ein dimensionsloses logarithmisches Zählmaß, deren "Zusammenhang" zur thermodynamischen Entropie lediglich darin besteht, daß in der Boltzmannschen Gleichung für die Entropie idealer Gase außer dem dimensionsbehafteten Faktor k auch ein logarithmisches Zählmaß ln w enthalten ist. Das Vorhandensein logarithmischer Zählmaße ist jedoch, wie auch Tabelle 4 des SI zeigt, weit verbreitet und stellt nur ein mathematisch formal ähnliches Verhalten dar. Keinesfalls ist jedoch daraus ein inhaltlicher Zusammenhang begründbar. Somit ergibt sich, daß auch die informationelle Entropie, genau wie das Bit, ein Zählmaß ist.

4. Information
Offensichtlich haben die Shannonschen Arbeiten die Information zum Inhalt. Da aber Bit und Entropie Zählmaße sind, d.h. abstrakte Definitionen der Zahlentheorie,stellt sich die Frage, was denn dann die Information ist, die offenbar mit diesen Zählmaßen gemessen wird, d.h. welche Dimension bzw. Maßeinheit ergänzt diese Zählmaße zu einem vollständigen Meßwert. Die Antwort kann nur lauten: Maßeinheiten sind die Objekte der Mengen, deren Mächtigkeit mit den Zählmaßen bestimmt wird. Insofern gleichen alle Zählmaße den natürlichen Zahlen, deren Anwendung auch in Verbindung mit Objekten erfolgt, z.B. indem ich zähle: 3 Mark, 4 Häuser, 5 Planeten usw. Rényi [Rén 82, S. 18] stellt z.B. fest, daß die Antworten eines Mädchens auf die Fragen "Fräulein, mögen Sie Käse?" oder "Fräulein, wollen Sie meine Frau werden?", seien sie nun "ja" oder "nein", jeweils 1 Bit Information enthalten, obwohl Gehalt und Bedeutung beider Fragen völlig verschieden seien. Hier handelt es sich um die Objekte "mich heiraten" und "Käse essen", die jeweils Mengen mit der Mächtigkeit 2 bilden (0 x "mich heiraten" und 1 x "mich heiraten" bzw. 0 x "Käse essen" und 1 x "Käse essen")
Wie bei den Zahlen ist es offensichtlich auch beim Bit oder der Entropie möglich, von den konkreten Objekten zu abstrahieren, indem ich z.B. die Aufgabe $3 + 5 = 8$ rechne. Trotzdem steht hinter jeder Zahl (ungeschrieben) das abstrakte Objekt, so auch hinter dem Bit und der informationellen Entropie.
Ausgehend von der Erkenntnis, daß wir mit Bit und informationeller Entropie letztendlich Objekte zählen läßt sich die Schlußfolgerung ziehen, daß Objekte als Elemente von Men-gen und Information identisch sind. Dabei ist zu berücksichtigen, daß immer Zusammen-hänge in der Form bestehen, daß jedes Objekt Eigenschaften aufweist, wie z.B. geometrische Ausdehung, Objekt-Masse und -Energie, Teil eines größeren Objektes ist (z.B. das Atom als Teil eines Moleküls) und aus untergeordneten Objekten besteht (ein Atom aus Elektronen und Nukleonen).

Als allgemeiner Begriff für die Vielfalt von Objekten. ihrer Eigenschaften und gegenseitigen Beziehungen hat sich die Bezeichnung Struktur als zweckmäßig erwiesen. Wir sehen daher die Gesamtheit aller Strukturen der Materie (bzw. der realen Welt) als Gesamtheit der Primärinformationen an. wobei wir Materie im Sinne von C.F. Weizsäcker [Wei 75] und reale Welt gemäß H. Noltemeier [Nol 82.S.12] verstehen. Als Sekundärinformation verstehen wir kodierte Information. Insgesamt ergibt sich somit das in Bild 1 gezeigte Schema des Aufbaus der Materie

Materie

Es gelten Erhaltungssätze	Es gelten keine Erhaltungssätze
Energie	**Information (Struktur)**
- mechanische (potentielle, kinet.)	Primärinformation
- innere (thermische)	(= Primärstruktur)
- elektromagnetische	- geometrische
- chemische	- zeitliche
-Kern-	- energetische
-Masse-	Sekundärinformation
	- Kodes

Abb. 1 : Aufbau der Materie

Materie besteht demnach generell aus den beiden Komponenten Energie und Information. Heisenberg [Hei 76] führte einmal aus: "Energie ist der graue Beton, aus dem alles geschaffen wurde. ... Die Ursubstanz Energie, wenn sie in Form von Elementarteilchen vorliegt, wird zur Materie." In Erweiterung zu Heisenberg fassen wir [Dub 88] jedoch die Materie nicht nur als Teilchen. d.h. stofflich auf, sondern einschließlich aller anderen energetischen Zustände, wie Lichtquanten, Feldern oder Temperaturen. Alle Erscheinungen der realen Welt sind aus Energie aufgebaut (nur Energie ist nachweisbar) und haben jeweils eine spezifische räumlich-zeitliche Struktur und Energiedichte, die ihre Information ausmachen..

Weiterhin ist allgemein anerkannt, daß die Materie in der Form strukturiert ist, daß sich Objekte definieren lassen, d.h. Teile der realen Welt, die sich von der Umgebung unterscheiden. Eine Liste von Beispielen für Objekte zeigt Tabelle 2.

Alle Objekte sind in mehr oder weniger starkem Maße der Bewegung und Weiterentwicklung im Verlaufe der Zeit unterworfen. Dabei gelten für die Energie in ihrer Gesamtheit (siehe Abb.1) Erhaltungssätze. Nicht wenige Autoren versuchen derartige Erhaltungssätze auch für die Information zu finden. So schreibt Rényi [Rén 82, S.42]: "Anscheinend gibt es eine gewisse Verwandschaft, eine Analogie zwischen den Begriffen Information und Energie!. Ferner sieht es so aus, als gäbe es ein Gesetz, das man als Gesetz der Informationserhaltung bezeichnen könnte." Unserer Meinung nach ist grade das Gegenteil richtig, ja wir

Objekte:

- Energiequanten	- Elementarteilchen
- Atome	- Moleküle
- chemische Strukturen	- mechanische Strukturen (Stein, Gebirge, Gewässer, Planet, Fixstern, Galaxis)
- biologische Objekte (Virus, Zelle, Bakterium, Pflanze, Tier, Mensch)	- gesellschaftliche Strukturen (Persönlichkeit, Familie, Gemeinde, Organisation, Land, Staat)

Tabelle 2: Beispiele für Objekte

sind zu der Schlußfolgerung gelangt, daß, wie in Abb. 1 angegeben, alle Erscheinungen, für die Erhaltungssätze gelten, der Energie zugerechnet werden müssen, und alle Erscheinungen, für die keine Erhaltungssätze gelten, Information sind.

5. Eigenschaften der Information

Natürlich genügt die eben angeführte Eigenschaft, daß für die Information keine Erhaltungssätze gelten, nicht zu deren grundlegender Beschreibung. Daher seien hier die wichtigsten Beschreibungsmerkmale von Objekten/Strukturen angegeben:

- Objektname
- Objekthierarchie (Teilobjekte, Zugehörigkeit zu übergeordneten Objekten)
- geometrische Form (z.B. im x,y,z-System)
- raum-zeitliches Verhalten (Bewegung, Entwicklung)
- Energieinhalt (Ruhemasse, Temperatur, kinetische, potentielle, elektro-magnetische und Bindungsenergie)
- Wechselwirkungen mit der Umwelt (energetisch und informationell)
- Intelligenz

Haupeigenschaften der Information (sowohl von Primär- als auch Sekundärinformation) sind:

- keine Erhaltungssätze
- neu erschaffbar
- zerstörbar
- kopierbar
- wandelbar räumlich <---> zeitlich
  parallel <---> seriell

Die Tatsache, daß Information neu erschaffbar ist, ist die Grundlage jeder Weiterentwicklung. Ständig entstehen in der unbelebten und belebten Welt ohne Einfluß des Menschen und auch durch dessen Tätigkeit neue Strukturen, die es vorher nicht gab. Die Umkehrung dazu ist die Zerstörung von Information infolge der Entwicklung, da jede neue Struktur Energie zu ihrem Aufbau benötigt, die ja vorher in einer anderen Struktur vorlag, die zerstört werden muß um die Energie für die neue Struktur verfügbar zu machen. Besonders beeindruckend ist es für den Menschen, wenn so wertvolle und unwiederbring-

Objekte zerstört werden, wie es durch Tod eines Menschen erfolgt. Sein gesamtes Wissen, seine gesamte Persönlichkeit sind in wenigen Minuten unwiderbringlich verloren. Gleiches gilt auch für andere Strukturen, wie etwa alte Schriften oder Städte. Wenn sie zerstört wurden, verbrannten, dann kündet manchmal noch die Überlieferung davon, daß es sie einmal gab, aber sicher oft auch nicht einmal diese. Was in diesen Schriften stand, ist unwiderbringlich verloren, wenn nicht Teile als Kopien oder kodiert in Form von Abbildungen die Zeiten überstanden.
Hinsichtlich der kodierten Information gibt es das interessante Phänomen, daß oft z.B. zeitliche Veränderungen als geometrische Strukturen dargestellt werden, aus der t-Achse eines Diagrammes wird die x-Achse. Hier ergibt sich ein Ansatz für die Ableitung einer Speichertheorie aus der Kommunikationstheorie |Dub 88, S.60| So entsteht aus der Kanalkapazität $C = 2\,\Delta f\, H_{max}(x_E;x_A)$ die lineare Speicherdichte $S_L = \frac{2}{\Delta s}\, H_{max}(x_E;x_A)$
Ähnliche Beziehungen gelten für die Flächen- und Raumspeicherdichte.

Allgemein ist festzustellen, daß das Kopieren von Information die beiden Gebiete Informationsübertragung und Informationsspeicherung umfaßt. Eine Kopie einer Struktur kann nur an außerhalb der zu kopierenden Struktur erfolgen, da sonst andere Energiedichten und damit eine andere Struktur entstehen. Demzufolge muß für eine Informationsspeicherung die Information an den Speicherort übertragen werden. Da andererseits Information an einen energetischen Träger gebunden ist, der sich nur mit eine endlichen Geschwindigkeit fortbewegen kann, ist die Information auch für die Übertragungszeit im Kanal gespeichert. Oft werden z.B. Übertragunsleitungen als Speicher mit zyklischem Umlauf der Information genutzt.
Ein allgemeines Schema einer Informationsausbreitung ist in Abb. 2 dargestellt.

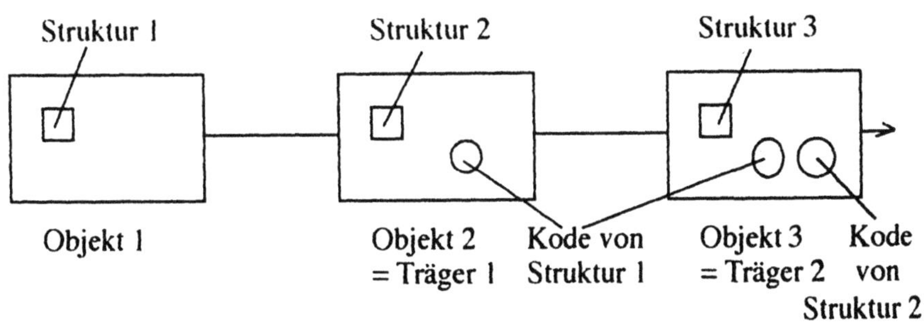

Abb. 2: Schema einer Informationsausbreitung

Jedes Objekt besitzt eine Struktur, die als kleines Quadrat im Objektsymbol angegeben ist. Kodierung heißt, daß ein (reduziertes) Abbild dieser Struktur auf einem anderen Objekt erzeugt wird,welches wir nun als Träger dieses Kodes bezeichnen, wobei auch mehrere Kodes auf einem Träger erzeugt werden können (kreisförmige Symbole in der Abbildung 2). Kode bedeutet dabei, daß ein (komplexes) Objekt existiert, das den Kode lesen kann. Allgemeine Wechselwirkungen der Objekte aufeinander fallen unter den Begriff "Entwicklung"

6.Literaturverzeichnis

Autl 84 Autorenkollektiv: Fachlexikon Meßtechnik.. Leipzig: VEB Fachbuchverlag 1984

Dub 88 Hans-Joachim Dubrau: Messen und Information. In: Wissenschaftliche Beiträge zur Informatik. 2(1988)H.4.S.52-61, hrsg.:Informatikzentrum des Hochschulwesens an der TU Dresden

Hei 67 Werner Heisenberg: Das Naturgesetz und die Struktur der Materie - Meilensteine des Denkens und Forschens.Vortrag Stuttgart 1967 In:W. Heisenberg: Tradition in der Wissenschaft - Reden und Aufsätze. München: 1977

Nol 82 Hartmut Noltemeier: Informatik III - Einführung in Datenstrukturen. München, Wien: Carl Hanser Verlag 1982

Rén 82 Alfréd Rényi: Tagebuch über die Informationstheorie. Berlin, Budapest: Gemeinschaftsausgabe der Verlage Akadémiai Kiado Budapest und VEB Deutscher Verlag der Wissenschaften Berlin, 1982

Sch1 90 Hans-Jochen Schneider: Information. In: Lexikon Informations- und Kommunikationstechnik.S.300, hrsg. v. Fritz Krückeberg; Otto Spaniol, Düsseldorf: VDI-Verl. 1990

Sch2 90 Hans-Jürgen Schneider: Bit. In: Lexikon Informations- und Kommunikationstechnik.S. 77, hrsg. v. Fritz Krückeberg; Otto Spaniol. Düsseldorf: VDI-Verl. 1990

Sch3 89 Alexej W. Schilejk o; Tamara I. Schilejko: Bezedi ob informatike. Moskau: Verlag Molodaja Gwardia 1989. (Deutsche Übersetzung des Titels: Gespräche über die Informatik)

Sha 48 Claude Elwood Shannon: A Mathematical Theory of Communication, Bell Syst.Techn.J. **27**(1948)S.379-423,623-656

Sha 49 Claude Elwood Shannon: W. Weaver: The Matematical Theory of Communication. Urbana: University of Illinois Press 1949 und 1963 (dt.Übersetzung: Mathematische Grundlagen der Informationstheorie. München, Wien: R.Oldenburg Verlag 1976)

Sie 91 Richard Sietmann: Schöpfer der Kommunikationstheorie - C. Shannons Schrift sorgt auch 50 Jahre nach der Veröffentlichung für Diskussionen. In: VDI-Nachrichten,45(1991)Nr.27, 5.7.91,Düsseldorf: VDI-Verlag

Völ 83 Horst Völz: Möglichkeiten und Grenzen einer Theorie der Informationstechnik (Informationstheorie). In: Nachrichtentechnik-Elektronik. **33**(1983)H.10, S.400-403. Berlin: VEB Verlag Technik

Wei 75 Carl-Friedrich Frhr. v. Weizsäcker: Die philosophische Interpretation in der modernen Physik. Zwei Vorlesungen..11.Aufl. NOVA ACTA LEOPOLDINA Neue Folge, Nr.207, Bd.37/2, Halle(Saale): Hrsg. Deutsche Akademie der Naturforscher Leopoldina 1987

Verfasser

Prof.Dr.-Ing. Hans-Joachim Dubrau

TU Dresden, Fak. Informatik, Inst.für Angewandte Informatik,

Lehrstuhl Messen und Informationsgewinnung

Hans-Grundig-Str.25, O-8036 Dresden

Physik und Informatik

—

Synergie für Forschung, Lehre und Entwicklung

Thomas Beth

Institut für Algorithmen und Kognitive Systeme
Universität Karlsruhe
Am Fasanengarten 5
7500 Karlsruhe

Zusammenfassung

Nicht erst durch das Aufkommen des massiven Parallelismus sind die von alters her gewachsenen engen Bindungen zwischen diesen beiden Gebieten so nachhaltig deutlich geworden. War es früher eher die Rolle des wissenschaftlichen Rechnens als Werkzeug der Physik (und die Physik als Lieferant von Hardware-Komponenten für den Bau von Computern), so setzt sich durch die neuen Entwicklungen großer Systeme die systematische Betrachtungsweise einer Naturwissenschaft als Grundvoraussetzung für die Beherrschbarkeit von Zusammenhängen über das Verständnis von Naturgesetzen zunehmend durch.

Die Synergie zwischen Physik und Informatik erstreckt sich in drei Richtungen:

- Informatik als Instrument für physikalische Experimente in Vorhersage, Simulation, Versuchsplanung, Messung, Auswertung.
- Physik als Entwicklungswissenschaft für fortgeschrittene Informationstechnologien wie Elektronik, Optik, Quantenschalter, Dünnfilmtechniken, Wellenmechanik, ...
- physikalische und informatische Modellierung im Zusammenspiel: Hier zeichnet sich die größte Neuerung ab, in der nur die gesamtheitliche systemtheoretische, auf mathematischen Methoden aufsetzende Arbeitsweise Erfolg verspricht.

Daher sind wesentliche Einflüsse aus diesen Überlegungen auf Forschungs- und Entwicklungsplanungen sowie auf die Lehre im Grund- und Hauptstudium der Informatik zu erwarten.

Mit der Planung eines neuen Curriculums, wie es z. Zt. an der Universität Karlsruhe entsteht, wird dieser Herausforderung zu begegnen sein!

Ein Konstruktionsalgorithmus für Mustererkennungssysteme

C.Cruse, M.Bode, H.-G. Purwins

Institut für Angewandte Physik
Westfälische Wilhelms-Universität Münster
Correnssstr. 2/4
4400 Münster

Durch iterative Anwendung weniger einfacher Regeln können komplexe Strukturen, etwa Fraktale, erzeugt werden. In analoger Weise gelingt es durch Iteration einfacher Konstruktionsregeln komplexe Neuronale Netze zu generieren, die endliche Klassifikationsprobleme sicher lösen. Benchmark-Tests ergeben eine hohe Konvergenzgeschwindigkeit des Konstruktionsprozesses im Vergleich mit optimierten backprop-Algorithmen.

1. Einleitung

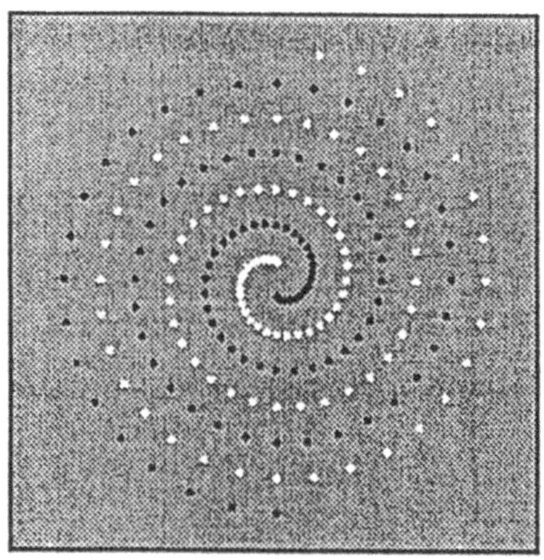

Bild 1) Das Spiralproblem: In einem 2-dimensionalen Musterraum befinden sich zwei spiralförmig ineinander verschlungene Musterklassen, bestehend aus je 97 Mustern, die voneinander getrennt werden müssen. [nach: Lang 88]

Durch iterative Anwendung von einfachen Regeln lassen sich komplexe Strukturen erzeugen, wie beispielsweise fraktale Farnblätter [Barnsley 86]. Hier wird ein Algorithmus vorgestellt, der die Struktur eines zur Informationsverarbeitung fähigen Systems entwickelt. Im Gegensatz zu den Ansätzen von Mezard und Nadal u.a. [Mezard&Nadal 89] oder den backprop-Algorithmen werden in diesem Verfahren den drei hinreichenden Netzebenen spezielle Aufgaben zugeordnet. Dadurch können die Ebenen sukzessive konstruiert werden. Das entwickelte System ist direkt einer Optimierung, etwa durch einen backprop-Algorithmus, zugänglich. Mit diesem Ansatz ist das XOR-Problem [Rummelhart 86] in 8 Iterationsschritten, das Spiral-Problem [Lang 88] in weniger als 300 Iterationsschritten lösbar. Ein verwandter Algorithmus von Marchand [Marchand 90] setzt binäre Musterräume voraus und benötigt daher auch nur zwei Netzebenen.

2. Das Verfahren

Sei eine endliche, zu klassifizierende Menge von Beispielmustern gegeben. Die, sämtlich eindeutigen, Klassifikationsklassen der Muster seien bekannt. Die Aufgabe bestehe nun darin, den Musterraum so zu gliedern, daß zunächst diese Beispielmuster korrekt klassifiziert werden, und daß darüber hinaus "ähnliche" Muster jeweils auf identische Klassen abgebildet werden. Da aber im Musterraum nahe beieinanderliegende Muster nicht unbedingt auch als "ähnlich" definiert sind, ist es bisher schwierig, solche Gliederungen zu finden.
Der hier vorgestellte Konstruktionsalgorithmus arbeitet in zwei Stufen:

I) Codierung:

Der Musterraum wird durch Separatrizes in Sektoren eingeteilt [Cruse 91]. Die Einteilung wird durch weitere Separatrizes solange verfeinert bis sämtliche Sektoren nur noch Muster aus höchstens einer Klasse enthalten. Die Separatrizes werden numeriert. Dadurch läßt sich jedem Sektor eindeutig ein Tupel zuordnen. Dieses Tupel wird als Muster in dem durch die Separatrizes aufgespannten "Coderaum" aufgefaßt, in dem die weitere Verarbeitung stattfindet.

Beispiel: Die Mustermenge (Bild: 2a) wird vom Algorithmus in der ersten Stufe durch Separatrizes in Gebiete gegliedert, die nachfolgend anhand der Beispielmenge klassifiziert werden. Der sich ergebende Generalisationsbereich ist für Klasse 2 der punktierte Bereich, für Klasse 1 dessen Komplement.

Bild 2)

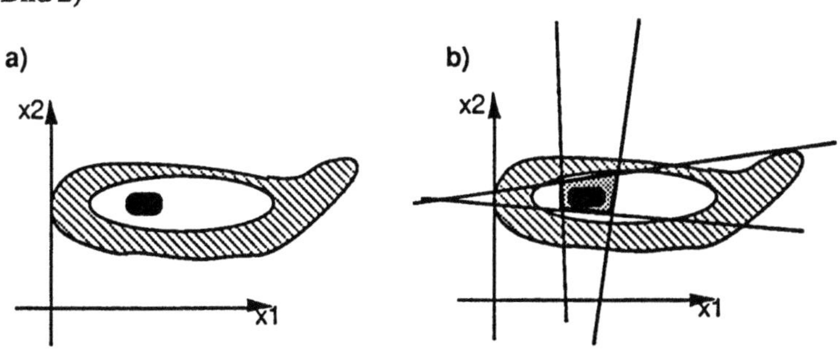

Musterklasse 1 : ▧
Musterklasse 2 : ■
Generalisationsbereich der Klasse 2 : ▩

Die Mustermengen bestehen aus einzelnen Mustern. Der Übersicht wegen werden nur die Umrisse gezeigt.
Musterraumkoordinaten: x1, x2

II) Klassifizierung:

Die, eventuell nicht zusammenhängenden, Sektoren mit identischer Klassenzugehörigkeit werden passend zusammengefaßt. Diese Zusammenfassung erfolgt in dem durch die im Musterraum liegenden Separatrizes erzeugten Coderaum, dessen besondere Eigenschaften ausgenutzt werden.

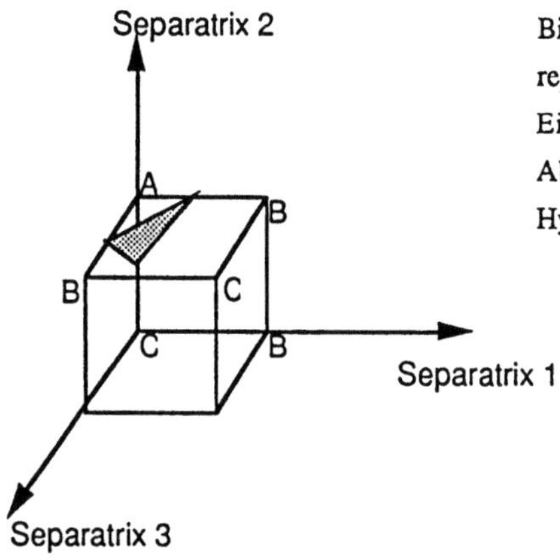

Bild 3) Die Transformation der die Sektoren repräsentierenden Muster im Coderaum auf kanonische Einheitsvektoren im Klassifizierungsraum durch Abtrennung der entsprechenden Eckpunkte des Hyperwürfels.

Unter Verwendung von binären Neuronen entsteht im Coderaum ein Hyperwürfel, auf dessen Eckpunkten die Sektoren repräsentiert sind. Der Abstand der Sektoren kann in diesem Raum durch einen Hamming-Abstand gemessen werden:Benachbarte Muster im Coderaum entsprechen benachbarten Sektoren im Musterraum. Die Muster können durch Separatrizes einzeln erfaßt werden, so daß als Ergebnis eine Transformation in den durch die Separatrizes des Coderaumes aufgespannten Klassifikationsraum erfolgt. In diesem entsprechen nun sämtliche Muster kanonischen Einheitsvektoren. Damit sind beliebige disjunkte Teilmengen im Klassifikationsraum linear separabel und können somit immer geeignet zusammengefaßt werden.

3. Anwendungen

Für den Standard-benchmark-Test, das XOR-Problem, benötigt dieser Algorithmus 8 Epochen, d.h. die Beispielmuster wurden durchschnittlich achtmal zur Berechnung herangezogen. Zum Vergleich [Jacobs 87] benötigt der Standard-backprop im Mittel über die erfolgreichen Versuche 538 Epochen, und 250 Epochen mit der "delta-bar-delta"-Regel. Fahlmann [Fahlmann 88] berichtet von 24 Epochen mit dem Quickprop-Algorithmus.

Eine weitere als benchmark-Test für Klassifikations-Algorithmen geeignete Mustermenge ist das Spiral-Problem (Bild 1): Im Musterraum befinden sich zwei auf ineinander verwundenen Archimedischen Spiralen angeordnete Musterklassen, bestehend aus je N Mustern, die mit konstantem Winkelinkrement von außen nach innen auf den Spiralen plaziert werden [Lang 88]. Die Schwierigkeit der Aufgabe wächst mit der Anzahl der Muster, da sich die besetzten Spiralbereiche immer mehr ineinander verschlingen. Bei N=10 Mustern ist das Problem noch linear separabel, deshalb ist in Bild 4) die Neuronenzahl anfänglich noch sehr gering. Mit zunehmender Komplexität wächst diese zunächst stark, dann aber für die Neuronen der ersten Schicht nur noch schwach. Dies läßt sich dadurch erklären, daß eine neue Separatrix mehr Sektoren erzeugt, wenn schon viele Separatrizes vorhanden sind.

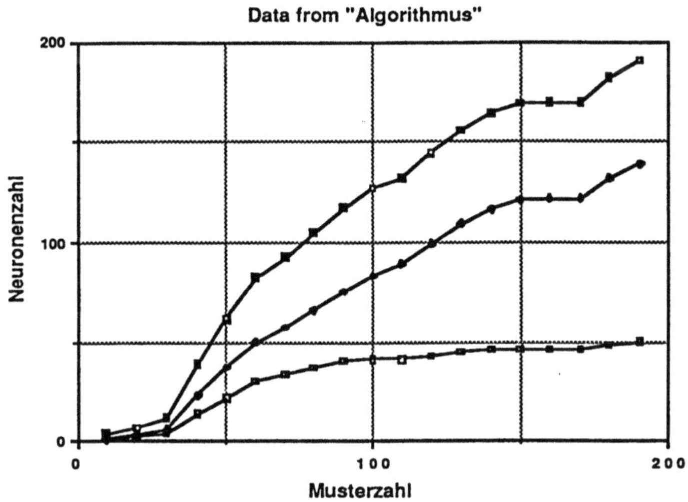

Bild 4) Die unterste Kurve stellt die Neuronenzahl der ersten Schicht dar, die darüberliegenden die der zweiten Schicht und die oberste die Gesamtzahl der Neuronen.

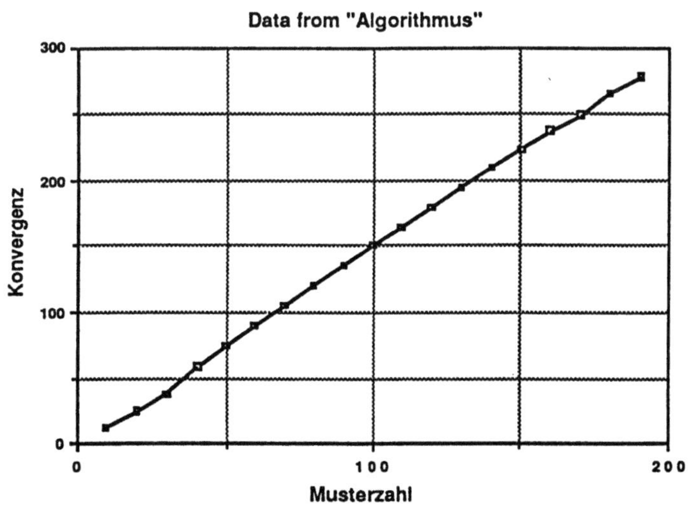

Bild 5) Die Konvergenzzeit in Musterzyklen aufgetragen gegen die Komplexität des Problems, gemessen in der Anzahl 2N der Muster.

Der Anstieg der Neuronenzahl der zweiten Schicht spiegelt die Anzahl der von den Mustern in Anspruch genommenen Sektoren wider (Bild 5).

Die mittleren Konvergenzzeiten für ein Problem mit insgesamt 194 Mustern (N=97) wurden [Lang 88] beim Standard-backprop mit 20.000 Epochen und beim Quickprop mit 8.000 Epochen angegeben. Der hier vorgestellte Algorithmus benötigt 292 Epochen, um die Aufgabe vollständig zu lösen.

4. Zusammenfassung

Es wurde ein Algorithmus vorgestellt, der ein auf ein gegebenes Klassifikationsproblem abgestimmtes Neuronen-Netz entwirft. Das Netz benötigt genau zwei "hidden"-Netzebenen. Die Konvergenzzeit liegt für die hier verwendeten benchmark-Tests erheblich unter den Vergleichswerten.

Literatur:

Barnsley, M.F.,Ervi,V.,Hardin,D.,Lancaster,J.(1986): Solution of an inverse problem for fractals and other sets, Proc.Natl.Acad.Sci.USA,Vol.83, S.1975-1977, April 1986.

Cruse, C.(1991): Die Entwicklung eines Klassifikationsalgorithmus aus Untersuchungen zur Wissensdarstellung in Neuronalen Netzen, Diplomarbeit an der WWU Münster

Fahlmann, S.E.(1988): An Empirical Study of Learning Speed in Backpropagation Networks, Tech. Report CMU-CS-88-162, Carnegie Mellon University.

Jacobs, R. A.(1987): Increased Rates of Convergence Through Learning Rate Adaptation, Technical Report COINS TR 87-117, University of Massachusetts at Amherst, Dept. of Computer and Information Science, Amherst, MA.

Lang, K., Witbrock, M.(1988): Learning to Tell Two Spirals Apart,Proceedings, 1988 Connectionist Summer School, Morgan-Kaufmann.

Marchand, M.,Golea, M., Rujan, P.(1990): A Convergence Theorem for Sequential Learning in Two-Layer Perceptrons, Europhysics Letters, Vol.11, S.487-492.

Mezard,M.,Nadal,J.-P.(1989): Learning in Feedforward Layered Networks: The Tiling Algorithm, Journal of Physics A, Vol.22,S.2191-2204.

Rummelhart, D.E. and McClelland,J.L.(1986): Parallel Distributed Processing: Explorations in the Microstucture of Cognition, MIT Press.

Relaxation im Orts/Geschwindigkeits-Raum

Berechnung objektbezogener Geschwindigkeits-Vektorfelder aus lokalen Messungen

Helmut Glünder‡

Institut für Medizinische Psychologie, Ludwig-Maximilians-Universität
Gœthestraße 31, D-8000 München 2, Germany

Es wird ein in der Physik übliches (Energie-)Minimierungsprinzip auf ein Grundproblem der Bewegtbildanalyse angewendet. Abweichend von den geläufigen mathematisch formalen Optimierungsansätzen und Lösungsalgorithmen, bzw. den dazu vielfach eingesetzten weitgehend universellen Hopfield- und Boltzmann-Netzen, wird hier eine anschauliche Netzwerk-Architektur vorgeschlagen, die nicht nur funktional sondern auch strukturell problemspezifisch ausgelegt ist.

Einführung

Geschwindigkeit ist ihrer differentiellen Definition gemäß, ebenso wie zum Beispiel der Helligkeitskontrast, ein lokales Maß. Nun ergibt sich aber keinesfalls an jedem Objektpunkt eines sich beispielsweise starr verschiebenden ausgedehnten Bildobjekts das gleiche, geschweige denn das wahre Maß, d.h. der wahre Geschwindigkeitsvektor. Es stellt sich somit die Frage, ob und wie aus der Gesamtheit der in einer Bildebene zu messenden Vektoren durch deren Verrechnung brauchbare Schätzungen wahrer Bewegungsvektoren zu ermitteln sind. Man weiß heute, daß dieses Problem unter Annahme wenig restriktiver Randbedingungen als Optimierungsproblem darzustellen und mit bekannten Methoden zu lösen ist [1–6]. Die bislang dazu vorgeschlagenen Berechnungsmethoden unterscheiden sich im wesentlichen in der Wahl dieser Randbedingungen und darin, ob flächig oder entlang von Konturen verrechnet wird sowie, welcher Optimierungs-Algorithmus angewendet wird. Abgesehen von universellen optimierenden Widerstandsnetzwerken [5, 7] sind derzeit keine Verfahren bekannt, die ausschließlich auf – möglichst biologisch plausiblen – hochparallelen Netzstrukturen beruhen oder zumindest nicht auf anderen weitgehend universellen Architekturen, also zum Beispiel Hopfield- oder Boltzmann-Netzwerken. Derart motiviert wird im folgenden ein Konzept vorgestellt, das schon durch seine physikalische Rechnerstruktur den Eigenheiten des konkreten Problems anschaulich Rechnung trägt.

Elementare Geschwindigkeitsmessung

Ohne Einschränkung der Allgemeinheit werden die elementaren Geschwindigkeitsmessungen in Form von Zuordnungen zwischen Bildpunkten zeitlich aufeinanderfolgender Momentaufnahmen angenommen, mithin also als Differenzmessungen. Jeder Punkt des zeitlich vorangehenden Bildes korrespondiert natürlich prinzipiell mit (sehr) vielen Punkten des nach der Zeitspanne Δt folgenden Bildes, so daß sich für jeden Punkt ein ganzes Spektrum vorerst gleichberechtigter Geschwindigkeitsvektoren $\vec{v}(\vec{r}, t) = \vec{\rho}(\vec{r}, t)/\Delta t$ ergibt (mit den Korrespondenzvektoren $\vec{\rho}(\vec{r}, t)$). Sicher weiß man nun von jedem Spektren nur, daß es den wahren Vektor einschließt. Keineswegs sicher, aber recht wahrscheinlich ist es, daß benachbarte Bildpunkte zum selben Objekt gehören und somit zumindest ähnliche Geschwindigkeitsvektoren aufweisen (Standard-Randbedingung). Überdies kann für hinreichend kleine Zeitspannen Δt die Annahme einer oberen Grenzgeschwindigkeit sinnvoll und hilfreich sein, so daß die Korrespondenzspektren dann a priori begrenzt sind.

---

‡ gefördert durch die "Volkswagen-Stiftung" im Rahmen des 'Synergetik'-Programms, Projekt I/65 914

Relaxation

Durch geeignete Operationen im soeben definierten – und im einfachsten Fall binär besetzten – vierdimensionalen Orts/Geschwindigkeits-Zustandsraum soll jedem Ortspunkt eine möglichst gute Schätzung seines wahren Geschwindigkeitsvektors zugeordnet werden. Dazu wird eine – für den Fall starrer Verschiebungen nach Geschwindigkeit und Ort aufspaltbare – vierdimensionale Korrelation der Korrespondenzsignale $g(\vec{r},\vec{v},t)$ mit einer im allgemeinen orts- und geschwindigkeitsvarianten, jedoch zeitinvarianten Funktion $w(\vec{r},\vec{v})$ vorgeschlagen, deren Resultat nichtlinear bewertet wird. Diese Verarbeitung wird solange wiederholt, bis ein zeitliches oder qualitatives Abbruchkriterium erreicht ist. Für sich verschiebende starre Objekte folgt damit in der einfachsten Form zum Zeitpunkt t und unter Vernachlässigung der Relaxationsdauer:

$$g_{i+1}(\vec{r},\vec{v})=\begin{cases} g_i(\vec{r},\vec{v})\ \underline{\text{corr}}\ w(\vec{r},\vec{v}) \\ 0 \quad \text{wenn} \quad g_i(\vec{r},\vec{v})<1 \end{cases}$$

- Die Korrelation in den *Orts*dimensionen mit $w_{exc}(\vec{r})$ sorgt für eine (geringfügige) Erhöhung der Werte in einer (isotropen) *Orts*nachbarschaft jedes Punktes des Zustandraums. Die Schnittmenge der Geschwindigkeitsspektren benachbarter Bildpunkte erhält dadurch erhöhte Werte.

- Durch die Korrelation in den *Geschwindigkeits*dimensionen mit $w_{inh}(\vec{v})$ werden in einer (isotropen) *Geschwindigkeits*nachbarschaft jedes Punkts des Zustandraums die Werte (geringfügig) erniedrigt und folglich die nicht zur Schnittmenge gehörenden Geschwindigkeiten an jedem Bildpunkt geschwächt.

Aufgrund der wiederholten Anwendung relaxiert die Repräsentation im vierdimensionalen Zustandsraum auf für jeden Ort sinnvolle Geschwindigkeitsvektoren. Gilt $w_{exc}(\vec{r})=-w_{inh}(\vec{v})$, so läßt sich für Verschiebungsanalysen entsprechend obiger Gleichung überdies zeigen, daß der wahre Vektor im Verlauf der Relaxation notwendigerweise erhalten bleibt.

Netzwerkarchitektur

Zur Analyse ortsdiskreter Bildsignale läßt sich eine einfache und global betrachtet hochrepetitive zweidimensionale Netzstruktur angeben, die jedoch eine regional inhomogene Verschaltung aufweist. Als Rechenelemente dienen Summierer deren Ausgangssignale nichtlinear bewertet werden. Größen, die die örtlich-zeitliche Korrespondenz von Bildpunkten und damit das Vorhandensein potentieller Geschwindigkeitsvektoren ausdrücken, bilden die externen Eingangssignale des Netzwerks. Netzinterne Eingangssignale der Rechenelemente stammen jeweils von den Ausgängen zweier Gruppen umliegender Elemente. Ordnet man jedem Ort der Bildebene einen Netzwerkbereich zu, in dem alle möglichen Geschwindigkeitsvektoren durch Rechenelemente vertreten sind (zweidimensionale Darstellung des vierdimensionalen Zustandraums mit erhaltener örtlicher Topographie), so muß jedes dieser Elemente mit den vektoriell gleich definierten Elementen benachbarter Bildpunkten in Verbindungen stehen (Korrelation in den Ortsdimensionen). Neben dieser für die gewählte Topographie 'weitreichenden' Erregung stehen die Rechenelemente an jedem Ort mit ihren vektoriell benachbarten Elementen in negativ bewerteter Verbindung (Korrelation in den Geschwindigkeitsdimensionen), so daß 'kurzreichweitige' Hemmung vorliegt. Das gesamte Netzwerk ist zeitinvariant.

Schlußbemerkungen

Verglichen mit einer langwierigen seriellen Abarbeitung sowohl der einzelnen Korrelationen als auch des übergeordneten Relaxationsprozesses, relaxiert ein geeignet ausgelegtes Netzwerk der geschilderten Architektur unmittelbar, nur begrenzt durch Latenzen in Verbindungen und Rechenelementen. Bei zumindest kurzzeitig gleichförmiger Bewegung beschleunigt sich dieser Vorgang, da sich dann Objektpunkte in bezüglich der vorangehend berechneten Geschwindigkeit sensibilisierte Bereiche hineinbewegen [8, 9].

Die örtliche Zuordnung von Geschwindigkeiten gelingt auch für solche – meist innere – Objektbereiche, in denen isoliert betrachtet keinerlei Bewegung zu messen ist. Das heißt, das Verfahren propagiert zuverlässige Geschwindigkeitsvektoren in derartige Bereiche. Außerdem können sich an jedem Bildpunkt mehrere – allerdings nur hinreichend unterschiedliche – Geschwindigkeitsvektoren ergeben, was nicht zuletzt in Hinblick auf die Analyse sogenannter transparenter Bewegung zu fordern ist [10, 11].

Im Gegensatz zur Mehrzahl der bekannten Vorschlägen zeigt das hier vorgeschlagene Netzwerk-Konzept funktional wie strukturell gute Übereinstimmung mit entsprechenden zentralnervösen Verarbeitungsstrukturen ([12] Kapitel 35). Da der belebten Natur die formal-deduktive mathematische Behandlung von Problemen der Informationsverarbeitung nicht geläufig ist, sie aber allem Anschein nach mit mathematisch formulierbaren physikalisch-induktiven Prinzipien in Einklang stehen muß, wird man in biologischen Systemen vergeblich nach Implementierungen von beispielsweise algorithmischen Lösungen formal konstruierter Optimierungsintegrale Ausschau halten.

Literatur

[1] Ullman S. (1979) *The Interpretation of Visual Motion.* The MIT Press, Cambridge/MA.

[2] Horn B. K. P. und Schunck B. G. (1981) Determining optical flow. *Artif. Intell.* **17**: 185-203.

[3] Nagel H.-H. (1983) Displacement vectors derived from second-order intensity variations in image sequences. *Comput. Vision Graphics Image Process.* **21**: 85-117.

[4] Hildreth E. C. (1984) *The Measurement of Visual Motion.* MIT-Press, Cambridge/MA.

[5] Poggio T., Torre V. und Koch C. (1985) Computational vision and regularization theory. *Nature* **317**: 314-319.

[6] Yuille A. L. und Grzywacz N. M. (1988) A computational theory for the perception of coherent visual motion. *Nature* **333**: 71-74.

[7] Wang H. T., Mathur B. und Koch C. (1989) I thought I saw it move: computing optical flow in the primate visual system. In: Gluck M. A. und Rumelhart D. E. (Hrsg.) *Neuroscience and Connectionist Theory.* Erlbaum, Hillsdale/NJ.

[8] Anstis S. und Ramachandran V. S. (1987) Visual inertia in apparent motion. *Vision Res.* **27**: 755-764.

[9] Snowden R. J. (1989) Motion in orthogonal directions are mutually suppressive. *J. Opt. Soc. Am. A* **6**: 1096-1101.

[10] Doorn A. J. van und Koenderink J. J. (1982) Temporal properties of the visual detectability of moving spatial white noise. *Exp. Brain Res.* **45**: 179-188.

[11] Doorn A. J. van, Koenderink J. J. und Grind W. A. van de (1985) Perception of movement and correlations in stroboscopically presented noise patterns. *Perception* **14**: 209-224.

[12] Braitenberg V. und Schüz A. (1991) *Anatomy of the Cortex. Statistics and Geometry.* Band 18 der Reihe *Studies of Brain Function.* Springer, Berlin.

VLSI für die Spracherkennung-
Chip-Architekturen für Suchalgorithmen
in der automatischen Spracherkennung

Alfred Hauenstein
Lehrstuhl für Integrierte Schaltungen
TU München
Arcisstr. 21
8000 München 2
Tel.: 089 / 23 18 17 - 47

1. Einleitung

Die automatische Erkennung von natürlicher fließend gesprochener Sprache, insbesondere bei großen Wortschätzen, stellt eine wichtige Voraussetzung der anwendergerechten Mensch-Maschine-Kommunikation dar. Dies ist insbesondere bei der Entwicklung zukünftiger Informations- und Kommunikationsendgeräte (Stichwort: MultiMedia) zu beachten. Von zentraler Bedeutung für die Akzeptanz ist es hierbei, daß folgende Anforderungen erfüllt werden: Großer Wortschatz (>1000 Worte) bei hoher Erkennungsrate, Sprecherunabhängigkeit und Durchführung der Erkennung in Echtzeit. Gerade letzteres stellt hohe Anforderungen an die Performance der sprachverarbeitenden Einheit.

2.Anforderungen der Spracherkennung an die verarbeitende Hardware

Ausgangspunkt für die folgende Darstellung sind die Erfahrungen und Ergebnisse die bei der Siemens AG mit den Projekten SPICOS und SPICOS II (eine Zusammenarbeit von Siemens, Philips und IPO-Eindhoven) gesammelt wurden.

Aus den Untersuchungen ergab sich, daß für ein Spracherkennungssystem mit einem Wortschatz von ca. 1000 Worten bei einer Realisierung mit allgemeiner Hardware (Workstations, DSPs) ohne Berücksichtigung der Signalvorverarbeitung eine Rechenleistung von über 200 MIPS (Millionen Instruktionen pro Sekunde) nötig ist, um Echtzeitbetrieb zu gewährleisten [1].

Daher liegt es nahe, die zeitintensivsten Komponenten des Erkennungsprozesses statt durch Software (Programmierung in C) besser mittels aufgabenspezifischer Hardware (ASIC) durchzuführen.

3. Modellierung des Sprachsignals

Die Modellierung der Referenzsignale erfolgt über Hidden Markov Modelle (HMMs) [2]. Sie ist phonembasiert, so besteht das Wort 'drei' z.B. aus drei Phonemen (*d, r, ai*). Jedes Phonem selber wiederum besteht aus drei kontextabhängigen Teilen (z.B. das 'r' in "drei": ^{d}r, r, r^{ai}), die mit mehreren sog. 'states' (Zuständen) modelliert werden. Die Anzahl der Zustände ist dabei von der Länge des Phonems abhängig. Die Reihenfolge der Zustände in einem Phonem kann dabei mit fortschreitender Zeit nur von links nach rechts durchlaufen werden. Dabei ist es nur möglich, daß das modellierte Sprachsignal im selben Zustand bleibt, in das nächste überwechselt oder eines überspringt. Dies wird mit den Zustands-Übergangswahrscheinlichkeiten $\mathbf{a}_{ii}$, $\mathbf{a}_{i(i-1)}$ und $\mathbf{a}_{i(i-2)}$ beschrieben.

Um die Erkennung des Eingangs-Sprachsignals durchzuführen (s.Bild 1), wird dieses zuerst vorverarbeitet (A/D-Wandlung, Filterung) und in sog. 'frames' mit Abständen von 10 ms aufgeteilt. Sodann werden in einem Vektorquantisierer (VQ) die frames mit den oben beschriebenen Modellen, die in einem Codebuch stehen, verglichen und ihre Emissionswahrscheinlichkeiten $\mathbf{b}_i(\mathbf{t})$, also der Grad der Übereinstimmung mit den Zuständen **i**, berechnet.

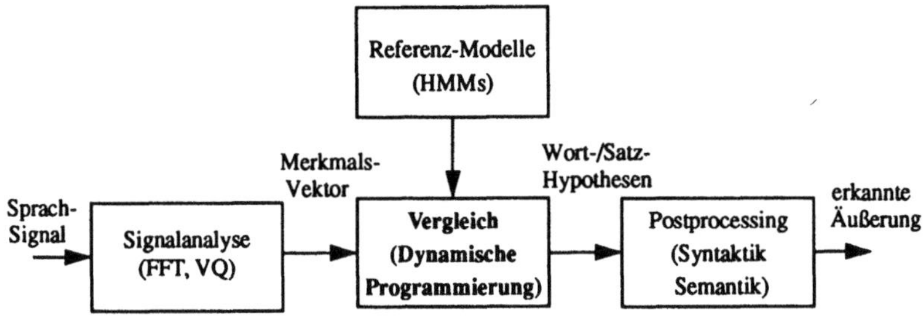

Bild 1: Signalfluß in einem Spracherkennungssystem

Als nächstes erfolgt nun der eigentliche Prozess der Erkennung. Dazu ist es zum ersten nötig, die einzelnen Zustände wieder zu Phonemen und weiter zu Worten zusammenzusetzen. Dies geschieht mittels des Viterbi-Algorithmus [3]. Es handelt sich hierbei um einen Spezialfall der Dynamischen Programmierung, deren Aufgabe es hier ist, über den Zeitverlauf (im 10 ms-Raster) die Aneinanderreihung von Zuständen zu finden, die am besten mit dem Modell übereinstimmt (Suche nach dem besten Pfad), also eine sog. Kostenfunktion zu minimieren. Mit den oben beschriebenen Phonem-Modellen kommt man dabei zu folgendem Berechnungsschema für die aktuellen Vorwärtswahrscheinlichkeiten $\alpha_i(\mathbf{w},\mathbf{t})$, die die Kosten eines Pfades beinhalten:

$$\log\alpha_i(wort,t) = \log b_i(wort,t) + min\begin{cases}\log\alpha_i(wort,t-1) + \log a_{ii}(wort)\\ \log\alpha_{i-1}(wort,t-1) + \log a_{i(i-1)}(wort)\\ \log\alpha_{i-2}(wort,t-1) + \log a_{i(i-2)}(wort)\end{cases}$$

Alle Werte liegen bereits logarithmiert vor, so daß die Berechnungen auf einfache Additionen und Minimumbildungen von Integer-Zahlen beschränkt werden können. Zu jedem Zeitpunkt werden die Pfade mit den geringsten Kosten, die im nächsten Zeitschlitz weiter verfolgt werden, ausgesucht. Die Information, in welchem Zustand sich die Pfade befinden werden in einem 'backtracking'-Feld abgelegt, um am Ende den Verlauf des oder der besten Pfade nachvollziehen zu können.

Um die Erkennung fließender Sprache zu erreichen, müssen die Worte zu Satzhypothesen verknüpft werden. Dazu ist es nötig, eine Grammatik zu definieren, nach der aus den Worten Sätze gebildet werden. Dafür werden i. A. Finite-State- oder stochastische Grammatiken verwendet. Für solche Grammatiken ist der Rechenaufwand für das Parsing der Wortfolge proportional zur Länge der Äußerung. Von sog. kontextfreien Grammatiken erwartet man sich eine höhere Erkennungsrate. Allerdings ist bei diesen der Rechenaufwand proportional zur dritten Potenz der Länge der Wortfolge [4], so daß eine praktische Anwendung erschwert wird. Um bei der Auswahl der Grammatiken zukünftige Ergebnisse einfließen lassen zu können, soll eine Architektur entwickelt werden, die möglichst unabhängig von der syntaktischen Modellierung ist.

4. Die Chip-Architektur

Die für die geschilderten Aufgaben nötige Chip-Architektur soll im weiteren vorgestellt werden. Dazu sollen vor allem die kritischen Komponenten aufgezeigt werden.

4.1 Der Viterbi-Prozessor

Als erstes ist die Ausführung des Viterbi-Algorithmus zu optimieren. Für die Implementierung auf einem DSP werden ca. 25 Instruktionen für einen Durchlauf des oben angegebenen Berechnungsschemas benötigt. Rechnet man dies mit der Anzahl der Worte (1000), den Zuständen pro Wort (ca. 50) und der Frequenz der ankommenden frames (1/10 ms) hoch, erhält man eine benötigte Rechenleistung von ca. 125 MIPS ausschließlich um den Viterbi-Algorithmus, ohne weiteren Auswertungen der Ergebnisse, durchzuführen. Da zudem für die Zukunft noch größere Wortschätze angestrebt werden, ist auf diese Weise eine Erkennung in Echtzeit nicht möglich.

Daher wurde die Architektur eines "Viterbi-Prozessors" entwickelt, die sich sowohl der Methoden der Parallelisierung als auch der Fließbandverarbeitung (Pipelining) bedient (s. Bild 2). Die Komponente Viterbi-Prozessor wird im Augenblick mit einer 1,0 µm Standardzellen-Bibliothek designt. Ihr Vorteil ist, daß sie sowohl für die Einzelworterkennung, bei der keine grammatischen Einschränkungen berücksichtigt werden müssen, als auch für die Erkennung fließender Sprache mit einer entsprechend erweiteren Architektur eingesetzt werden kann.

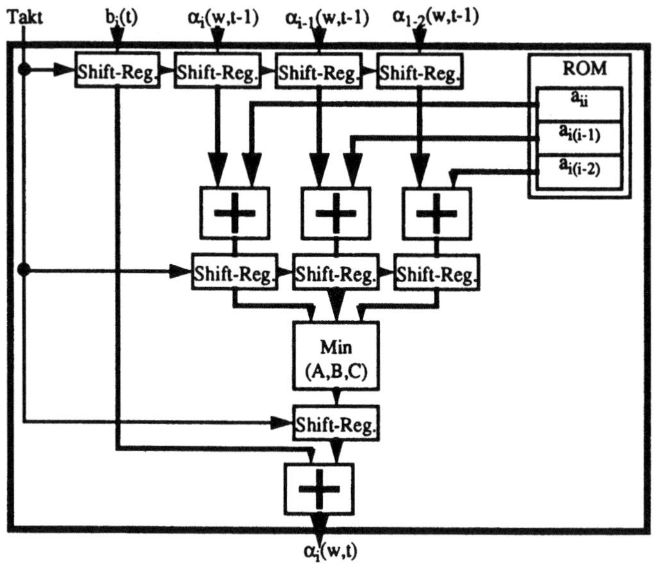

Bild 2: Viterbi-Prozessor

4.2 Die Speicherarchitektur

Betrachtet man die Systemarchitektur nach Bild 3, so wird deutlich, daß der Aufbau der Speicher großen Einfluß auf die System-Performance hat. In der vorliegenden Architektur sind drei RAM-Blöcke vorgesehen. Dies ist zum ersten der Speicher für die Emissionswahrscheinlichkeiten $\mathbf{b_i(t)}$ der Phoneme. Da auf diese während des Erkennungsprozesses wahlfrei zugegriffen werden muß, ist es wichtig, daß es sich um schnelle SRAMs handelt, die die Systemfrequenz nicht durch lange Zugriffszeiten drücken. Da sie außerdem nur 24 kbit groß sein müssen, ist es möglich, sie als on-Chip-RAMs auszuführen.

Die RAMs für die Abspeicherung der Vorwärtswahrscheinlichkeiten $\alpha_i(\mathbf{w,t})$ und $\alpha_i(\mathbf{w,t-1})$ sind je 1 Mbit groß. Diese müssen als externe Speicher ausgeführt werden. Normale DRAMs kommen nicht in Frage, da sie zu langsam sind. Im Augenblick wird der Einsatz von Video-RAMs (VRAMs) diskutiert. Dies sind von der Technologie DRAMs, die jedoch so organisiert sind, daß nach Anlegen der Adresse nicht ein einzelnes Wort, sondern hintereinander ein ganzer Block von Worten ausgegeben wird, wodurch sich die effektive Zugriffszeit drastisch verkürzt. Diese Art der Adressierung ist möglich, da auf die Vorwärtswahrscheinlichkeiten nicht wahlfrei, sondern in einer stetigen Abfolge zugegriffen wird. Für einen variierten Erkennungs-Algorithmus, der unwahrscheinliche Pfade ausschließt (sog. 'Pruning'), wäre es jedoch wünschenswert, daß die VRAMs nicht nur am blockweise, sondern wahlfrei adressiert werden könnten.

Das externe RAM für das backtracking-Feld (Größe: 10 Mbit) kann ebenfalls als VRAM ausgeführt werden. Desweiteren treffen die Randbedingungen zum VRAM für die Vorwärtswahrscheinlichkeiten hier genauso zu.

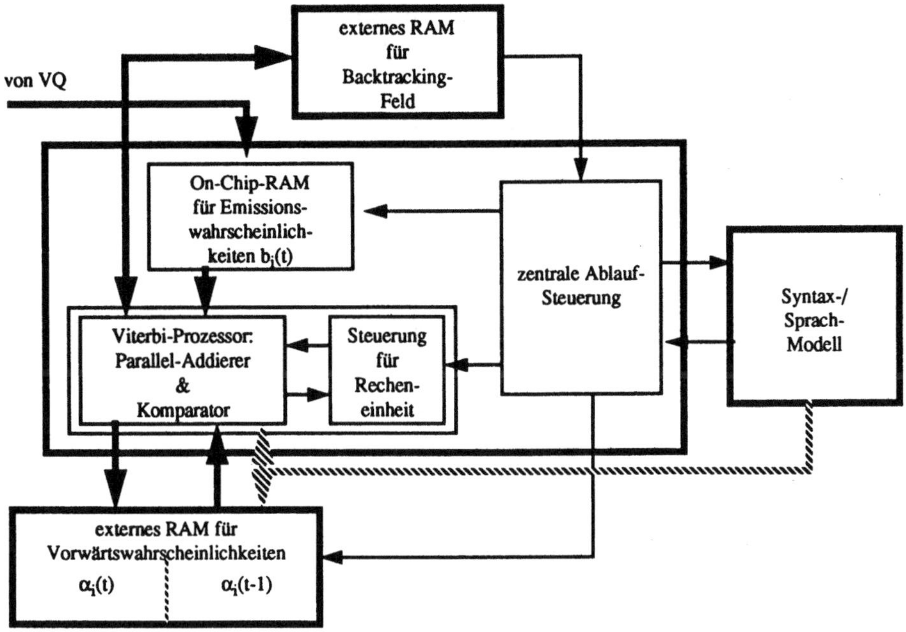

Bild 3: Systemarchitektur

4.3 Das Sprachmodell

Da für das Sprach-Modell (die Grammatik) noch Freiheitsgrade offen gehalten werden sollen, wird dieses als externe Komponente Einfluß nehmen. Dafür ist eine Schnittstelle zur zentralen Ablaufsteuerung vorgesehen. Diese Ablaufsteuerung nimmt Einfluß auf den Viterbi-Prozessor und gibt ihm dabei den Suchraum vor. Ebenso ist sie für die Adressierung der Speicher zuständig, um für einen korrekten Ablauf die richtigen Daten zur Verfügung zu stellen. Optional können weitere Daten des Sprachmodells (wie Wortübergangswahrscheinlichkeiten) direkt in den Prozessor geladen werden.

5. Ausblick

Die vorgestellten Chip-Architekturen sollen eine flexible Basis zur Hardware-Realisierung von Spracherkennungssystemen, die Echtzeitanforderungen genügen, bieten. Dabei sind noch einige Fragen zu klären bevor das Gesamtsystem designt werden kann:

Welche VRAMs mit flexibler (wahlfreier ?) Adressierung werden als externe RAMs eingesetzt ?

Steht ein Parsing-Algorithmus für kontextfreie Grammatiken für den Spezialfall der Erkennung natürlicher Sprache zur Verfügung, dessen Rechenaufwand nicht kubisch mit der Satzlänge wächst ?

Literatur

[1] J. Jann, A. Aktas, "Multiprozessor- und ASIC-Architektur für den Echtzeitbetrieb in der Spracherkennung", GME-Fachtagung 1991, Baden-Baden, S. 53-58.

[2] Lawrence L. Rabiner, "A Tutorial on Hidden Markov Models and Selected Applications in Speech Recognition", Proc. IEEE, Feb. 89, S. 257-286.

[3] G. David Forney jr., "The Viterbi Algorithm", Proc. IEEE, vol. 61, March 1973, S. 268 ff.

[4] H. Ney, "Dynamic Programming Speech Recognition Using Context-Free Grammar", Proc. ICASSP 87, S. 69 ff.

Wie chaotisch verhält sich ein neuronaler adaptiver Regler, und ist er im Netzwerkmanagement einsetzbar?

H. Nikolaus Schaller
Lehrstuhl für Datenverarbeitung
Technische Universität München
Arcisstr. 21
8000 München 2
hns@ldv.e-technik.tu-muenchen.de

Zusammenfassung

Für einige Probleme aus dem Bereich des Netzwerkmanagements bietet es sich an, adaptive Regler zu verwenden. Dazu ist es wichtig, das Verhalten adaptiver Regler zu verstehen. Ein bestimmtes Modell zeigt komplexes, vielleicht chaotisches Verhalten, das durch die Theorie chaotischer Systeme beschrieben werden könnte. Entsprechende Fragen werden formuliert.

1. Problemstellung

Im Bereich des Managements von Rechner- und Kommunikationsnetzen [1] gibt es Aufgaben, die sich als Optimierungsproblem beschreiben lassen (Abbildung 1). So soll z.B. allen Netzbenutzern für eine Datenübertragung eine gewisse minimale Datenrate garantiert werden. Der klassische Ansatz sieht vor, daß der Netzbetreiber eine entsprechende Bandbreite fest installiert und reserviert. Bei WANs (Wide Area Networks) stellt sich aber das Problem, daß eine Koordinierung vieler Wünsche kompliziert wird und eine feste Reservierung ineffizient ist. Eine Automatisierung ist daher erstrebenswert.

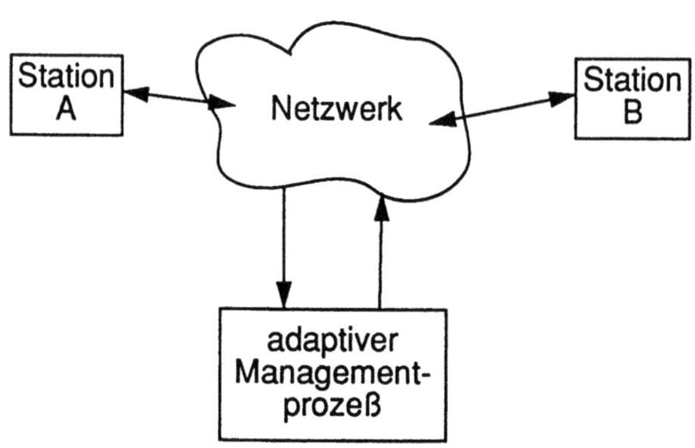

Abbildung 1: Optimierung in Kommunikationsnetzen

Daher bieten sich lokale adaptive Entscheider an, die lernen, wie sie in ihrem Zuständigkeitsbereich (Private Management Domain) eingreifen müssen, indem sie die Netzressourcen (Managed Objects) dynamisch beeinflussen. Mit herkömmlichen Reglern ist hier kein Erfolg zu erwarten, da die Entscheidungen in diskreter, nichtlinearer und sich zeitlich ändernder Weise erfolgen müssen. Außerdem ist es kaum möglich, das Streckenverhalten zu bestimmen und dann einen angepaßten Regler zu konstruieren. Der Vorteil selbstorganisierender Systeme wäre, daß sie auf Änderungen in der Netzstruktur und Ausfälle flexibel und automatisch reagieren können. Ein Beispiel für die Dienstgüteregelung (Quality of Service) wird in [2] für das ATM-Netzwerk untersucht.

Barto et. al. [3] haben einen adaptiven Regler[1] auf neuronaler Basis beschrieben, der lernen kann, einen Stab auf einem Wagen zu balancieren (inverses Pendel), wobei der Wagen nur durch eine konstante Beschleunigung nach links oder rechts beeinflußt werden kann (diskrete Entscheidung). Der Regler (ein Neuron, ASE) wählt regelmäßig für den aktuellen (quantisierten) Streckenzustand eine Aktion aus (Abbildung 2). Die Adaption des Regelverhaltens wird durch einen Überwacher gesteuert, der grobes Fehlverhalten feststellt (Reinforcement) und durch Veränderung der Parameter des ASE in Zukunft zu vermeiden sucht. Eine weitere Komponente im Regler ist ein Prädiktor (ACE), der das drohende Fehlverhalten abhängig vom Streckenzustand kontinuierlich zwischen kritisch und unkritisch einteilt. Ein verwandtes System wird z.B. in [4] beschrieben.

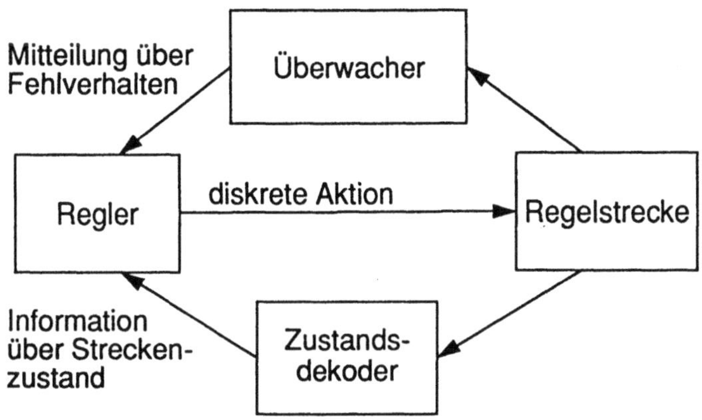

Abbildung 2: Struktur eines adaptiv geregelten Systems

Der Einsatz von Verfahren dieser Art bietet sich für das Managementproblem an, weil

- die Strecke ungenau bekannt und nichtlinear ist (d.h. es ist keine Arbeitspunktsfestlegung und Linearisierung z.B. für einen PID-Regler möglich),
- sich das Streckenverhalten gelegentlich gravierend ändern kann (Ausfall oder Umkonfigurierung im Rechnernetz),
- nur eine diskrete Menge von Aktionen möglich ist (z.B. Leitung benutzen oder nicht), und
- ein gewisser Spielraum für die zulässigen Streckenzustände vorhanden ist und damit keine genaue Sollwertbeschreibung vorliegt (z.B. garantierte Mindestübertragungsrate).

Allerdings gibt es einige Unterschiede zum inversen Pendel:

- Die Zahl der Aktionsalternativen kann hoch sein,
- mehrere Regler können/müssen zusammenspielen,
- die Strecke kann nicht durch nichtlineare Differentialgleichungen beschrieben werden, sondern eher als stochastischer Prozeß (abhängig vom Verkehrsaufkommen), und
- die Strecke kann eine größere Totzeit aufweisen (Übertragungsdauer von Aktionen, Messung von Parametern in hohem Zeitabstand).

1. Regler im Sinne von Wirkungskreis, nicht Sollwertnachführung.

2. Verhalten adaptiver Regler

Experimente an einer Simulation des inversen Pendels [5] haben gezeigt, daß das Adaptionsverfahren bei einer gewissen, allerdings problemangepaßten, Einstellung der Parameter (Lernraten etc.) nach ca. 60 Versuchen konvergiert. Das System zeigt allerdings im eingeschwungenen Zustand Grenzzyklen, die vermutlich chaotisch sind. Grenzzyklen wären für die beabsichtigte Anwendung unkritisch, sofern bekannt ist, in welchem Bereich sie sich bewegen.

3. Fragestellungen

Für einen Entwicklungsingenieur, der solche adaptiven Systeme einsetzen möchte, wären daher unter anderem die Antworten auf die folgenden Fragen interessant, die vermutlich durch die physikalische Chaostheorie beantwortet werden können:

- ist das Grenzverhalten wirklich chaotisch?
- wie hängt es von Parametern der Strecke ab?
- wie hängt es von den Parametern des Reglers/Adaptionsverfahrens ab?
- wie kann die Reglerstruktur verbessert werden?

4. Literatur

[1] H.-G. Hegering, R. Valta, "Netzmanagement - Aufgaben und Architekturkonzepte", Tutorium der 6. GI/ITG-Fachtagung "Kommunikation in verteilten Systemen", Stuttgart, 20.-24. Februar 1989

[2] A. Hiramatsu, "ATM Communications Network Control by Neural Networks", IEEE Tr. on Neural Networks, Bd. 1, Nr. 1, S122-130, 1990

[3] A. G. Barto, R. S. Sutton, C. W. Anderson, "Neuronlike Adaptive Elements That Can Solve Difficult Learning Control Problems", IEEE Tr. on System, Man, and Cybernetics, Bd. SMC-13, Nr. 5, S834-846, 1983

[4] D. Nguyen, B. Widrow, "The Truck Backer-Upper: An Example of Self-Learning in Neural Networks", in: Advanced Neural Computers, Ed.: R. Eckmiller, Elsevier, North-Holland, 1990

[5] C. W. Mack, "Analyse adaptiver Verfahren am Beispiel eines instabilen nichtlinearen Reglersystems", Diplomarbeit LSt. f. DV, TU München, 1991

Selbstlernende nicht-algorithmische Prozeßregelung in einem strukturierten konnektionistischen System

Dieter Butz
Kratzer Automatisierung GmbH
Maxfeldhof 5–6
8044 Unterschleißheim

1. Abstract

Es wird eine konnektionistische Methode vorgestellt, die eine nicht-parametrische Regelung eines physikalischen Prozesses durch graduelle Verbesserung einer Güteschätzung im on-line-Betrieb ermöglicht. Wichtig ist dabei die Möglichkeit der assoziativen Speicherung von beliebigen Stützstellen. Die beschriebene Regelaufgabe zeigt an einem Beispiel wie ein Regler mit einem Konnektionistischen System (KS) realisiert werden kann.

2. Aufgabenstellung und Lösungsstrategie

Wir gehen von einem unbekannten Prozeß P aus, von dem zur Zeit t_i ein 1-dim., wertequantisiertes Signal x_n seines inneren Zustandes gemessen wird; x_n dient zudem als Trigger für den folgenden Regelzyklus. Abhängig von dem Wert x_n produziert ein Regler ein 1-dim., wertequantisiertes Stellsignal $u_n(x_n)$, welches durch die Transformationsfunktion von P auf eine Gütefunktion $Q(u_n(x_n))$ abgebildet wird, die die Qualität von u_n in impliziter Abhängigkeit von der Zustandsgröße x_n beurteilt (Abb. 1).

Es werde nun angenommen, daß es für jedes x_n genau eine optimale Stellgröße $u_n = u_n^{opt}(x_n)$ gibt, für die die Güte maximal wird. Die Aufgabe besteht somit darin, durch schrittweise Verbesserung der Funktion $u_n(x_n)$ die Güte der Prozeßregelung zu erhöhen.
Das Problem ist gelöst, wenn es zu jedem Zustandswert x_n innerhalb der Meßgenauigkeit ein eindeutiges Stellsignal $u_n^{opt}(x_n)$ gibt, für das die Güte maximal wird; d.h. wenn es ein $\epsilon > 0$ gibt mit

$$|u_n(x_n) - u_n^{opt}(x_n)| < \epsilon \quad \text{für alle } x_n \text{ und } \epsilon \rightarrow 0 \tag{1}$$

Damit läßt sich das Extremwertkriterium für den Gradienten der Güte wie folgt formulieren:

$$(\Delta Q / \Delta u) \rightarrow 0 \quad \text{für alle } u_n(x_n) \tag{2}$$

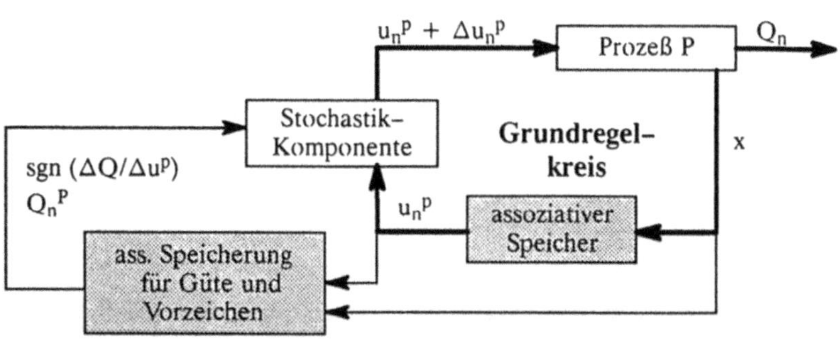

Abb. 1. *Blockschema des Gesamtsystems: Grundregelkreis und Steuereinheit für den Stochastikanteil. Nicht eingezeichnet: Entscheidungsinstanz zur Steuerung der Lernvorgänge in den assoziativen Speichern im Anschluß an einen Regelzyklus.*

Zur graduellen Verbesserung der Regelgüte beginnt das KS, eine geschätzte Stellgröße $u_n^p(x_n)$ zu liefern (p: "predictive"). Das Erreichen der optimalen Regelgüte geschieht durch zusätzliche Information über den Gradienten $(\Delta Q/\Delta u^p)$. Dazu schätzt das KS in zwei weiteren Teilkomponenten die zu u_n^p gehörigen Güte $Q^p(u_n^p,x_n)$ und das Vorzeichen des Gradienten $sgn(\Delta Q/\Delta u^p)$. Die Schätzwerte stammen implizit aus dem Datenmaterial der Vergangenheit (s.u.) und dienen einer gerichteten stochastischen Beeinflussung von u_n^p. Die Vorzeichenschätzung erlaubt die Richtung des Stochastikanteils von u_n^p zu bestimmen, für die der Gradient positiv angenommen wird. Die Schätzung der Güte Q^p dient zum späteren Vergleich mit der tatsächlichen Güte, die aus dem realen Prozeßablauf mit der zufallsbehafteten Stellgröße resultiert.

3. Assoziative Informationsspeicherung

Die Informationsspeicherung geschieht in einem strukturierten KS mit mehreren, sich im Aufbau gleichenden Einheiten, von denen jede zur Schätzung einer bestimmten Input-Output-Relation dient. Jede funktionelle Einheit besitzt zwei Neuronenschichten mit einer dazwischenliegenden Kopplungsschicht. Entscheidend für eine Speicherung von nicht-linearen, funktionalen Abhängigkeiten (z.B. $u(x) = \sin x$) in einer einzigen Kopplungsschicht sind drei Umstände:

a. Die Neuronen einer Schicht bilden eine wohlgeordnete Menge auf dem Zahlenstrahl; jedem Neuron s ist genau eine Zahl x_s zugeordnet. Dadurch besitzt es genau zwei nächste Nachbarn x_{s-1} und x_{s+1}.

b. Für die Inputneuronen wird zur Kodierung numerischer Werte die von Geiger [1] vorgeschlagene topologische Kodierung verwendet, die zur Zahlendarstellung ein verteiltes Aktivitätsmuster erzeugt. Die Aktivität a_s eines Neurons s beinhaltet somit den Abstand der Zahl x zu dem Neuronenstandort x_s. Dies ist Ausdruck der topologischen Ordnung der reellen Zahlen auf dem Zahlenstrahl.

c. Die Interpretation der Aktivitäten einer Ausgangsschicht geschieht über Berechnung des Schwerpunkts. Die Verbindungsstärken können den Schwerpunkt direkt durch ihre Stärke und Richtung beeinflussen.

Zur iterativen Speicherung der Information im KS wird eine modifizierte Δ-Regel [2,3] verwendet, welche die am postsynaptischen Neuron vorhandene Differenz zwischen Soll- und Istaktivität und die momentane Kopplungsstärke benützt.

Wichtig ist die Tatsache, daß sich das KS selbständig die impliziten Abhängigkeiten der Variablen durch schrittweises Lernen erschließt. Deshalb wird diese Informationsgewinnung als "selbst-lernend" bezeichnet.

4. Simulationsergebnisse

In der durchgeführten Simulation wurde eine sinusförmige Abhängigkeit der optimalen Stellgröße $u^{opt}(x)$ von der Zustandsgröße x gewählt.

$$u^{opt}(x) = 500 + 100 \bullet \sin(2\pi x/200) \quad \text{mit } x \in [0,1000] \pm 1 \tag{3}$$

Der Regler wurde mit einer willkürlich gewählten Funktion $u^{init}(x) = x$ vorinitialisiert (siehe Abb. 2).

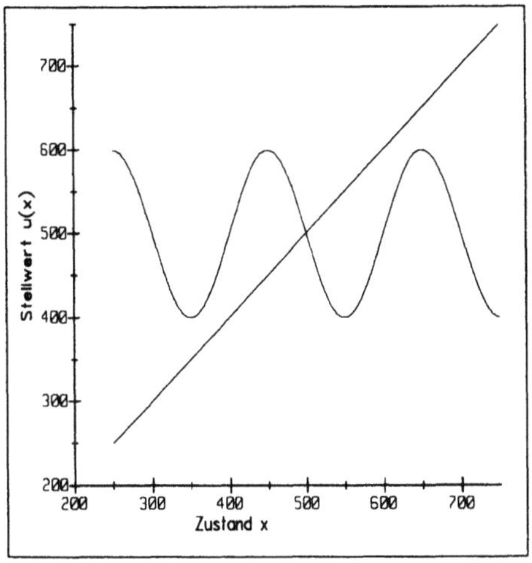

Abb. 2. *Die Funktionen* $u^{init}(x)$ *und* $u^{opt}(x)$

Die Gütefunktion zur Beurteilung des Reglerverhaltens wurde wie folgt ermittelt:

$$Q(u, u^{opt}) = \exp\left[-(u-u^{opt})^2/0.005\right] \qquad (4)$$

Die Größe des Netzwerks als assoziativer Speicher wird anhand der Reglerkomponente verdeutlicht: Die Abbildung eines auf 0.1% genau zu kodierenden Inputintervalls auf ein gleich genau zu kodierendes Outputintervall erfolgt mit je 101 topologisch angeordneten Neuronen in den 2 Neuronenschichten, wobei jedes Outputneuron mit jedem Inputneuron unidirektional gekoppelt ist (10101 Kopplungen).

Um Randeffekte zu vermeiden, wurden die Prozeßzustandswerte x nur aus dem Intervall von [250,750] gewählt. Abb.3 zeigt die sukzessive Verbesserung des Reglers nach 100 und 500 Regelzyklen.

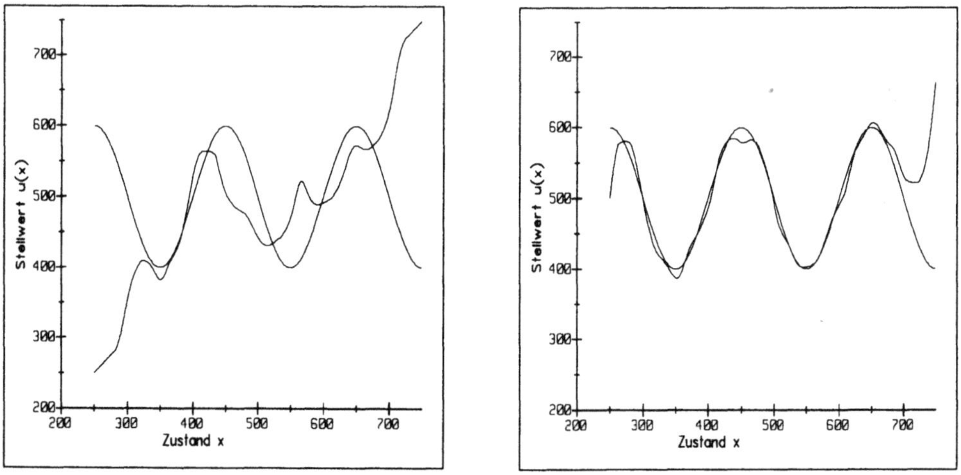

Abb. 3. *Reglerverhalten nach 100 (a) bzw. 500 (b) Regelzyklen im Vergleich mit der optimalen Stellkurve*

5. Diskussion

Aufgrund der sich überlappenden rezeptiven Bereiche der Inputneuronen und damit der Ausprägung der Kopplungen durch Lernen an benachbarten Stellen, ist es möglich, einen assoziativen Speicher mit dynamischer Stützstellenzahl zu realisieren.
Es wurde gezeigt, daß diese Art der on-line-Speicherung sinnvoll in Regelaufgaben zu verwenden ist. An der Regelung multivariabler Strecken ohne explizites mathematisches Modell mit dieser Methodik wird momentan gearbeitet.

6. Literatur

[1] Geiger, H.; Storing and Processing Information in Connectionist Systems. In: R. Eckmiller (Hrsg.). Advanced Neural Computers, International Symposium on Neural Networks for Sensory and Motor Systems (NSMS 1990), Proceedings, Elsevier Science Publishers, Amsterdam, 1990.

[2] Waschulzik, T., Geiger, H.; Theorie und Anwendung strukturierter Konnektionistischer Systeme. In: G. Dorffner (Hrsg.). Konnektionismus in Artificial Intelligence und Kognitionsforschung. 6. Österreichische Artificial-Intelligece-Tagung (KONNAI), Proceedings, Springer-Verlag, Berlin, 1990.

[3] Widrow, B., Hoff, M.E.; Adaptive Switching Circuits, Western Electronic Show and Convention, Convention Record IV, S. 96-104, 1960.

Neue Rechnerprinzipien und Materialien

G. Mahler
Universität Stuttgart

B. Schürmann
Siemens AG, München

Der Themenkreis "Neue Rechnerprinzipien und Materialien" versucht, eine Brücke zu schlagen zwischen physikalischen Anforderungen und Bedingungen spezifischer Materialien und darauf zugeschnittenen Anwenderstrategien. Hintergrund solcher Überlegungen ist der zu beobachtende Trend, daß ein Grundtheorem der Informatik, nämlich die Unabhängigkeit von Rechnerprinzipien und deren Realisierung, gerade durch fortschrittliche Technologien zunehmend in Frage gestellt wird: Strukturell homogene Netzwerke, wie etwa neuronale- oder Quanten-Netze, erfordern als verteilte Rechner sowohl andersartige Benutzer- wie auch Design-Prinzipien.

Die Nähe zu biologischen Systemen (Gehirnmodelle) wird in den beiden Übersichtsbeiträgen Görz und Kinzel deutlich. Während Herr Görz Beobachtungen über menschliche Gedächtnis-Leistungen zur Optimierung assoziativer Speichermodelle auszunutzen sucht, diskutiert Herr Kinzel eine statistische Mechanik des Lernens. Im Beitrag von Herrn van der Meer kommt unter anderem die Grundproblematik, die Beherrschbarkeit komplexer Systeme (im Sinne einer Verifikation des postulierten Verhaltens) zur Sprache. Hierzu wird von Herrn Ramacher eine Analyse neuronaler Algorithmen im Rahmen einer Hamilton'schen Dynamik und eines Extremalprinzips vorgeschlagen. Herr Schuster diskutiert eine aus der Nachrichtentechnik bekannte Größe, die Transinformation, die den Informationsaustausch zwischen Neuronengruppen mit hoher Auflösung erlaubt. Der Beitrag von Herrn Hofmann beschäftigt sich mit optischer Datenverarbeitung am Beispiel von Matrixmultiplikationen. Herr Körner schließlich schlägt ein Hybrid-Konzept zur optischen Steuerung eines (zukünftigen) VLSI-Netzes vor, welches auf logische Charakterisierung verzichtet, zugunsten direkter (physikalischer) Input-Output-Relationen.

In der sehr lebhaften Diskussion wird die Frage aufgeworfen, ob konventionelle optische Datenverarbeitung gegenüber der sich rasch weiterentwickelnden Elektronik noch eine echte Durchbruchchance haben wird (Folberth). Die Rolle der Grundlagenforschung, insbesondere der Theorie, wird von mehreren Diskussionsrednern aufgegriffen (u.a. Kornwachs, Mahler, Mühlenbein, Hübler). Einerseits wird von einer Preisgabe analytischen Grundverständnisses im Zuge bloßer Beobachtungen an Computerexperimenten gewarnt, andererseits werden Funktionsaspekte als willkommenes Ordnungsschema zur Charakterisierung der komplexen Dynamik synthetischer Nanostrukturen gesehen.

Verteilte Repräsentation

G. Görz, Universität Erlangen-Nürnberg

Erweiterte Zusammenfassung

Einleitende Bemerkungen. Wir stellen ein Speichermodell vor, das hochgradig parallel, verteilt und dünn besetzt ist. Dabei konzentrieren wir uns auf einen Ansatz von Kanerva [1]. Ein Grund für diese Wahl aus einer Vielfalt von Ansätzen, die ähnliche Eigenschaften haben, liegt in seiner Ableitung aus konventionellen Rechnerarchitekturmodellen. Es besteht die Erwartung, zugleich neue Ansatzpunkte zu gewinnen für die Diskussion über die scheinbare (?) Kluft zwischen subsymbolischer und symbolischer bzw. diskreter und analoger Repräsentation und Verarbeitung.

Repräsentationen und die kognitivistische Hypothese. Die Repräsentationsdebatte in der Kognitionswissenschaft und Künstlichen Intelligenz (KI) ist gekennzeichnet durch zwei kontroverse Positionen. Auf der einen Seite behauptet die kognitivistische Hypothese, daß das Gehirn eine Repräsentation der Welt erzeugt, die kompositional ist. Den Extremfall stellt Fodors "Sprache des Geistes" dar. Auch der gegenwärtige Konnektionismus ist eindeutig auf der repräsentationalistischen Seite. Dem steht die Auffassung gegenüber, daß es zwar mentale Repräsentation, aber keine Repräsentationen gebe. Diese wird etwa vom radikalen Konstruktivismus vertreten und von einigen Neurophysiologen, indem behauptet wird, daß es kein Abbild der Welt im Gehirn gebe, sondern nur Veränderungen von Verhaltensdispositionen. Da in beiden Fällen *etwas* zu repräsentieren ist, werden wir am Ende die Frage stellen, ob das vorgestellte Speichermodell im Sinne einer Implementationstheorie Mittel bereitstellt, um Simulationen in jedem der beiden "Paradigmen" durchzuführen. Die Einordnung als Implementationstheorie geht auf einen Vorschlag von Marr für den Entwurf und die Realisierung komplexer informationsverarbeitender Systeme zurück. Demnach sind drei Ebenen zu unterscheiden: die berechnungstheoretische (was?, warum?) zur Spezifikation des Ziels, die algorithmische (wie?) zur Spezifikation von Repräsentationsformalismus und Algorithmus und die implementierungstechnische zur Realisierung einer abstrakten Maschine, die die Aufgabe löst.

Speicher- und Gedächtnismodelle. Vor allem durch Performanzaspekte bestimmt gibt es eine Reige von Desiderata, die ein Speicher- oder Gedächtnismodell erfüllen soll: Assoziativität, Unschärfe, Approximativität, Robustheit, Verteiltheit und Parallelismus. Die von Kanerva entwickelte Theorie hat das Ziel, ein Speichermodell aus neuronenartigen Komponenten aufzubauen in der Absicht, eine *physiologisch plausible* Theorie der Repräsentation zu finden. Eine wichtige Rolle dabei spielt die Frage, ob das menschliche Gedächtnis ein inhalts-adressierbarer Speicher ist und was es heißt, zu sagen, daß es assoziativ ist.

Kanervas Speichermodell: Verteilter, dünn besetzter Speicher. Im folgenden wird die von Kanerva entwickelte Theorie eines verteilten, dünn besetzten Speichers (SDM) skizziert.

- Als mathematische Grundlage dient eine Untersuchung des n-dimensionalen Raums binärer Vektoren. Der Raum $\{0,1\}^n$ für große n $(100 < n < 10000)$ hat Eigenschaften, die intuitiven Vorstellungen über Relationen zwischen Begriffen entsprechen, sodaß es geeignet erscheint, Gedächtnis-Elemente als Punkte des Raums $\{0,1\}^n$ zu repräsentieren. Die gespeicherten und zugegriffenen Daten sind Muster aus Nullen und Einsen mit n Komponenten.

- Aus einer Darstellung der Prinzipien der Speicherarchitektur des klassischen Universalrechners, insbesondere der Adreßdekodierung und ihres inhärenten Parallelismus, ergeben sich Anhaltspunkte für eine Verallgemeinerung zu einem Speichermodell, das die gewünschten Eigenschaften (s.o.) aufweist. Aus der Theorie der formalen Neuronen werden werden für das Speichermodell relevante Eigenschaften untersucht, insbesondere das Konzept der linearen Schwellenfunktion.

- Daraus ergibt sich eine neue Interpretation der Funktion formaler Neuronen. Eine lineare Schwellenfunktion — ein formales Neuron — ist eine "Pattern-Matching"-Einheit für eine Menge von Mustern in der Umgebung eines spezifischen Musters, der sog. *Adresse* des Neurons. In der Tat ist die Adressen-Dekodierung in einem konventionellen Speicher ein Spezialfall der Adressen-Dekodierung durch ein Neuron. Formale Neuronen mit n Eingaben sind somit ideal als Adreßdekodierer eines Direktzugriffs-Speichers mit dem Adreßraum $\{0,1\}^n$.

- Bei der Konstruktion eines Speichers mit einem riesigen Adreßraum (z.B. 2^{1000} Adressen), ohne daß eine sehr große Zahl an Speicherlokationen bereitgestellt werden muß, tritt eine Reihe praktischer Probleme auf. Eine kleine zufällige Stichprobe (2^{10}) von Speicherlokationen aus den möglichen 2^{1000} leistet die Speicherfunktion, wenn Kopien jedes Elements in vielen Lokationen gespeichert werden. In diesem Fall gibt es eine kritische Distanz von 450 bit um eine Adresse, innerhalb der alle Lokationen bedient werden.

Funktionalität des SDM. Sie ergibt sich aus einer Untersuchung der Lese- und Schreiboperationen. Das Hauptergebnis besteht im Aufweis der Durchführbarkeit einer mathematischen Behandlung dieses verteilten dünn belegten Speichers (SDM). Es besteht eine Entsprechung der formalen Eigenschaften des Modells zu dem "Wissen über das, was man weiß" und zu der approximativen Erreichbarkeit von Elementen.

- Das *Schreiben* eines Worts (1000 bit) in multiplen Kopien innerhalb der kritischen Distanz erfolgt durch paralleles Inkrementieren bzw. Dekrementieren der (1000) Zähler der adressierten Lokation.

- Zum *Lesen* eines Worts werden die Inhalte der durch die Leseadresse aktivierten Lokationen (innerhalb der kritischen Distanz) gesammelt und der Inhalt des Worts nach der Majoritätsregel statistisch rekonstruiert. Das gelesene Wort ist also eine statistische Approximation; das Speichermodell ist sensitiv für Ähnlichkeit. Iteriertes Lesen (innerhalb der kritischen Distanz) *konvergiert* schnell (< 10) zum best-passenden Wort.

- Die *Kapazität* des SDM ist im Beispiel ca. 10^5 Worte bei 10^6 Lokationen.
- Wie soll Information gespeichert werden, um später, wenn die Situation es erfordert, auf sie schnell zugreifen zu können? Eine Lösung besteht darin, Sequenzen als verzeigerte Listen zu speichern. Für den Zugriff auf diese besteht ebenfalls die genannte Konvergenzeigenschaft.
- Der SDM kann als neuronales Netz mit drei Schichten interpretiert werden.

Neurophysiologische Analogien. Die Konstruktion eines SDM aus neuronenartigen Komponenten führt zu einer Methode der Speicherung der Daten: Eine Speicherlokation kann als Menge von n Zählern realisiert werden, wobei für jede bit-Position einer vorhanden ist. Die resultierende Struktur ähnelt überraschenderweise Strukturen, die man in der Rinde des Kleinhirns findet. Daher liegt es nahe, nach Analogien zu suchen. In den dort vorhandenen Zelltypen kann aufgrund struktureller und quantitativer Ähnlichkeiten eine klare Zuordnung von Teilaufgaben vermutet werden: Ein- und Ausgabeleitungen, Adreßleitungen, Adreßdekodierer, bit-Lokationen, Datenspeicher, Aufsummation zur statistischen Rekonstruktion von Ausgaben und Einstellung der Schwellwerte. Dies legt die Hypothese nahe, daß die Kleinhirnrinde ein verteilter Direktzugriffs-Speicher ist.

Aspekte der Realisierung und Anwendung. Die (Re-)konstruktion autonomer lernender Systeme, wofür Menschen und Tiere natürliche Vorbilder sind, dient als Anwendungsrahmen für das Speichermodell. Die im repräsentationalistischen Sinn entscheidende Überlegung besteht darin, daß die Operationalität solcher Systeme auf einem internen Modell der Welt beruht, das sie durch Erfahrung aufbauen, und daß ein SDM in idealer Weise zur Speicherung eines prädiktiven Weltmodells geeignet ist. Die zeitliche Dimension ist in Zugriffspfaden eingeschlossen über den Ort, an dem Daten gespeichert werden. Das interne Modell erzeugt Erwartungen und vervollständigt partielle Wahrnehmung der Außenwelt. Auch komplexere Strukturen als bloße Vektoren von binären Merkmalen können auf das Speichermodell abgebildet werden. Die Adressierung des Speichers muß nicht exakt sein; sie besitzt eine Attraktoreigenschaft. Die prädiktive Kraft des Modells resultiert aus seiner Fähigkeit, Sequenzen zu speichern und zu generalisieren. Doch auch Vertreter der Richtung, die behaupten, daß es kein Abbild der Welt im Gehirn gibt, müssen Voraussetzungen für die Repräsentation der Dynamik von Verhaltensdispositionen annehmen. Es ist also zu vermuten, daß in beiden Fällen "Erfahrung" durch Folgen von Mustern im Fokus der Aufmerksamkeit darzustellen ist. Der Fokus ist — technisch gesprochen — ein kombiniertes Adressen-Daten-Register. Die Objekte werden nicht durch "direkte Abbilder" repräsentiert, sondern durch *Aktivierungsmuster,* die von den *Objekten verursacht* werden. Lernen wird ermöglicht durch eine skalare *Präferenzfunktion* über Zuständen. Die *Codierung* als Transformation von Eingabesignalen in eine für die Repräsentation geeignete Form wird vom sensorischen System durchgeführt. Nach Marr exisitiert ein massiv paralleles abgeschlossenes System zur Gestalterkennung, das keine Referenzmuster oder Wissen über bestimmte Objekte, die Welt, benutzt. Zusammenfassend können wir festhalten:

- Der Fokus ist assoziiert mit einer Menge von Zuständen von Neuronen. Er ist sehr klein im Vergleich zur Gesamtzahl der Neuronen.
- Die Information im Fokus ist zu jedem Zeitpunkt ein kleiner Ausschnitt aus der gesamten gespeicherten Information.
- Der Inhalt des Fokus dient als Adresse für den Speicher.
- Der Zugriff auf Information erfolgt iterativ über den Fokus durch approximative Schlüssel.
- Wahrnehmung und Aufmerksamkeit hängen vom Fokus ab.
- Assoziationen sind inhärent (Formähnlichkeit im Modell) oder gelernt (Invarianzen).
- Der Speicherinhalt ist hochgradig verteilt, Elemente werden statistisch rekonstruiert.
- Schreiben in und Lesen aus Speicher erfolgen massiv parallel.

Literatur

[1] Kanerva, P.: *Sparse Distributed Memory.* Cambridge, Mass.: The MIT Press, 1988

Physik lernender Netzwerke

Wolfgang Kinzel
Lehrstuhl Theoretische Physik III
Universität Würzburg
Am Hubland
8700 Würzburg

1. Einleitung

Seit etwa 6 Jahren beschäftigen sich weltweit theoretische Physiker mit der Theorie neuronaler Netzwerke. Es geht dabei um die Erforschung der kooperativen Informationsverarbeitung, also um die Frage, ob und wie ein großes Netzwerk von einfachen Einzelteilen und mit einfachen Mechanismen Informationen verarbeiten kann. Es wird untersucht, mit welchen Gesetzmäßigkeiten unendlich große Systeme fehlertolerant und inhalt-adressierbar speichern, selbsttätig lernen und verallgemeinern können. Es zeigte sich, daß man mit Modellen und Methoden der theoretischen Vielteilchenphysik allgemeine mathematische Gesetze zu den Eigenschaften neuronaler Netzwerke erhält.

2. Das Modell

Die Physiker benutzen dabei Modelle, die aus der Festkörperphysik kommen. Zum Beispiel ist ein vieluntersuchtes Modell zu magnetischen Legierungen (Spingläser) das Edwards Anderson Modell (1975), das mit folgender Energie definiert ist.

$$H = -\frac{1}{2} \sum_{i,j} J_{ij} S_i S_j \qquad (1)$$

Dabei bedeuten $S_i \epsilon \{+1,-1\}$ die Richtung der magnetischen Momente (Spins), und J_{ij} kennzeichnet die Stärke der magnetischen Wechselwirkung zwischen den Spins S_i und S_j. Die besonderen Eigenschaften der Spingläser entstehen durch den Wettbewerb zwischen positiven und negativen J-Werten.

Aber Gl.(1) kann auch als ein Problem der kombinatorischen Optimierung angesehen werden. H ist dabei die Kostenfunktion, zu der bei festen J-Werten die Konfiguration $\{S_i\}$ mit den geringsten Kosten zu finden ist. Falls die J-Werte zufällig verteilt sind, geben die Methoden der Festkörperphysik exakte Werte für die geringsten Kosten, aber auch Aussagen über die Zahl der stabilen Zustände als Funktion der Kosten.

Ferner kann Gl.(1) auch als neuronales Netzwerk verstanden werden. Dabei beschreiben die S_i–Variablen den Zustand des Neurons i: $S_i=+1(-1)$ bedeutet, daß Neuron i feuert (ruht). Die J_{ij}–Werte beschreiben die Stärke der synaptischen Kopplung, dabei ist wiederum der Wettbewerb von erregenden ($J>0$) und hemmenden ($J<0$) Synapsen wichtig für die kollektiven Eigenschaften. Für symmetrische Synapsen wird die Dynamik der Neuronen durch Gl.(1) beschrieben.

Als letzte Anwendung des Modells (1) soll hier ein weiterer Aspekt der neuronalen Netzwerke genannt werden: Lernen als selbstorganisierende synaptische Plastizität. Jetzt sind die S_i–Variablen die Stärken der Synapsen, und die J–Werte enthalten die zu lernende Information. Lernen ist dabei die Bewegung in einem komplexen Energie–Gebirge der Funktion $H(\{S_i\})$, mit sehr vielen lokalen Minima.

3. Statistische Mechanik

Die mathematischen Methoden zur Beschreibung unendlich großer wechselwirkender Systeme wurden schon am Anfang dieses Jahrhunderts von Boltzmann und Gibbs entwickelt. Seit 1975 wurden diese Methoden auf ungeordnete Systeme mit konkurrierenden Kopplungen (Spingläser) erweitert, und damit stehen der Physik Werkzeuge zur Verfügung, um neuronale Netzwerke zu berechnen.

Betrachte z.B. folgende Dynamik des unendlich großen Netzwerkes (1):

$$S_i(t+1) = \text{sign}\left[\sum_j J_{ij}\, S_j(t)\right] \tag{2}$$

Jedes Neuron S_i richtet sich nach seinem lokalen Feld aus, das von der Aktivität sämtlicher verkoppelter Neuronen S_j abhängt. Gl.(2) ist ein nichtlineares rückgekoppeltes System mit unendlich vielen Freiheitsgraden, also ein scheinbar unlösbares Problem. Da aber (bei symmetrischen $J_{ij}=J_{ji}$) die Dynamik (2) in jedem Zeitschritt t die Funktion H, Gl.(1), ein wenig erniedrigt, kann der stationäre Endzustand mit den Methoden der statistischen Mechanik berechnet werden. Dabei summiert man über alle Konfigurationen $\{S_i\}$ (bei N Neuronen sind dies 2^N viele!) und wichtet jeden Zustand mit dem Boltzmann–Faktor $\exp(-H/T)$. T ist ein Parameter (Temperatur), der bei der deterministischen Dynamik (2) nach den Rechnungen auf den Wert null gesetzt wird.

Es wird also die Zustandssumme

$$Z = \sum_{\{S_i\}} \exp[-H(\{S_i\})/T] \tag{3}$$

berechnet. Dies geschieht durch Sattelpunkts-Methoden und Einführen von "Ordnungsparametern", also durch Festhalten makroskopischer Eigenschaften. Wichtig ist nun, daß Z von den festen Kopplungen J_{ij} abhängt. Diese enthalten z.B. die zu lernenden Muster. Mit den Methoden der Festkörperphysik kann man nun die typischen Eigenschaften der Netze berechnen, indem man über alle möglichen (Zufalls-) Muster mittelt. (Man muß dabei log Z mitteln, dies geschieht mit der "Replika-Methode"). Man erhält damit exakte Aussagen über typische (=mittlere) Probleme.

4. Attraktor-Netzwerke

Dieser Netzwerk-Typ hat die Funktion eines Assoziativspeichers. Dabei soll eine Menge von binär codierten Informationen (Muster) mit je N Bits im Netzwerk der synaptischen Kopplungen J_{ij} gespeichert werden. Im betrachteten Grenzfall $N \rightarrow \infty$ wächst die Zahl der gespeicherten Muster wie αN; α ist die Speicherkapazität.

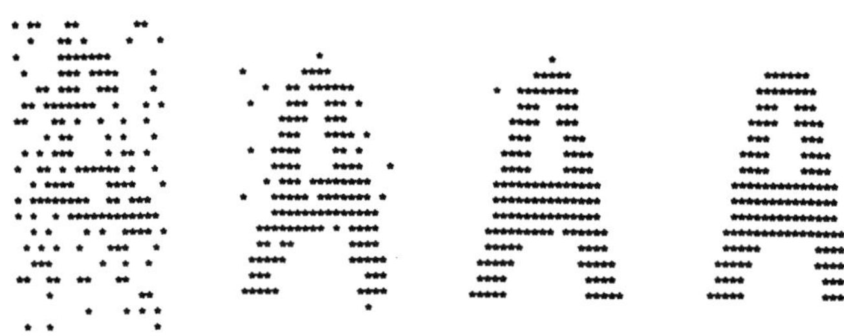

Abb. 1: Dieses Computermodell arbeitet mit 20 x 20 in einem Rechteck angeordneten "Neuronen". Die Sterne kennzeichnen die aktiven Neuronen. Nicht gezeigt sind die inaktiven Neuronen sowie das Netz, das die Neuronen koppelt. Es hat 30 Buchstabenmuster gelernt. Hier erkennt es in nur drei Schritten das aufgeprägte unscharfe Muster "A".

Die Kopplungen J_{ij} können entweder konstruiert werden, oder das Netzwerk kann selbsttätig lernen, d.h. die Kopplungen passen sich mit einfachen Mechanismen an die zu lernenden Muster an. Nach der Lernphase kann das Netzwerk dann mit der Dynamik Gl.(2) eine Teilinformation vollständig restaurieren (siehe Bild 1). Ein verrauschtes Bild oder ein fehlerhaft geschriebener Name werden wieder restauriert, zusätzlich nimmt das Netzwerk denjenigen Zustand an, der vorher als Muster gespeichert wurde. Die Muster sind Attraktoren der Vielneuronen–Dynamik.

Mit der statistischen Mechanik wurden viele Eigenschaften des unendlichen Netzwerkes exakt berechnet: Maximale Speicherkapazität für verschiedene Kopplungen, Wiedererkennungsfehler, Einfluß von Störungen, Einfluß von asymmetrischen Synapsen, etc.. Bemerkenswert ist, daß das Netzwerk extrem fehlertolerant und unabhängig von den Modell–Details funktioniert.

Ersetzt man z.B. die deterministische Dynamik Gl.(2) durch eine stochastische, so wird $T>0$ in Gl.(3) ein Maß für die Stärke des Rauschens. Die analytischen Rechnungen geben ein Gebiet in der T–α Ebene, in dem das Netzwerk als Assoziativspeicher arbeitet. An den Grenzen dieses Gebietes findet man oft die "Überladungs–Katastrophe": Erhöht man die Speicherkapazität ein wenig, so springt der Erkennungsfehler von einem kleinen zu dem maximalen Wert; das Netzwerk funktioniert nicht mehr. Dieser "Phasenübergang" ist analog zu Phasenumwandlungen in der Physik.

5. Perzeptron: Lernen und Verallgemeinern

Eine der bemerkenswertesten Eigenschaften von neuronalen Netzwerken ist ihre Fähigkeit. selbsttätig zu lernen. Ein Netzwerk lernt nicht durch Eingabe von Regeln und Programmen, sondern durch Anpassung an einen Satz von Trainings–Beispielen. Sowohl im biologischen Vorbild wie auch in den "Neurocomputern" passen sich die Stärken der synaptischen Kopplungen an die zu lernenden Muster an. Nach der Lernphase kann das Netzwerk aber nicht nur die gelernten Beispiele richtig reproduzieren, sondern es kann auch verallgemeinern, d.h. es kann auch neue Eingabemuster mit einer geringen Fehlerrate richtig abbilden.

Diese Eigenschaften kann man auch im Rahmen der statistischen Physik mit einfachen Modellen quantitativ studieren. Das einfachste Modell ist das wohlbekannte Perzeptron: Es besteht aus N Eingabe–Neuronen, einer Schicht von synaptischen Kopplungen und einem Ausgabe-Neuron S_0. Gl.(2) wird dann zu

$$S_0 = \operatorname{sign}\left[\sum_j J_j S_j\right] \qquad (4)$$

Lernregeln zu diesem Netzwerk sind schon seit 30 Jahren bekannt, z.B. ADALINE von Widrow und Hoff und der Perzeptron–Algorithmus von Rosenblatt. Aber erst jetzt kann man für typische (=Zufalls) Probleme die Speicherkapazitäten, die Verteilung der Lernzeiten und andere Eigenschaften mit den Methoden der Festkörperphysik exakt berechnen. Mit diesen Erkenntnissen konnten dann schnelle neue Lernalgorithmen entwickelt werden.

Auch die Fähigkeit zu verallgemeinern konnte für unendlich große Netzwerke exakt berechnet werden. Es zeigte sich, daß ein Netzwerk mit N Neuronen αN viele Beispiele lernen muß, um zu verallgemeinern. (Beachte, daß es 2^N viele Beispiele gibt). Dabei ist der Verallgemeinerungsfehler ϵ definiert als die Wahrscheinlichkeit, daß das Netzwerk (3) (= "Schüler") eine andere Antwort S_0 als die zu lernende Abbildung (= "Lehrer") gibt.

Für den Fall, daß der Lehrer ebenso wie der Schüler ein einfaches Perzeptron ist, erhielten wir z.B. folgende analytische Resultate:

1. Der ADALINE-Algorithmus versucht, ein lineares Problem perfekt zu lernen. Dies gelingt nur bis $\alpha=1$, aber auf Kosten der Verallgemeinerungsfähigkeit. Bei $\alpha=1$ kann der Netzwert nicht mehr verallgemeinern.
2. Der Perzeptron-Algorithmus lernt perfekt und verallgemeinert gut, der Fehler ϵ verschwindet wie $1/\alpha$ für $\alpha \to \infty$.
3. Das Netzwerk mit binären Synapsen $J_j \epsilon \{+1,-1\}$ verallgemeinert unstetig: Bei $\alpha=1.245$ springt der Verallgemeinerungsfehler auf $\epsilon=0$. Für $\alpha>1.245$ reproduziert der Schüler, der alle αN Beispiele perfekt lernt, den Lehrer vollkommen.
4. Ein Schüler, der entsprechend seinem Wissen selbst Fragen stellt, verallgemeinert sehr viel besser. Für den Perzeptron-Algorithmus verschwindet ϵ exponentiell schnell mit wachsendem α (statt $1/\alpha$ für Zufalls-Beispiele).

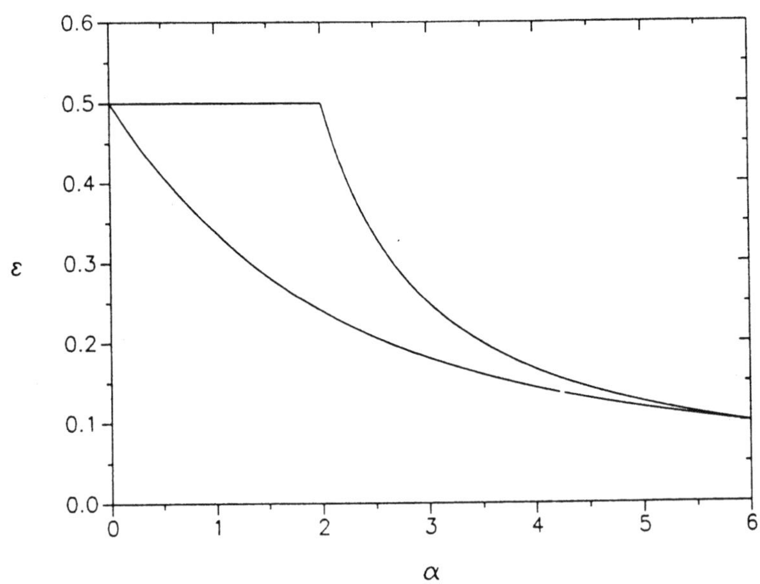

Abb. 2: Verallgemeinerungsfehler ϵ als Funktion der Anzahl αN der gelernten Beispiele für ein einfaches Perzeptron. Die obere Kurve zeigt eine Abschätzung von Cover während die untere Kurve das exakte Ergebnis für ein typisches Perzeptron wiedergibt.

5. Ein komplexer Schüler (hier ein Mehrschicht–Perzeptron) verallgemeinert schlechter als ein einfacher Schüler (hier ein Einschicht–Perzeptron, ebenso wie der Lehrer).

Diese Resultate stehen in direktem Zusammenhang mit Ergebnissen aus der Mathematik (Winder, Cover, Vapnik, Baum, Haussler...). Der Unterschied ist aber: Zum einen liefert die Informatik exakte Schranken für den schlechtesten Fall, während die Physik den typischen Fall berechnet. Zum anderen ergeben sich aus der Sicht der Mathematik ganz andere Abschätzungen als die Methoden der Physik. In Abb.2 ist dies am Beispiel des Verallgemeinerungsfehlers ϵ des Perzeptrons (2) als Funktion der Anzahl αN der gelernten Beispiele dargestellt. Dabei ist der Lehrer ebenfalls ein einfaches Perzeptron. Die obere Kurve ist eine Abschätzung von Cover während die untere Kurve mit den Methoden der Festkörperphysik berechnet wurde: Sie zeigt den Fehler für ein mittleres Perzeptron, das alle Beispiele exakt gelernt hat. Beides sind analytische Resultate.

6. Zusammenfassung

Die Funktionsweise neuronaler Netzwerke ist völlig verschieden von derjenigen heutiger Computer. Netzwerke benötigen keine Steuereinheit und kein Programm, sie verarbeiten die Information selbsttätig mit einfachen lokalen Mechanismen und als kooperativen Effekt. Netzwerke speichern die Information verteilt und inhalt–adressierbar, nicht in nummerierten Schubladen wie im Computer. Netzwerke sind extrem fehltertolerant, ihre Arbeitsweise ist ein mehr/weniger statt ja/nein, und ihre Eigenschaften sind robust gegenüber Modell–Einzelheiten. Lernen geschieht an Hand von Beispielen und nicht durch logische Regeln und Expertenwissen. Netzwerke können nach einer Lernphase verallgemeinern, ohne irgendwelche allgemeine Regeln einprogrammiert zu bekommen.

Zu allen diesen Eigenschaften kann man mit den Methoden der theoretischen Physik für einfache Netzwerke allgemeine Aussagen machen. In vielen Fällen findet man exakte mathematische Gesetze für unendlich große Netzwerke. Diese Relationen sind keine Abschätzungen für den schlechtesten (und unwahrscheinlichen) Fall, sondern sie betrachten den typischen (= wahrscheinlichsten) Fall.

Deshalb glauben wir, daß die Physik neuronaler Netzwerke wichtige Grundlagen liefert, die auch für die Neurobiologie und die Informatik von Interesse sind.

Einführende Literatur zu Physik neuronaler Netzwerke, der man auch die Referenzen zu Originalarbeiten entnehmen kann:

Neural Networks, An Introduction
B. Müller and J. Reinhardt, Springer Verlag 1990

Introduction to the Theory of Neural Computation
J. Hertz, A. Krogh and R.G. Palmer, Addison Wesley 1991

Dynamics of Learning
W. Kinzel and M. Opper, in: Models of Neural Networks,
ed. by E. Domany, J.L. van Hemmen and K. Schulten, Springer Verlag 1991.

Architektur künstlicher neuronaler Netze und die darauf modellierbaren Problemfelder der Fuzzy Logic Application

Marius van der Meer
Deutsche Forschungsanstalt für Luft- und Raumfahrt
PT Informationstechnik
Rudower Chaussee 5
O-1199 Berlin-Adlershof

0. Motivation

Die hier niedergelegten Gedanken erheben weder Anspruch auf Vollständigkeit noch auf "letzte Wahrheiten". Es soll vielmehr im Geiste der Veranstaltung ein Gespräch angeregt werden über die faszinierenden Berührungspunkte zwischen der Theorie Neuronaler Netze und den Systemen der unscharfen Logik. Beide gehören nach Auffasung des Autors noch viel enger zusammen, als durch die verschiedenen Ursprünge beider Gebiete heute allgemein akzeptiert ist. Diese "große Unifikation" wird mit Sicherheit zu wesentlich flexibleren Fuzzy-Systemen einerseits und besser verstehbaren künstlichen neuronalen Netzen andererseits führen.

1. Gehören neuronale Netze zur KI?

1.1. Konnektionismus versus KI

Ein aktueller Streitpunkt zwischen Künstlicher Intelligenz und Konnektionismus ist die Durchschaubarkeit, also Nachvollziehbarkeit dessen, was ein künstliches neuronales Netz - z.B. als Kern eines entsprechenden Expertensystems - an Schlüssen gezogen hat, durch den Menschen.

Das bedeutet, daß in letzter Konsequenz neuronale Computer mit drastisch höherer Leistungsfähigkeit als herkömmliche serielle oder parallele Computerarchitekturen in grober Näherung biologischen Nervensystemen sowohl in ihren Vor- wie auch ihren Nachteilen vergleichbar werden. Die Nachteile wären wohl neben der bekannten Unsicherheit oder Ungenauigkeit in komplexen formalen Rechenoperationen (ohne externe Speicher) vor allem in der Eigenheit zu sehen, daß die Wissensstrukturierung nicht in Form expliziter Regeln angebbar ist. Wahrscheinlich sind nicht einmal implizit stets Regeln vorhanden, die diese Bezeichnung verdienen, sondern eben aus einer Vielzahl von Beispielen gewonnene "Erfahrungen" in Gestalt einer synaptischen Matrix. Das sich daraus ergebende Paradoxon lautet, daß in dem Maß, wie die Verarbeitungsstrukturen sich dem biologischen Vorbild nähern, sie sich dem Verständnis für die zugrundeliegenden Strukturen dieser Wissensbasis entziehen [Cr]. Der Mensch wird daher oftmals die Gefahr einer unkontrollierbaren Eigendynamik argwöhnen (auch dort, wo sie tatsächlich unbegründet sein mag), wenn sich beispielsweise im Bereich der Expertensysteme Probleme aus der mangelhaften Transparenz ergeben. Neuronale Netze lassen nicht ohne weiteres ihre Argumente für eine vorgeschlagene Lösung explizit hinterfragen. Erst durch eine dedizierte Einbindung in hybride Überwachungs-Strukturen in Form "klassischer" Rechnerarchitekturen könnte solch eine Erklärungskomponente nachträglich dem Gewohnten "klassischer" Expertensysteme angeglichen werden. Ein Beispiel hierfür wird von [Pa] vorgeschlagen. Hier ist also durchaus ein offenes Aufeinanderzugehen von Konnektionisten und Künstlichen Informatikern geboten, das sich in der gegenwärtigen Phase der Neuroinformatikforschnung hierzulande ansatzweise entwickelt [Pa].

1.2. Warum sind NN undurchschaubar?

Künstliche neuronale Netze gehen vom Existenzbeweis der Natur für die biologische Realisierung von Wissensrepräsentation und analytischem Denken aus und versuchen, die in der Natur beobachteten mentalen Leistungen durch Elementaranalyse zu erklären, um sie nachzuvollziehen. Das geschieht also in modelltheoretischem Sinne gänzlich gegensätzlich zum Vorgehen der Künstlichen Intelligenz, die die Systemanalyse und Modellbildung geistiger Leistungen durch formalisierte und möglichst noch algorithmische Beschreibung zu vollziehen versucht. Erst nach dieser top-down Analyse wird die Implmentierung in geeignete technische Systeme möglich oder für möglich gehalten. Hierfür ist eigentlich das Verständnis der realen neuronalen Mechanismen, mit denen biologische Nervensysteme ihre Leistungen vollbringen, durchaus noch verzichtbar.

Katastrophal endet dieser Zugang, wenn es Probleme zu lösen gilt, für die entweder keine expliziten Algorithmen vorliegen oder die algorithmische Beschreibung alle Grenzen der Verhältnismäßigkeit überschreitet und letzlich auch die technische Leistungsfähigkeit verfügbarer Systeme überfordert.

Dagegen sind trotz oder wegen der Komplexität von gewöhnlichen Leistungen biologischer Nervensysteme (erst recht des menschlichen ZNS) wie Erkennen, Objektunterscheidung, Umweltanpassung, Erinnerung, situationsangemessenes Agieren diese Verhaltensweisen in ihrer Regelhaftigkeit weder dem Aktor noch dem Beobachter bekannt oder bewußt. Trotz der scheinbaren Einfachheit der verabeitenden Elemente wird diese Komplexität der Leistungen letztenendes hervorgebracht durch die massive Parallelität dieser Systeme: vom Ohr 30.000, vom Auge 10^6 abgehende Neuronensignale an die etwa tausendmal größere Anzahl von Neuronen in den primären Sinnesarealen der Hirnrinde; insgesamt im menschlichen ZNS etwa 10^{11} Neuronen (etwa soviel wie Sterne in unserer Milchstraße oder Galaxien im sichtbaren Universum) mit je etwa 10^4 synaptischen Eingängen, was 10^{15} synaptische Kopplungen insgesamt ergibt [Th].

Neuronale Netze leisten daher äußerst hochdimensionale Input-Output-Abbildungen, und zwar in äußerst flexibler, trainierbarer Weise. Dabei fehlt jegliches vorgegebenes mathematisches Modell dieser Abbildungsfunktion - sie arbeiten modellfrei. Außerdem leisten sie Erkennungsfunktionen für Sachverhalte, die nicht einmal analytisch vollständig beschrieben sind, leisten also eine Klassifikation ohne (explizite) Definition. Viele dieser Leistungen vollbringen in entsprechend niedrigdimensionalen Ansätzen auch künstliche neuronale Netzwerke als vieldimensionlae nichtlineare Systeme trotz der stark vereinfachten Architektur und der weitgehenden Uniformität der künstlichen Neuronen. Die langsame Dynamik eines solchen Systems besteht in der Verschiebung der Aktivierungsschwellen und synaptischen Kopplungsstärken in dem gesamten Netz, die schnelle Dynamik entspringt der Zustandsänderung infolge der von außen aufgeprägten sensorischen Reize. Beides dürfte sich schon aus Gründen der Dimensionalität einer letzten Analyse verschließen, was plausibel scheint, wenn man als Analogie das physikalische Drei-Körper-Problem der vollständigen Bahnanalyse aller Sterne der Milchstraße gegenüberstellt.

2. Können neuronale Netze fusselig sein?

2.1. Gibt es Schluß-Neuronen?

Mit der beginnenden Verbreitung der unscharfen Logik infolge der industriellen Nutzung theoretisch lange vorgedachter Verallgemeinerungen der klassischen Logik [Zi] verstärkt sich das Bemühen um das Verständnis einer intuitiv vermuteten Verwandschaft zu den neuronalen Netzen [Ko]. Auf den ersten Blick ist bereits durch die diametral entgegengesetzten Zugänge (die Verallgemeinerung der mathematischen Logik in der Fuzzy Logic einerseits und der bionischen Analyse natürlicher informationsverarbeitender Systeme andererseits) die Verschiedenheit oder sogar Unvereinbarkeit beider "Netze" viel eher zu vermuten. Dennoch soll hier der Gedanke weiter diskutiert werden, spezifische Eigenheiten des einen Modells im anderen fruchtbringend zu suchen oder zu implementieren.

Gemeinsam ist beiden Netzen die funktionale Zuordnung eines Inputs zu einem Output in einer aus Beispielen lernenden Weise. Dabei stellen diese Beispiele bei hinreichend komplexen Systemen sicher nur einen sehr kleinen Bruchteil der nach der jeweiligen Dimensionalität möglichen Vielfalt der Input-Muster (und entsprechend der Outputs) dar, sind also stark vorselektierte Proben.

Zwei wesentliche Unterschiede fallen sofort ins Auge: erstens in der Art der Proben für die Belehrung, zweitens in der Art der Schlußfolgerung eines Output-Signals aus einem gegebenen Input-Muster.

Während bei Fuzzy-Systemen die Input-Proben aus Werte-Tupeln geeigneter unscharfer (z.B. linguistischer) Variabler in Form von Ausprägungsgraden bestehen, erhalten neuronale Netze Werte-Tupel als Input, die entsprechend der Dimensionalität der Input-Schicht gewisse scharfe (oder verrauschte) Punkte in einem unstrukturierten Input-Raum darstellen. Während das Fuzzy-System aus seinen eingegebenen Ausprägungsgraden mittels eines linguistischen Regel-Netzes die Ausprägungsgrade der verschiedenen möglichen Schlußfolgerungen nachvollziehbar ermittelt, erzeugen die Inputs in einem neuronalen Netz infolge der nichtlinearen Übertragungsfunktionen der einzelnen Neuronen ein ganz spezifisches Aktivierungsmuster in den verdeckten Schichten und natürlich auch in der Outputschicht. Das vollzieht sich, ohne eine andere Begründung für Aktivierungen und synaptische Kopplungsstärken angeben zu können, als daß letztere bei der Belehrung des Systems sich so eingestellt haben und die reizabhängigen Aktivierungen damit festzulegen. Wenn auch letzendlich große Aktivierungs- Muster (räumlich oder auch zeitlich) strukturiert mit Erkennungsleistungen oder Assoziationen in neuronalen Netzen verbunden sein mögen, so ist doch für Output-Schichten künstlicher neuronaler Netze ebenso wie für Fuzzy-Aktor- Steuerung die Idee eines "Großmutter"-Neurons oder einer entsprechenden Schlußregel ein Mittel zur besseren Verständlichkeit der absolvierten Schlußprozesse.

3. Unscharfe Logik in neuronalen Netzen

3.1. Was ist fusselig an den Neuronen?

Die oft beklagte Unschärfe oder gar fehlende Logik menschlichen Denkens ist geschuldet der ungeheuren Komplexität der zu bewältigenden Aufgaben und des dafür herangezogenen Vorwissens. Dazu kommt noch eine oft ebenfalls schwer faßbare Unischerheit durch die Leistungsgrenzen der verfügbaren Sensorik oder durch eingeschränkte Beobachtbarkeit oder durch zu kleine Stichproben. Das alles führt zu Unsicherheiten, die wenig mit Wahrscheinlichkeiten zu tun haben, weil jene eigentlich mathematisch-statistisch wohlbegründbar für große Stichproben kreiert wurden. Zu dieser Unschärfe oder Unsicherheit der sensorischen Eingangsgrößen für neuronale Netze tritt außerdem noch die Tatsache, daß nach menschlichem Verständnis durchaus viele Elementareindrücke mit abgestuften Ausprägungsgraden sogenannter linguistischer Variablen beschreibbar wären. Interessanterweise ähnelt nun aber auch die in neuronalen Netzen zu beobachtende Output-Sigmoid-Funktion als Realisierung des Schwellenverhaltens des jeweiligen Neurons einer Flanke der Ausprägungsvariabilität von linguistischen Variablen. Das läßt darauf hoffen, durch geeignete Interpretation von Inputvariablen und inneren Variablen in neuronalen Netzen eine Analogie zwischen der Netzstruktur und der Struktur fuzzy-logischer Schlußregeln herstellen zu können. Wenn man einzelnen Neuronen-Inputs die Bedeutung fuzzifizierter Variabler zuweisen könnte, dann wäre mit der Interpretation des Neurons als Schlußregel und der Aktivierung des Neurons als Ausprägungsgrad des Schlusses eine Analogie vollzogen, die jedes einzelne künstliche Neuron als Inferenzglied eines Fuzzy-Netzes verstehen könnte.

Wohlgemerkt ist eine solche Analogiebildung jenseits von jeglicher Anwendbarkeit zur Interpretation biologischer Gegebenheiten. Für technische Realisierungen dagegen ließe sich auf diese Weise eine Brücke von theoretischen Modellen neuronaler Netze, ihrer Dynamik und Belehrbarkeit über die Konzepte unscharfer Logik bis hin zur Hardware-Realisierung für geeignete Anwendungsfälle schlagen. Dabei käme die Verwendung von vereinheitlichten Beschreibungshilfen für Hardware- wie Software-Knoten [Is] der jeweils angemessenen Realisierung als Schaltkreis oder Simulationssoftware zugute. Die unter dem Blickwinkel neuronaler Netze gewöhnliche Trainierbarkeit eines Netzes führt unter dem Blickwinkel des Fuzzy-Schlußregel-Netzes zu möglicherweise völlig neuen, vereinfachten und damit optimierten Regelsammlungen. Hierbei können topologische und evolutionstheoretische Zugänge, wie sie heute in der Theorie neuronaler Netze genutzt werden, der weiteren gezielten Optimierung und damit Verbreitung von Fuzzy- Steuerungen Auftrieb verleihen.

3.2. Ein Beispiel: Gray-Code

Ein anschauliches Beispiel für die möglichen Beziehungen zwischen künstlichen Neuronalen Netzen und Logischen Schaltungen, die als in Chips materialisierte Schlußregeln aufgefaßt werden können, soll im folgenden erläutert werden. Letztlich können diese Betrachtungen auch noch auf fuzzifizierte Variable und entsprechende Schlußregeln verallgemeinert werden. Es handelt sich um eine in der Form eines Umcodierers versteckte Realisierung der oft untersuchten logischen Funktion des exklusiven Disjunktion (XOR). Aufgabe des Netzes soll es sein, für eine gewisse Bitbreite normalen Binärcode in den sogenannten Gray-Code zu überführen. Die praktische Relevanz könnte in einer drastischen Minimierung und Homogenisierung von Bitwechseln bei Zählvorgängen sein: während beim Binärcode während des Zählens die niedrigste Bitposition bei jedem Inkrementieren mit Sicherheit ihre Bitbelegung und höhere Bitpositionen stets, wenn ein Übertrag auftrat ändern, wird beim Zählen im Gray-Code mit jedem Inkrement genau eine Bitposition geändert, die niedrigste dagegen stets genau nach zwei Inkrementierungen. Das bedeutet eine Verdoppelung der maximal möglichen Zählfrequenz bei gleicher physikalischer oberer Grenzfrequenz.

Table 1. Umkodieren von Binärcode in Gray-Code

x_3	x_2	x_1	x_0	$\longrightarrow$	y_3	y_2	y_1	y_0
0	0	0	0		0	0	0	0
0	0	0	1		0	0	0	1
0	0	1	0		0	0	1	1
0	0	1	1		0	0	1	0
0	1	0	0		0	1	1	0
0	1	0	1		0	1	1	1
0	1	1	0		0	1	0	1
0	1	1	1		0	1	0	0
1	0	0	0		1	1	0	0
1	0	0	1		1	1	0	1
1	0	1	0		1	1	1	1
1	0	1	1		1	1	1	0
1	1	0	0		1	0	1	0
1	1	0	1		1	0	1	1
1	1	1	0		1	0	0	1
1	1	1	1		1	0	0	0

Diese Umkodierung erfolgt logisch nach einer sehr einfachen Regel:

$$y_n := (x_{n+1} \vee x_n) \wedge \neg(x_{n+1} \wedge x_n)$$

Die Realisierung einer solchen Regel kann sowohl in Hardware mit XOR-Gattern als auch durch eine Netz logischer Schlußelemente erfolgen.

Nachdem mit den mehrschichtigen neuronalen Netzen die grundsätzliche Kritik von Minsky am Rosenblattschen Perceptron, nicht einmal die XOR-Funktion erlernen zu können, überwunden wurde, kann man auch ein neuronales Netz auf die Realisierung dieser Umkodierung durch Vorzeigen genügend vieler Beispiele trainieren. Man kann aber für derartig einfache Fälle auch explizit eine Lösung für die synaptische Matrix und die Schwellenwerte der beteiligten Neuronen angeben, um die lineare Separierung aller möglichen Inputs als Cluster der gewünschten Outputklassen zu erreichen. Dabei kann man in direkter Umsetzung der logischen Zuweisung je ein künstliches Neuron mit einer logischen Verknüpfung beauftragen, indem alle synaptischen Gewichte für nicht zu negierende Variable gleich 1, für zu negierende Variable gleich -1 gesetzt werden und man den Schwellenwerten für konjunktive, disjunktive und negierende Neuronen die Werte 1.5, 0.5 bzw. -0.5 zuweist.

Erstaunlicherweise können für dieses Problem durch analytische Überlegungen noch effektivere Lösungen mit weniger Neuronen (zwei Neuronen je XOR-Funktion) gefunden werden, wenn man nicht nur Kopplungen zwischen benachbarten Schichten von Neuronen zuläßt. Dies geht zurück auf das Verständnis für die Notwendigkeit der linearen Klassenseparierung durch Einführung von Hilfsvariablen. Andererseits können logische Verknüpfungen wie die oben angegebene natürlich auch in der Form von Fuzzy-Schluß- Figuren realisiert werden, wobei bedarfsweise auch verallgemeinerte "fuzzifizierte" Verknüpfungsoperationen anstelle der Konjunktion und Diskjunktion eingebracht werden könnten. Während man diesen Aspekt durch von 0.5/1.5 abweichende Schwellenwerte und modifizierte Aktivierungsfunktionen eines Neurons interpretieren kann, bietet sich durch Abschwächung der synaptischen Input-Gewichte die Möglichkeit, das Gültigkeitsmaß gewisser Fuzzy- Schluß-Regeln zu variieren.

4. Ausblick

Ein herkömmlich belehrtes Netz künstlicher Neurone muß also nicht unverständlich oder uninterpretierbar bleiben, sondern kann in übersichtlichen Fällen durchaus als Fuzzy-Schluß-Regel-Gebilde betrachtbar werden. Auch der umgekehrte Weg sollte oftmals zu vernünftiger Vorstrukturierung von neuronalen Netzen als Ausgangspunkt für eine endgültige Nachbelehrung gangbar sein.

Es muß in diesem Beitrag darauf verzichtet werden, die in der Fuzzy Logic üblicherweise verwendeten Zugehörigkeitsfunktionen eingehender mit den Sigmoidfunktionen neuronaler Netze und die verallgemeinerten logischen (Fuzzy-) Operatoren mit der Input-/Aktivierungs- /Output-Funktion künstlicher Neuronen in Relation zu setzen. Derartige Vergleiche sind aber mit Sicherheit fruchtbringend für die enge Verbindung des Konnektionismus mit der Fuzzy Logic.

Literaturverzeichnis

[Ko] Kosko, B.: Neural Networks and Fuzzy Systems, Prentice-Hall, Englewood Cliffs, 1992

[Cr] Creutzfeldt, O. et al.: Neurobiologie/Hirnforschung - Neuroinformatik, Künstliche Intelligenz, BMFT, Bonn, 1991

[Pa] Palm, G., Rückert, U., Ultsch, A.: Wissensverarbeitung in neuronaler Architektur. in: Brauer, W., Hernandez, D. (Hrsg.): Verteilte Künstliche Intelligenz und kooperatives Arbeiten, Informatik-Fachberichte 291, Springer- Verlag Berlin Heidelberg 1991

[Th] Thompson, R.F.:The Brain: An Introduction to Neurosience, W.H.Freeman & Co, New York, 1985

[Zi] Zimmermann, H.-J., Fuzzy Sets Theory - and its Application, Kluwer-Nijhoff Publishing, 1990

[Is] Issendorf, H.v., Ein Kalkül zur einheitlichen Behandlung von Software- und Hardware-Netzen, in: Technischer Bericht 2. Workshop über Parallele Systeme und Algorithmen (PASA), Heinz-Nixdorf-Institut (Hrsg), Uni-GH Paderborn, 1991

Die Transinformation: Werkzeug und Extremalprinzip bei Neuronalen Netzwerken

H. G. Schuster
Institut für Theoretische Physik
Universität Kiel, 2300 Kiel, FRG

1 Einleitung:

Die Transinformation ist eine aus der Nachrichtentechnik wohlbekannte Größe, die den Informationsfluß zwischen Sender und Empfänger durch einen gestörten Kanal mißt[1]. Im folgenden soll nun die doppelte Rolle, der Transinformation für Neuronale Netzwerke näher erläutert werden.

Zum einen läßt sich die zeitaufgelöste Transinformation dazu verwenden, um den Informationsaustausch zwischen Neuronengruppen mit hoher Zeitauflösung zu messen[2–3]. Zum anderen stellt sich heraus, daß sich die bekannte Hebb'sche Lernregel[4] (die besagt, daß sich eine Synapse dann ändert, wenn sie ein Signal empfängt und das nachgeschaltete Neuron gleichzeitig feuert) den Informationsfluß durch ein Neuron maximiert, d.h. die Hebb'sche Regel ist aus einem Maximalprinzip ableitbar[5–6].

2 Die zeitaufgelöste Transinformation

Betrachten wir zunächst den Fall der zeitabhängigen Transinformation zwischen Neuronengruppen. Dazu gehen wir von der Annahme aus, daß die zeitabhängigen Aktivitäten der Neuronengruppen (Feldpotentiale) wie in Abbildung 1 dargestellt als Zeitreihen x (t) und y (t) bekannt sind. Die zugehörigen Wahrscheinlichkeiten sind in Abbildung 1 definiert. Die Transinformation M berechnet sich daraus zu[3]:

$$M = \sum_{ij} p_{ij}^{xy} \, \mathrm{ld}(p_{ij}^{xy}/p_i^x p_j^y) \tag{1}$$

Falls Sender und Empfänger unabhängig von einander sind gilt:

$$p_{ij}^{xy} = p_i^x \cdot p_j^y \tag{2}$$

und die Transinformation verschwindet. Falls beim Sender und Empfänger die gleiche Zeitreihe vorliegt, gilt:

$$p_{ij}^{xy} = p_i^x = p_i^y \tag{3}$$

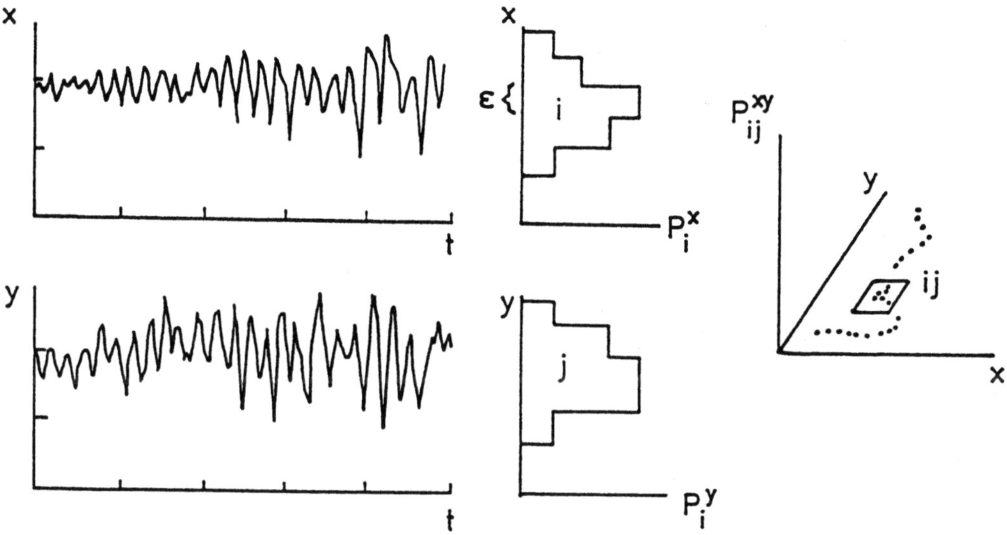

Abbildung 1: Definition der verschiedenen Wahrscheinlichkeitsdichten $p_i^x, p_j^x, p_{ij}^{xy}$ für die Zeitreihen x(t) und y(t). Die ein- und zweidimensionalen Intervalle haben die Größen ϵ bzw. ϵ^2 und werden mit i, j bzw.$_{ij}$ durchnummeriert.

d.h. $M = -\sum_i p_i^x \mathrm{ld} p_i^x$,die ganze ankommende Information wird übertragen.

Diese beiden Grenzfälle verdeutlichen die Rolle der Transinformation als Maß für den Informationsaustausch zwischen den beiden durch die Zeitreihen x(t) und y(t) beschriebenen Systemen. Um zur zeitaufgelösten Transinformation zu gelangen, schreiben wir M als Zeitsumme. Dies wird dadurch erreicht, daß wir, wie in der nichtlinearen Dynamik üblich, die Wahrscheinlichkeiten p_i^x, p_j^y und p_{ij}^{xy} durch die neuen Wahrscheinlichkeiten $\tilde{p}_n^x, \tilde{p}_n^y$ und $\tilde{p}_n^{xy}$ ersetzen. Dabei sind die letzteren wie folgt definiert:

$$\begin{aligned}
\tilde{p}_n^x &= \frac{1}{N}\sum_m \Theta(\epsilon - \mid x_n - x_m \mid), \\
\tilde{p}_n^y &= \frac{1}{N}\sum_m \Theta(\epsilon - \mid y_n - y_m \mid), \\
\tilde{p}_n^{xy} &= \frac{1}{N}\sum_m \Theta(\epsilon - \sqrt{(x_n - x_m)^2 + (y_n - y_m)^2}).
\end{aligned} \tag{4}$$

Wir haben dabei die Zeit die Schritte der Länge Δt unterteilt, d.h. $t = \Delta t \cdot n$. $\tilde{p}_n^x$ ist die Wahrscheinlichkeit die Variable x in einem ϵ-Intervall, um den Punkt x_n zu finden. $\Theta(x)$ ist die Heaviside-Funktion. Die Transinformation M läßt sich nur als Zeitsumme schreiben:

$$M = \frac{1}{N}\sum_n \mathrm{ld}(\tilde{p_n}^{xy}/p_n^x p_n^y) \tag{5}$$

Dabei ist $N = T/\Delta t$ und wir durften, wie beim Übergang vom Scharmittel zum Zeitmittel üblich die "Phasenraumdichte" p_{ij}^{xy} als Gewichtsfaktor weglassen[7]. Gleichung (5) zeigt, daß

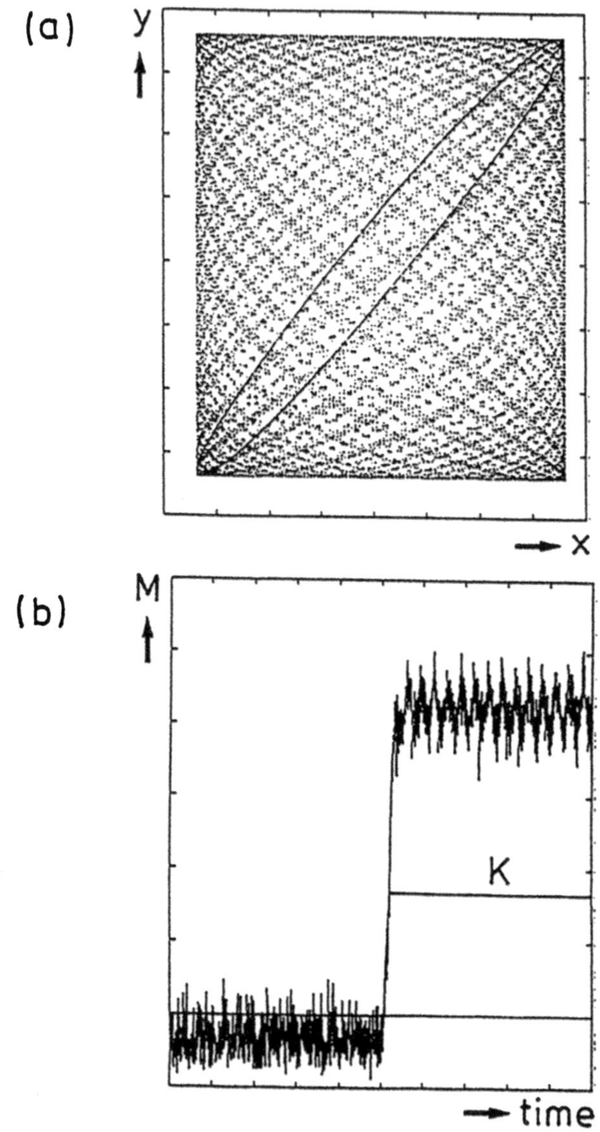

Abbildung 2: (a) Phasenportrait der Lösungen der Gleichungen (7a -c). Für nicht verschwindende Kopplung wird die Trajektorie des gekoppelten Systems eine Ellipse. (b) Die zeitabhängige Transinformation M_n und die zeitabhängige Kopplung K.

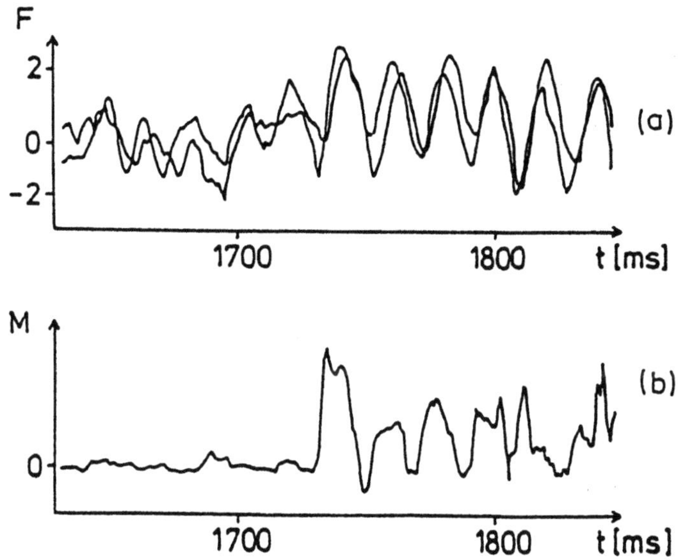

Abbildung 3: (a) Zeitabhängigkeit der Feldpotentiale F für neuronale Oszillatoren im visuellen Cortex einer Katze der ein langer Balken gezeigt wird. (b) Zeitabhängigkeit der Transinformation in gleichen Zeiträumen.

die Transinformation M als Zeitmittel dargestellt werden kann und es ist daher sinnvoll die zeitabhängige Transinformation

$$M_n = \mathrm{ld}(\tilde{p}_n^{xy} / \tilde{p}_n^x \tilde{p}_n^y) \tag{6}$$

einzuführen.

Wir betrachten nun zwei praktische Anwendungen dieser zeitaufgelösten Transinformation. Abbildung 2a zeigt die Phasenraumtrajektorien für ein System aus zwei gekoppelten Phaseoszillatoren. Die Zeitreihen x (t) und y (t) werden durch das Gleichungssystem:

$$\begin{aligned} x(t) &= cos\phi_1(t); \qquad y(t) = cos\phi_2(t) \\ \dot{\phi}_1(t) &= \omega_1 - Ksin(\phi_1(t) - \phi_2(t)) \\ \dot{\phi}_2(t) &= \omega_2 - Ksin(\phi_2(t) - \phi_1(t)) \end{aligned} \tag{7}$$

generiert. Im ungekoppelten Fall (K = 0) sind x (t) und y (t) unabhängig voneinander und die Trajektorie erfüllt den zweidimensionalen Phasenraum falls das Frequenzverhältnis ω_1/ω_2 irrational ist. Für nichtverschwindende Kopplungen ($K \neq 0$) ist die gemeinsame Trajektorie eine Ellipse. Abbildung 2b zeigt, daß die zeitabhängige Transinformation im wesentlichen die zeitabhängige Kopplung zwischen Oszillatoren mißt.

In Abbildung 3a ist die gemessene zeitliche Aktivität zweier Neuronengruppen im visuellen Cortex einer Katze dargestellt[8]. Es stellte sich die Frage, nach dem Informationsaustausch zwischen diesen Neuronengruppen. In Abbildung 3b sehen wir, daß die zeitaufgelöste Transinformation auf einer Zeitskala von etwa 5msec. um Größenordnungen varieren kann und ferner, daß sie mit der Periode der Schwingungen der Neuronengruppe fluktuiert. Die zeitaufgelöste Transinformation mißt den Informationsaustausch zwischen den Neuronengruppen und der wesentliche um Faktor 10 höheren Zeitauflösung als dies Kreuzkorrelationsfunktion tun kann (vergl. Ref.2,3).

Die biologische Bedeutung dieser hochaufgelösten Transinformationsmessung ist wie folgt: Die beiden Neuronengruppen deren Aktivitäten in Abbildung 3a dargestellt werden, sind im visuellen Cortex einer Katze lokalisiert. "Synchrones" Schwingen der Aktivitäten bedeutet, daß die Katze ein zusammenhängenden Gegenstand (hier einen langen Balken) sieht[9,10]. Eine nichtverschwindende Transinformation zwischen Neuronengruppen kodiert die Tatsache, daß die verschiedenen Bildpunkte des Balkens zu einer Einheit (Gesamtbalken) gehören. Der kurzfristige Informationsaustausch zwischen den Neuronengruppen erlaubt einer Neuronengruppe sich rasch mit einer anderen in Verbindung zu setzen. Damit wird das Erfassen rascher Szenenwechsel, etwa der Übergang von einem langen Balken (mit M $\neq 0$) zu einem kurzen Balken (mit M $= 0$) im Gehirn ermöglicht.

3 Das Prinzip der maximalen Transinformation

Im folgenden soll gezeigt werden, daß das folgende Extremalprinzip gilt:

"Während des Lernvorgang ohne Lehrer verändern sich die synaptischen Gewichte eines Neuronen derartig, daß die zwischen Eingang (Dendriten) und Ausgang (Axon) des Neurons übertragenden Information maximiert wird."

Der Beweis dieser Aussage erfolgt dadurch, daß wir erstens die Transinformation als Potential für zeitliche Änderung der Synapsenstärke $\vec{w} = (w_1 \ldots w_n)$ verwenden, d.h. es gilt:

$$\dot{\vec{\omega}} \propto \vec{\nabla}_\omega M \tag{8}$$

und zweitens M für ein nichtlineares Neuron berechnen [6]. Das Hauptproblem besteht darin M für ein nichtlineares Neuron mit der Input- Outputrelation

$$y = \sigma[\vec{\omega} \cdot \vec{x} - \Theta], \tag{9}$$

wobei $\sigma(z)$ die übliche Sigmoidfunktion ist, explizit zu bestimmen. In Referenz 6 wurde gezeigt, daß für den Grenzfall starken synaptischen Rauschens die Transinformation durch das mittlere Schwankungsquadrat des Outputs gegeben ist.

$$M =< \sigma^2 > - < \sigma >^2 \tag{10}$$

Dabei erstreckt sich die Mittelung $< \ldots >$ über alle Gaussverteilten Eingangsmuster. Aus Gleichung (9) ergibt sich dann mit Gleichung (10)

$$\dot{\vec{\omega}} =< \sigma(\vec{\omega} \cdot \vec{x} - \Theta)x > - < \sigma(\vec{\omega} \cdot \vec{x} - \Theta)(\sigma(\omega x - \Theta)\vec{x} > \tag{11}$$

oder

$$\dot{\vec{\omega}} = F(x)Q\vec{\omega} \text{ mit } x = \vec{\omega}^T Q\omega \text{ und } Q_{ij} =< x_i x_j > . \tag{12}$$

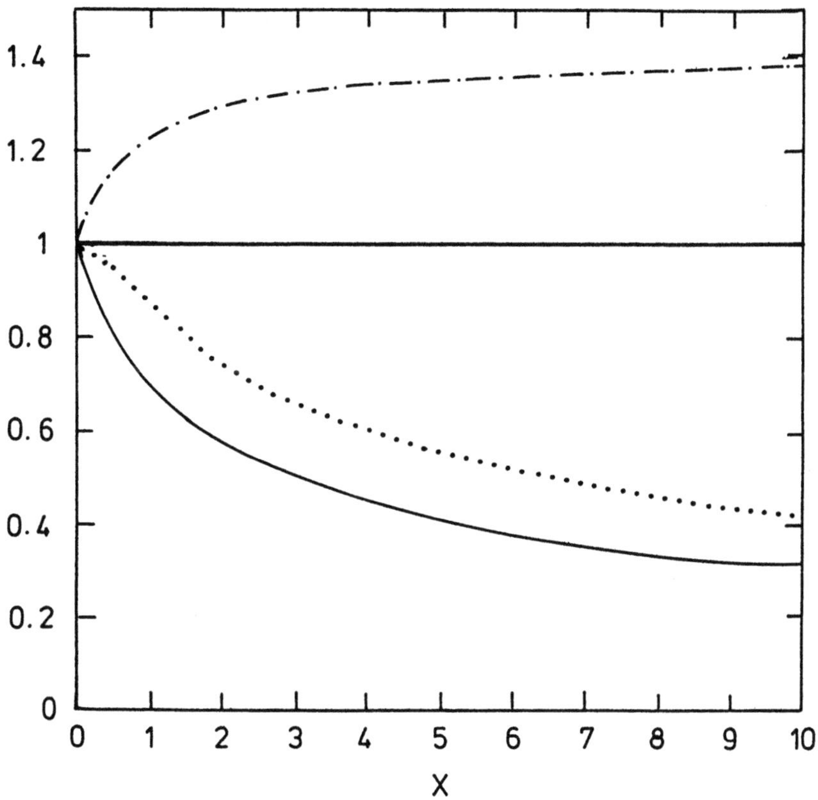

Abbildung 4: Die Filterfunktionen F(x) und G(x). F(x) entspricht einem nichtlinearen Neuron, welches durch Maximierung der Transinformation lernt (gestrichelte Linie). G(x) gilt für Hebb'sches Lernen (dünne Linie). Die dicke Linie entspricht dem Lernverhalten eines linearen Neurons. Die strichgepunktete Linie bezeichnet das Verhältnis F(x)/G(x).

Die Funktion F(x) ist in Abbildung 4 dargestellt. Die Resultate (11) und (12) können nun mit der Hebb'schen Lernregel

$$\dot{\vec{\omega}} = < \sigma(\omega x - \Theta) x >, \tag{13}$$

die nach Ausführung der Mittelung zu

$$\vec{\omega} = G(x)\, Q\vec{\omega} \tag{14}$$

wird, verglichen werden.

Abbildung 4 zeigt, daß die Funktionen F (x) und G (x) nun geringfügig voneinander abweichen, d.h. eine Maximierung der Transinformation führt (abgesehen von kleinen numerischen Abweichungen) auch für nichtlineare Neuronen zur Hebbschen Lernregel.

Damit erweist sich das Konzept der maximalen Transinformation, das zunächst nur für ein etwas künstlichen Fall linearer Neuronen bewiesen wurde[5], auch für den Fall nichtlineare Neuronen[6] als äußerst tragfähig und sein Anwendungspotential etwa für die Selbstorganisation in geschichteten Netzwerken ist und bleibt ein interessantes Forschungsgebiet. Dabei interessieren insbesondere folgende Fragen: Erstens, gilt dieses Extremalprinzip (unter Umständen mit geeigneten Nebenbedingungen) auch für schnelle zeitliche Prozesse? Zweitens, was passiert bei rückgekoppelten Systemen oder für den Fall das die Einzelneuronen selbst aktiv werden (zum Beispiel spontane Oszillationen zeigen)?

Wie auch immer die Antworten auf diese Fragen aussehen mögen, die Transinformation in ihrer zeitaufgelösten oder auch möglicherweise in ihrer raumaufgelösten Version ist und bleibt ein äußerst nützliches Instrument bei der Untersuchung realer und artifizieller neuronaler Netzwerke.

Literatur:

1. Shannon, C.E and Weaver, W.: "The Mathematical Theory of Information", Univ. Ill. Press, Urbana (1949).

2. Pawelzik, K. and Schuster, H.G.: "Detecting information-flows from multielectrode measurements", Submitted to Biol. Cybern. (1991).

3. Schuster, H. G. ed.: "Nonlinear Dynamics and Neuronal Networks", VCH Verlag Weinheim (1991).

4. Hebb, D. O.: "The Organization of Behavior", Wiley N.Y. (1949).

5. Linsker, R.: Computer **21**, 105 (1988).

6. Schuster, H. G.: "Learning by Maximierung the Information Transfer Through Nonlinear Noisy Neurons and Noise Breakdown", Phys. Rev. A. (accepted).

7. Schuster, H. G.: "Deterministic Chaos", second edition, VCH Weinheim (1988).

8. Gray, C.M. König, P., Engel, A.K. and Singer, W.: "Oscillatory responses in cat visual cortex exhibit inter-columnar synchronization which reflects global stimulus properties", Nature 338: 334-337 (1989).

9. Schuster, H. G. and Wagner, P.: Biol. Cybernetics **64**, 77-83 (1990).

10. Niebur, E., Schuster, H.G. and Kammen, D.: Phys. Rev. Lett. **67**, 2753 (1991).

Gemeinsamkeiten von Physik und Informatik am Beispiel eines systolischen optischen Prozessors

H. Hofmann, J. Enderlein
Technische Universität Dresden
Fakultät Informatik
Institut Technische Informatik
Mommsenstraße 13
O-8027 Dresden

1. Einleitung

Innovative Aspekte zur Entwicklung der Informatik werden zukünftig insbesondere auch durch Nutzung neuer Rechnerarchitekturen, beispielsweise Prinzipien der Parallelverarbeitung /1/, erwartet. Dabei rücken optische Computer /2/ ins Blickfeld der Grundlagenforschung, weil sich das Licht mit einer Reihe physikalischer Eigenschaften als Informationsträger bei der Signalverarbeitung im Rahmen neuer Rechnerarchitekturen anbietet. Solche Eigenschaften sind die Invarianz der gegenseitigen Beeinflussung von Lichtstrahlen, die gebündelte Ausbreitung, die Ausbreitungsgeschwindigkeit, die Vielzahl von Informationsparametern und die Modulierbarkeit. Analysiert man vorhandene Rechnerarchitekturen, so kommt man zu dem Ergebnis, daß solche Operationen wie Matrizenmultiplikation, Faltung und Korrelation, bei denen die Addition und Multiplikation bei gleichzeitiger Verschiebung der Komponenten vorkommen, geeignet sind, mit optischen Computern vorteilhaft realisiert zu werden, weil sie eine hohe Parallelität der Bearbeitung der Teilprozesse zulassen. Dabei sind für die Entwicklung optischer Computer interdisziplinäre Kenntnisse zur Rechnerarchitektur, physikalischen Optik und wegen des Vergleiches mit elektronischen Rechnern zur Computer-Hardware erforderlich, um die Effizienz optischer Lösungen zu optimieren. In diesem Beitrag werden die Anforderungen an das Physik- und Informatikwissen am Beispiel des Entwurfs eines integriert-optischen Braggmodulators für die Vektor-Matrix-Modulation vorgestellt.

2. Physikalische Aspekte des Braggmodulators

Der Grundgedanke optischer Informationsverarbeitung mittels Beugung des Lichtes wurde von TSAI /3/ verwirklicht, indem er mit dem Bragg-Modulator ein planar-optisches Bauelement auf Lithiumniobat entwickelte, bei dem geführte optische Wellen (GOW) mit akustischen Oberflächenwellen (AOW) wechselwirken. Das Ziel dieses Modulators war zunächst die Einseitenmodulation des Monomodenlichtes, eine Weiterentwicklung stellte der akusto-optische Braggmodulator mit Hornlichtwellenleitern /4/ dar. Die Grundstruktur eines integriert optischen Braggmodulators für die Vektor-Matrix-Multiplikation nach TSAI /5/, der an der TU Dresden auf einem X-Schnitt-Lithiumniobat-Substrat mit den Abmessungen 6 * 22 * 1 mm entwickelt wurde, ist in Bild 1 dargestellt. Die Matrixelemente a(ij) werden als optische Amplitudensignale in die Streifenwellenleiter eingekoppelt und gelangen über die TIPE-Mikrolinsen (Titandiffundiert, Protonenausgetauscht) in den Wechselwirkungsbereich AOW/GOW. Dort treffen sie auf eine im Braggwinkel einlaufende akustische Oberflächenwelle, die die Vektorkodierung (Elemente b(j)) trägt. Durch die entstehende Braggbeugung wird alle Lichtenergie in die 1. Beugungsordnung gebeugt und von der großen TIPE-Linse im Detektorpunkt D gesammelt. Die zeitliche Folge der Intensitäten des Lichtes im Punkt D entspricht den Elementen der Ergebnismatrix C. Dabei stellt die Wechselwirkung von AOW mit GOW die Multiplikation der einzelnen Vektor- und Matrixelemente dar, während die große TIPE-Linse die gleichzeitig gebildeten Produkte optisch summiert. Für den Layout-Entwurf ist die Optimierung der Interdigitalwandler (IDW) zur Erzeugung der AOW besonders wichtig. Durch den reziproken piezoelektrischen Effekt wird unter den Interdigitalwandlerfingern auf der Oberfläche des Lithiumniobats eine akustische Welle erzeugt, die in Form und Frequenz von der Wandlerwichtung und dem Fingerabstand abhängt /6/. Weitere physikalische Probleme stecken in der optischen Wellenleitung und der Lichtein- und auskopplung, insbesondere auch in der Herstellung der Wellenleiter und Linsen.

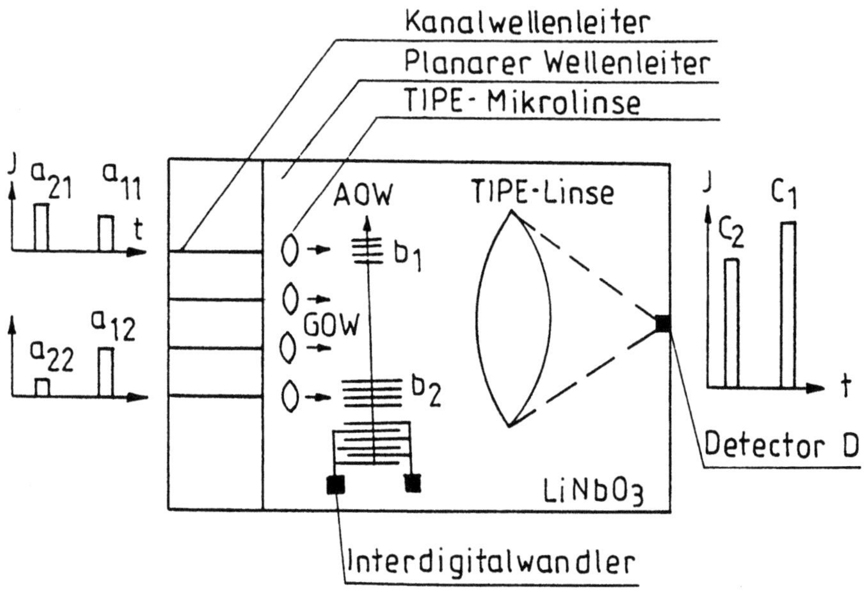

Bild 1 : Grundstruktur eines akusto-optischen Prozessors

Der Vektor-Matrix-Multiplikator kann sowohl für Analogwert-Matrizen (Elemente werden durch Intensitäten der GOW und AOW kodiert) als auch für Binärmatrizen angewendet werden, letzteres ist einfacher zu realisieren und bietet eine höhere Genauigkeit bei der Ausführung der mathematischen Operationen. Für einen in Entwicklung befindlichen Braggmodulator mit 14 GOW-Kanälen wurden mittels Simulation die Operatioszeiten der elektronischen mit der optischen Lösung verglichen. Die Operationszeit für den elektronischen Rechner (20 MHz-Takt) wurde für die Vektor-Matrix-Multiplikation mittels Produktzerlegung nach Kronecker, Außenproduktmultiplikation und herkömmlichen Verfahren berechnet und liegt für das letztere mit 14 Elementen bei etwa 7800 µs. Für die analoge optische Variante ergibt sich unter Berücksichtigung von Signallaufzeiten eine Operationszeit von 32 µs, das ist ein Verhältnis 2000:1 zugunsten der Optik (bei größeren Matrizen wird das Verhältnis noch größer). Eine vergleichbare technologische Herstellbarkeit in Zukunft vorausgesetzt, besitzt der optische Prozessor aufgrund der teilweise parallelen Verarbeitung der Matrix merkliche Vorteile.

3. Informatik-Aspekte des optischen Prozessors

Bereits von Kung /7/ wurde darauf hingewiesen, daß die Entwicklung systolischer Prozessoren Möglichkeiten zur Steigerung der Rechengeschwindigkeit von rechenintensiven middle-level Operationen, wie die Vektor-Matrix-Multiplikation, erschließt. Ein systolisches System besteht aus einer Anzahl untereinander verbundener Zellen, von denen jede Zelle eine Einzeloperation ausführen kann. Ein solcherart aufgebautes System läßt sich beliebig zu einem systolischen Feld oder Baum erweitern. Im Gegensatz zu Pipeline-Prozessoren, in denen sich nur die Ergebnisse bewegen, fließen in systolischen Systemen sowohl die Eingabedaten als auch die Teilergebnisse. Durch Aufteilung der Gesamtoperation in streng abgegrenzte Teilaufgaben lassen sich einfache genormte Prozessormodule aufbauen, aus denen komplexe Strukturen für unterschiedliche Aufgaben entwickelt werden können.

Bild 2 zeigt eine Zusammenstellung verschiedener Rechnerarchitekturen, wobei der Braggmodulator seinen Platz als Parallelrechner neben dem Pipelinerechner und dem Rechner mit Nebenläufigkeit findet. Im konkreten Fall des Braggmodulators bewegen sich die Eingabedaten in Form der modulierten AOW und GOW von einem Wechselwirkungsbereich (Zelle) zum anderen und gleichzeitig bewegen sich die Beugungsprodukte in Richtung der TIPE-Linse und werden dort zu Teilergebnissen, den Elementen der Ergebnismatrix aufsummiert /9/. Damit entspricht die Arbeitsweise der eines systolischen optischen Prozessors im Ensemble der in Bild 2 dargestellten Rechnerarchitekturen. Weiterhin ermöglicht die Anwendung des systolischen Prinzips eine Erweiterung und Optimierung des Vektor-Matrix-Multiplikators, z.B. durch Anwendung der Convolution /10/. Die Rechnerarchitektur ist damit die Quelle von Anregungen für die Layoutgestaltung des optischen Prozessors.

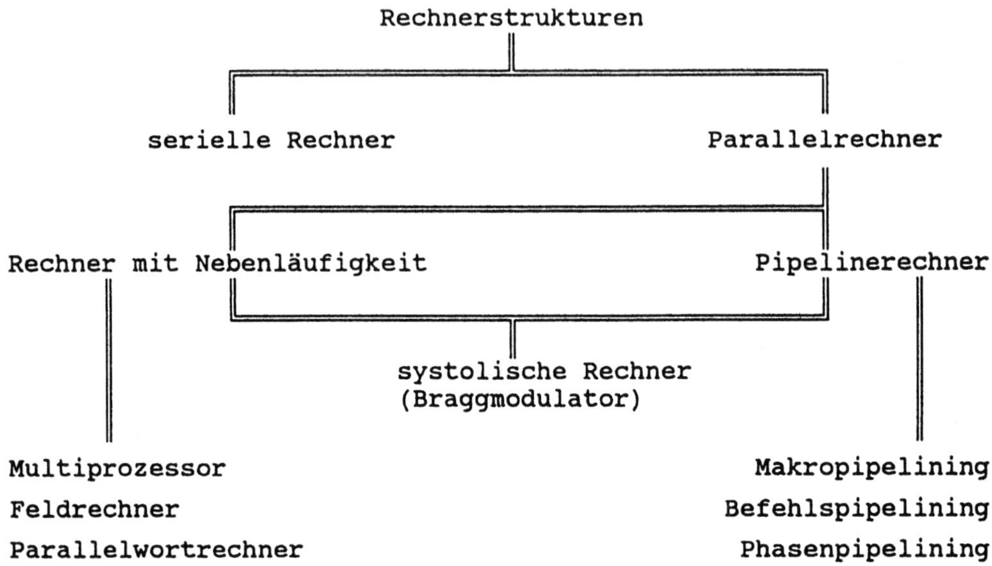

Bild 2 : Überblick Rechnerarchitekturen nach Bode,Händler /8/

4. Schlußfolgerungen

Anhand des vorgestellten systolischen optischen Prozessors ist ersichtlich, daß die interdisziplinäre Zusammenarbeit von Physikern, Informatikern und Informationstechnikern unerläßlich ist, wenn es um die Findung innovativer Lösungen bei der Weiterentwicklung von Computern geht. Damit die genannten Wissenschaftler eine gemeinsame Sprache und das gegenseitige Verständnis finden, wird gegenwärtig bereits in den Studienplänen der Informatikstudenten mit dem Fach Technische Informatik ein umfangreiches Hardware-Verständnis (optische Prinzipien eingeschlossen) ausgebildet. Beim Physikstudium sollte hingegen neben der im Stundenplan befindlichen Grundlagenausbildung Informatik das Systemdenken und die Computerarchitektur stärker gefördert werden, damit der Physiker neueste physikalische Erkenntnisse auf ihre Anwendbarkeit in der Computertechnik überprüfen und einsetzen kann. Aber nicht nur die Ausbildung von

Studenten, sondern auch die Weiterbildung tätiger Physiker und Informatiker sollte unter diesem Aspekt organisiert werden, wie es bereits auf der Frühjahrsschule Optik der DPG/DGAO 1991 in Oberbärenburg diskutiert wurde. Das Münchener Arbeitsgespräch "Physik und Informatik - Informatik und Physik" ist dazu ebenfalls ein wertvoller Beitrag.

Für die Bearbeitung von Teilproblemen sei cand.-inf. Katina Hampf und cand.-inf. Birgit Wernicke gedankt. Besonderer Dank gilt Herrn Prof. Dr. W. Sohler (FB Physik der Universität Paderborn) für die technologische Unterstützung.

Literatur :

/1/ Wherret, A. : Workshop on Photonic Logik and Information Processing, Proc. Edingbourgh 1988

/2/ Arrathon, R. : Optical computing digital and symbolic, Marcel Dekker Inc. New York 1989

/3/ Tsai, C. S. : Guided-Wave Acoustooptics, Springer Series in Electronics and Photonics, Vol. 23, Springer-Verlag Berlin 1990

/4/ Schubert, H.; Hofmann, H. : Der akusto-optische Modulator mit Hornlichtwellenleitern, Poster Jahreshaupttagung der Phys. Ges. 1990 in Leipzig

/5/ Tsai, C. S.; Zang, D. Y.; Lee, P. : Acusto-optic Bragg diffraction in a LiNbO3 channel-planar composite waveguide with application to optical computing, Appl.Phys. Lett. 47 (1985)6., S. 549-551

/6/ Hofmann, H. : Möglichkeiten der Signalverarbeitung mit akustischen Oberflächenwellenbauelementen. Elektron. Inf. verarb. Kybern. EIK 24 (1985) 7/8, S. 363-370

/7/ Kung, H. T. : Why systolic architectures ?, Computer 15 (1982), S. 34-46

/8/ Bode, A.; Händler, W. : Rechnerarchitektur II-Strukturen, Springer-Verlag, Berlin Heidelberg New York Tokyo 1983, S. 18

/9/ Enderlein, J.; Hofmann, H. : Systolische Architektur von optischen Prozessoren zur Realisierung von Matrix-Vektor-Multiplikationen mittels Faltung. Wiss. Beitr. Inform. Fakultät Informatik a. d. TU Dresden, 6(1991)H6, S. 80-85

/10/ Cartwright, S.; Gustfson, S. C. : Convolver-based optical systolic processing architectures, Optical Engineering, Vol. 24 (1985), S. 59-64

Optische Informationsverarbeitung durch Quanten-Netzwerke

H. Körner, W. G. Teich und G. Mahler
Institut für Theoretische Physik, Universität Stuttgart,
Pfaffenwaldring 57, 7000 Stuttgart 80

Zusammenfassung

Komplexe synthetische Nanostrukturen („Quanten-Netzwerke") besitzen ungewöhnliche optische Eigenschaften: Durch sogenannte „*pump and probe*"-Experimente können typische bildverarbeitende Funktionen realisiert werden. Dabei läuft die Informationsverarbeitung in diesem „Quantencomputer" auf molekularer Ebene ab und kann durch das optische Steuerfeld in gewissem Rahmen programmiert werden.

1. Einleitung

Inwieweit ein physikalisches System ein Computer darstellt, wird letztlich erst durch Konvention eines Benutzers festgelegt: Er untersucht in physikalischen Experimenten, wie Meßgrößen von äußeren Steuerparametern abhängen; diesen Zusammenhang versucht er als „Computerfunktion" verfügbar zu machen. Die Art des physikalischen Experiments legt dabei die relevante Beschreibungsebene fest.

In der herkömmlichen Mikroelektronik ist diese Beschreibungsebene klassisch (typischerweise hydrodynamisch). Sie wird bei einer weiteren Miniaturisierung, d. h. beim Vordringen in atomare Dimensionen, nicht mehr anwendbar sein: Es müssen neue Computerkonzepte entwickelt werden, bei denen die quantenmechanische Beschreibungsebene wichtig wird.

Wir untersuchen hier „Quanten-Netzwerke" als mögliche Realisierung eines „Quantencomputers". Die einzelnen Subeinheiten (Knoten) des Netzwerks bestehen aus räumlich lokalisierten Wenig-Niveau-Quantensystemen. Durch eine Hierarchie der Energieskalen kann gesichert werden, daß die einzelnen Subeinheiten zwar ihre Individualität behalten, andererseits aber die „Vernetzung" durch eine Wechselwirkung untereinander gegeben ist.

Die zur physikalischen Realisierung solcher Quantennetzwerke notwendige Technologie setzt eine Strukturkontrolle auf atomarer Ebene voraus. Vielversprechend ist hier der Ansatz über eine dreidimensionale Halbleiterstrukturierung („Quanten-Dot-Gitter") [1]. Als alternative Möglichkeit wird die chemische Synthese und Selbstorganisation (z. B. durch epitaktisches Aufwachsen) von funktionellen (Makro-)Molekülen diskutiert [2].

Das dynamische Verhalten des Netzwerks zeigt in Abhängigkeit von seiner Struktur ein adaptives Verhalten an seine Nichtgleichgewichts-Umgebung (optische Felder): Wir werden zeigen, daß diese Umgebung zum Teil als „Input", zum Teil als „Programm" funktionalisiert werden kann. Dazu untersuchen wir in sogenannten „*pump and probe*"-Szenarien die optische Antwort des Netzwerks auf ein Pump-Lichtfeld, das in dem Netzwerk eine komplexe Dynamik induziert. Diesen Zusammenhang kann man als „Bildverarbeitung" interpretieren und ihm damit eine Computerfunktion zuordnen. Informationsverarbeitung auf molekularer Ebene steht hier im Vordergrund. Unser Ansatz unterscheidet sich somit von den üblichen Konzepten, in denen optische Bildverarbeitung durch lokale Filterung in der Fourierebene mit optisch nichtlinearen Materialien geschieht [3].

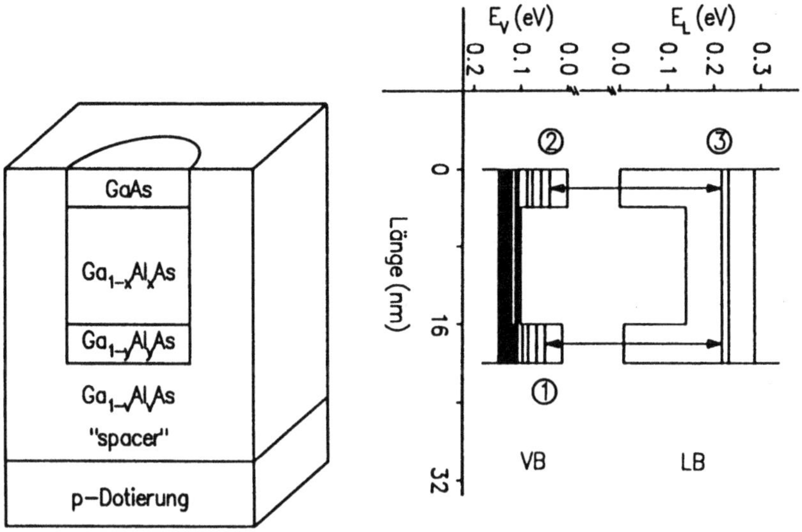

Abb. 1: Ladungstransfer-Quanten-Dot und zugehöriges Spektrum ($v = 0.4, x = 0.2, y = 0.01$).

2. Netzwerkmodell

Das Netzwerkmodell wird hier anhand einer möglichen Realisierung der Subeinheiten als Quanten-Dots dargestellt. In einem Quanten-Dot (Abb. 1) sind das Spektrum und über die Lokalisierung der Wellenfunktionen auch Übergangsraten und stationäre Dipolmomente im Prinzip durch die Strukturierung von außen einstellbar. Die hier relevante Dynamik erlaubt die Beschränkung auf ein Drei-Niveau System. Die Skalentrennung („Lokalisierungsauswahlregel") in den Übergangsraten $A_{12} \ll A_{13}, A_{23}$ stabilisiert den Ladungstransfer zwischen dem Grundzustand 1 und dem metastabilen Zustand 2: Auf der Zeitskala $A_{12}t \ll 1$ ist die Subeinheit bistabil, sie kann durch optische Anregung vom Zustand 1 in den transienten Zustand 3 und nachfolgender Relaxation in den Zustand 2 „geschaltet" werden, und umgekehrt.

Als Netzwerk wird nun eine zweidimensionale Anordnung von Quanten-Dots auf einem quadratischen Gitter betrachtet. Die einzelnen Subeinheiten behalten hier ihre Individualität in dem Sinn, daß der Charakter der Wellenfunktionen erhalten bleibt. Die unterschiedlichen Dipolmomente der Zustände ($-d_1 = d_2 = d$, $d_3 = 0$) führen aber über die Dipol-Dipol-Wechselwirkung zu einer Kommunikation der Subeinheiten: Die Lage der Energieniveaus und damit die optischen Übergangsfrequenzen einer Subeinheit hängen vom Zustand der anderen Subeinheiten ab:

$$\omega_{i_m j_m}(\{i_n\}) = \omega^0_{i_m j_m} + \sum_{n \neq m} \frac{d_{i_n}(d_{j_m} - d_{i_m})}{4\pi\epsilon_0 |\mathbf{R}_n - \mathbf{R}_m|^3}. \tag{1}$$

Dabei ist $\mathbf{R}_m$ der Ort der Subeinheit m und $\hbar\omega^0_{i_m j_m} = E^0_{j_m} - E^0_{i_m}$ die „ungestörte" Übergangsenergie zwischen den Zuständen i_m und j_m.

Die Kopplung des Netzwerks an ein inkohärentes Pump-Lichtfeld führt zu optischen Übergängen mit einer Rate

$$R_{i_m j_m}(\{i_n\}) = A_{i_m j_m} + B_{i_m j_m} U_m(\,\omega_{i_m j_m}(\{i_n\})\,), \tag{2}$$

die wegen der Frequenzrenormierung (1) vom Zustand der anderen Subeinheiten abhängt. A und B sind die Einsteinkoeffizienten für den spontanen bzw. induzierten Übergang, $U_m(\omega)$ ist die spektrale Energiedichte des Pump-Lichtfeldes am Ort m.

Die Absorption eines Probe-Lichtfeldes der Frequenz ω am Ort m ist durch

$$\alpha(m,\omega)=\frac{1}{\hbar\epsilon_0}\sum_{i_m,j_m}\sum_{\{i_n,n\neq m\}}\frac{\gamma_{i_mj_m}|d_{i_mj_m}|^2\left(P(i_1,\ldots,i_m,\ldots,i_N)-P(i_1,\ldots,j_m,\ldots,i_N)\right)}{(\omega-\omega_{i_mj_m}(\{i_n\}))^2+\gamma_{i_mj_m}^2} \tag{3}$$

gegeben. Dabei ist $P(\{i_n\})$ die stationäre Wahrscheinlichkeit, daß sich das Netzwerk im Zustand $\{i_n\}=\{i_1,i_2,\ldots,i_N\}$ befindet; sie wird hier durch eine Simulation der zu (2) gehörenden stochastischen Dynamik bestimmt. $d_{i_mj_m}$ ist das Übergangsdipolmoment und $\gamma_{i_mj_m}=\frac{1}{2}\sum_{k_m}(A_{k_mi_m}+A_{k_mj_m})$ die homogene Linienbreite. Die Absorption am Ort m hängt wegen der Kopplung der Subeinheiten nicht nur lokal von der Pumpenergiedichte U_m ab: Die optischen Eigenschaften des Netzwerks unterscheiden sich qualitativ von denen konventioneller Quantensysteme [1].

3. Optische Informationsverarbeitung

Ziel ist nun, dem Zusammenhang zwischen Pump-Lichtfeld und Absorption des Probe-Lichtfeldes eine Computerfunktion zuzuordnen. Das Pump-Lichtfeld bestehe aus zwei inkohärenten Gaußschen Moden, die jeweils selektiv die Übergänge $1\leftrightarrow 3$ und $2\leftrightarrow 3$ induzieren:

$$U_m(\omega)=\sum_{k=1}^{2}U_m^k\exp\left(-\left(\frac{\omega-\omega_{k3}^0+\delta_k}{\Delta_k}\right)^2\right). \tag{4}$$

Gemessen wird die Absorption eines Probe-Lichtfeldes mit der Spektralverteilung $g(\omega)$:

$$\alpha(m)=\int d\omega\, g(\omega)\,\alpha(m,\omega). \tag{5}$$

Die räumlichen Intensitätsverteilungen U_m^k werden nun normiert als „Input“, das räumliche Absorptionsmuster $\alpha(m)$ als „Output“ definiert. Das „Programm“ ist in der Funktion $g(\omega)$, der absoluten Intensität des Pump-Lichtfeldes und seinem Spektrum, d. h. in der Verstimmung δ_k gegenüber der ungestörten Übergangsfrequenz ω_{k3}^0 und der Bandbreite Δ_k „codiert“.

In Abb. 2 ist der zu dem Input U_m^1 (Abb. 2a) gehörende Output (Abb. 2b-f) für verschiedene Programmparameter dargestellt. Der zweite Inputkanal U_m^2 wird räumlich konstant gewählt. Für $\Delta_k\to\infty$ (Abb. 2b) ist die Dynamik der Subeinheiten entkoppelt (siehe Gl. (2) und (4)): Die Antwort bleibt rein lokal. Eine endlich Bandbreite Δ_k führt zu einer „Ferro-Kopplung“ der Subeinheiten [1]: Die kooperative Dynamik glättet das Inputmuster auf einer umso gröberen Skala, je kleiner Δ_k ist (Abb. 2c, d). Bei hoher Pumpenergie (Abb. 2e) kommen Sättigungseffekte ins Spiel: Das Muster U_m^1 wird mit dem räumlich konstanten Muster U_m^2 als Schwellwert „verglichen“. In Abb. 2b-e wurde eine breitbandige Spektralverteilung $g(\omega)$ gewählt, die Absorption $2\to 3$ wird unabhängig von der lokalen Umgebung gemessen. Die schmalbandige (selektive) Messung in Abb. 2f führt zu einer „Kantenerkennung“.

Da aus Rechenzeitgründen bisher nur ein kleines Netz (20×20 Zellen) simuliert werden konnte, wurde die (unphysikalische) Annahme gemacht, daß das Lichtfeld ($\lambda\approx 10^{-6}m$) auf der Skala des Dot-Abstandes ($\approx 10^{-7}m$) variieren kann. Erste Simulationen zeigen, daß für ein realistischeres Modell nach entsprechender Parameterskalierung ähnliche Eigenschaften erwartet werden können.

4. Diskussion

Den optischen Eigenschaften unseres Quanten-Netzwerk Modells lassen sich hochgradig-parallele bildverarbeitende Funktionen zuordnen. Ein derartiger „Quantencomputer“ wäre aber keine Implementierung eines herkömmlichen Computers auf molekularer Ebene, er setzte vielmehr völlig

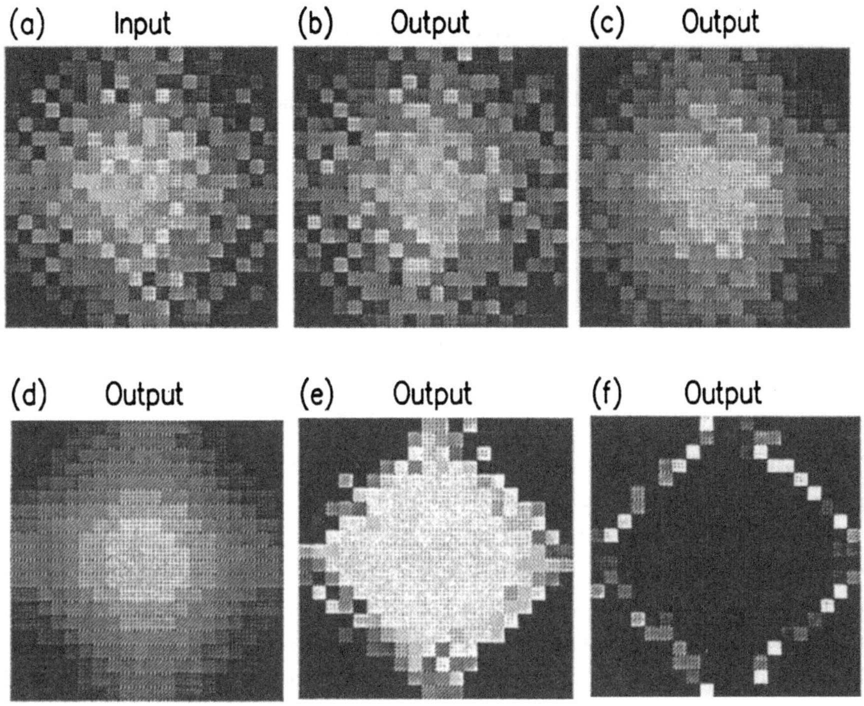

Abb. 2: Simulierte *„pump and probe"*-Experimente an einem Quanten-Netzwerk (20 × 20 Zellen): Pump-Intensitätsverteilung (a) als Input und zugehörige Probe-Absorptionsmuster (b-f) als Output bei verschiedenen Steuerparametern. Glättung des Inputmusters (b-d), Schwellwertentscheidung (e) und Kantenerkennung (f). hell: hohe Pump-Intensität bzw. starke Absorption; dunkel: niedrige Pump-Intensität bzw. schwache Absorption.

neue Anwendungsstrategien voraus. So bestimmt zum Beispiel die interne Dynamik des Quantennetzwerks nicht das Rechenergebnis allein, wie es bei herkömmlichen Computern der Fall ist, sondern die Funktion wird zusätzlich durch Art und Beschreibungsebene des physikalischen Experiments (z. B. durch die Spektralverteilung $g(\omega)$) an dem Netzwerk ausgewählt („Schnittstelle"). In solchen neuen Konzepten („programmierbare Materie") stecken aber auch Schwierigkeiten: Einem vorgegebenen Satz von Steuerparametern läßt sich zwar unter Umständen eine Funktion zuordnen, offen ist aber die Frage, inwieweit die Umkehrung, zu einer gegebenen Funktion die passenden Steuerparameter zu finden, durchführbar ist: Dies wäre die Voraussetzung für eine breitere technische Anwendung.

Literatur

[1] G. Mahler, H. Körner, W. Teich, in: Festkörperprobleme **31**, Hrsg. U. Rössler (Vieweg, Braunschweig, 1991), S. 357

[2] F. L. Carter, *Proceedings of the Second International Workshop on Molecular Electronic Devices* (Marcel Dekker Inc., New York, 1987)

[3] R. Thoma, N. Hampp, C. Bräuchle, D. Oesterhelt, Opt. Lett., **16**, 651 (1991)

Optisches Rechnen

Thomas Beth Volker Hatz Stephan Teiwes

Institut für Algorithmen und Kognitive Systeme
Universität Karlsruhe
Am Fasanengarten 5
7500 Karlsruhe

Zusammenfassung

Die Aussicht auf eine erfolgreiche Entwicklung schneller elektrooptischer Hybrid-Computer in den nächsten Jahren ist ausgesprochen gut. Eine wachsende Anzahl von Herstellern führt heute bereits optische und elektrische Bauelemente auf der IC- oder Schaltkreis-Ebene zusammen. Zukünftig denkbar sind außerdem spezielle Arithmetik-Prozessoren auf rein optischer Basis, die die elektronischen Standard-Prozessoren unterstützen. Hierbei können Methoden der Algorithmentechnik eine bedeutende Rolle spielen, um die Entwicklung einer neuen hybriden Rechnergeneration voranzutreiben. Vergleichbares ist bereits in der Chip-Technologie während der letzten Jahrzehnte geschehen. Anhand einiger markanter Beispiele zeigen wir auf, daß die Algorithmentechnik verschiedene Themen im Bereich *Optical Computing* aufgreift. Wie schnell und in welcher Qualität die Entwicklung elektrooptischer Hardware vorangeht, hängt entscheidend von den vorhandenen Entwurfswerkzeugen ab. Diesen Gesichtspunkt behandeln wir zum Abschluß .

Einleitung

Die Vorteile voll-optischer Rechnersysteme liegen auf der Hand. Sie besitzen eine hohe zeitliche und teilweise auch räumliche Bandbreite und arbeiten vielfach inhärent massiv parallel. Optische Signale können sich räumlich überkreuzen und dies bei sehr geringer Störanfälligkeit. Die Fan-Out-Beschränkungen liegen dabei wesentlich günstiger als bei elektrischen Signalen. Aus wirtschaftlichen Gesichtspunkten ist es zudem interessant, daß optische Implementierungen bestimmter Anwendungen durchaus preiswerter sein können als ihre elektronischen Pendants ([13]).

Die zur Zeit verfügbare Technologie ist noch nicht genügend weit fortgeschritten, um einen universell einsetzbaren optischen Rechner zu konstruieren. In naher Zukunft realisierbar erscheinen dagegen Hybrid-Rechner, die nur in Teilbereichen von optischen Bauelementen bzw. von optoelektronischen ICs profitieren. Der Markt bietet heute schon eine Reihe solcher optoelektronischer Chips; weiterhin auch akustisch oder elektrisch ansteuerbare Lichtmodulatoren und optische Speichermedien. So liegt es nahe, in Hochleistungsrechnern (Supercomputern) optische Bauteile dort einzusetzen, wo es auf höchste Geschwindigkeit und Parallelität ankommt. Ansatzpunkte bieten sich in den folgenden Bereichen.

Arithmetik Von Interesse ist die Konstruktion einer optischen arithmetisch logischen Einheit. Daneben gilt es, besonders zeitaufwendige, vielverwendete Operationen durch optische Elemente zu realisieren.

Kommunikation Kommunikation zwischen Teilschaltungen auf einem Chip oder über Chipgrenzen hinweg sowie Interprozessor-Kommunikation in Parallelrechnern kann man durch transparente Substrate (wave-guides) oder in Form einer Freiraum-Kommunikation (räumliche Lichtmodulation, Holographie) bewerkstelligen.

Speicher Optische Speicher packen große Datenmengen auf engstem Raum und ermöglichen einen schnellen Zugriff.

In allen drei Bereichen kann die Algorithmentechnik zur Zeit bekannte Verfahren anbieten oder zumindest helfen, neue Ansätze zu finden. Beispiele hierzu behandeln wir in den folgenden Abschnitten.

Optische Arithmetik

Die algorithmischen Verfahren, eine digitale arithmetisch logische Einheit (ALU) optisch zu realisieren, können sehr unterschiedlich ausfallen. Eine naheliegende Möglichkeit besteht darin, boolesche Funktionen durch nichtlineare optische Elemente nachzubilden, um damit in bekannter Weise Halb- bzw. Volladdierer oder andere Gatter aufzubauen. Daneben gibt es Ansätze, die die natürlichen Eigenschaften optischer Apparaturen, wie massive Parallelität und hohe Konnektivität, besser nutzen. Bekannt geworden sind in diesem Zusammenhang die *symbolische Substitution* und die *chinesische Restetechnik*. Wir betrachten nun diese drei Ansätze zum Aufbau einer optischen digitalen ALU genauer.

Optische Implementierung serieller Schaltkreise

Die Funktionalität einer ALU läßt sich präzise durch einen endlichen Automaten beschreiben, dessen Übergangsdiagramm allerdings recht umfangreich ist. Betrachtet man die Teilfunktionen der ALU getrennt, so korrespondieren damit mehrere, aber einfachere endliche Automaten.

In [11] zeigen die Autoren am Beispiel eines einfachen seriellen Addierers, daß die optische Realisierung endlicher Automaten möglich ist. Die Methodik dafür greift altbekannte Algorithmen auf.

Ein serieller Addierer wird durch seine Übergangs- und Ausgabefunktion sowie durch deren Komplemente beschrieben. Selbige lassen sich durch die zugehörigen (rekursiven) booleschen Funktionen ausdrücken. Diese booleschen Funktionen sind in mehrstufigen Netzwerken aus AND/OR-Gattern berechenbar, wobei jede Stufe eine festgewählte Permutation, z. B. einen *perfect-shuffle*, durchführt. Ein Netzwerk, das nur gleichartige perfect-shuffle-Permutationen enthält, legt eine einfache, aber effiziente optische Implementierung nahe. Es genügt, eine Stufe zu realisieren und selbige rückzukoppeln.

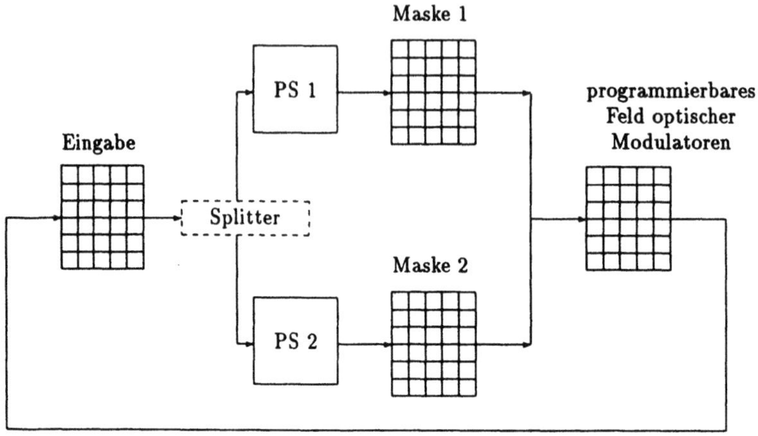

Abbildung 1: Schema eines optisch realisierten endlichen Automaten

Abbildung 1 zeigt das Prinzip eines optischen endlichen Automaten, gemäß dem ein serielles Addierwerk aufgebaut werden kann. Die zweidimensionale Eingabe ist ein binäres Lichtmuster und wird spaltenweise verarbeitet, d. h. der Aufbau kann mehrere Additionen (eine pro Spalte) ausführen. Ein Strahlenteiler konvertiert die Eingabe in zwei identische Muster, die dann getrennt in zwei verschiedene optische Perfect-Shuffle-Stufen [10] gelangen. Binäre Masken lassen jeweils nur die Lichtstrahlen durch, die anschließend durch ein Array nichtlinearer optischer Elemente logisch (AND, OR) verknüpft werden müssen, woraus sich die Funktion eines Volladdierers ergibt.

Diese Version der optischen Implementierung führt zu einer sehr schnellen seriellen ALU. Um die hohe räumliche Bandbreite des Lichts besser zu nutzen, bedarf es eines anderen Ansatzes, beispielsweise den der symbolischen Substitution.

Symbolische Substitution

Zwei elementare Verarbeitungsschritte kennzeichnen eine symbolische Substitutionslogik. Zunächst gilt es, ein digitales Eingabemuster zu erkennen, um dieses gemäß einer vorgegebenen Substitutionsregel durch ein anderes Muster zu ersetzen. Dabei können auch logische Operationen ausgeführt werden. Gemäß dem SIMD-Prinzip wird zellenweise auf dem Eingabe-Muster eine logische Operation durchgeführt. Es gibt diverse Möglichkeiten, Daten optisch in zweidimensionalen Feldern zu kodieren. Entsprechend unterschiedlich fallen auch die Realisierungen der logischen Operationen in der Substitutionsebene aus.

Die symbolische Substitution erlaubt nicht nur eine massiv parallele optische Umsetzung aller logischen Operationen, sondern auch die Regelung des Datentransports auf optischen Freiraumkanälen. Definiert man also in zweidimensionalen, optisch kodierten Feldern Nachrichten- und Steuersignale, dann ist die Konstruktion einer massiv parallelen optischen ALU möglich. Eine völlig andere, vielleicht etwas exotische Methode, arithmetische Funktionen umzusetzen, bietet die *chinesische Reste-*

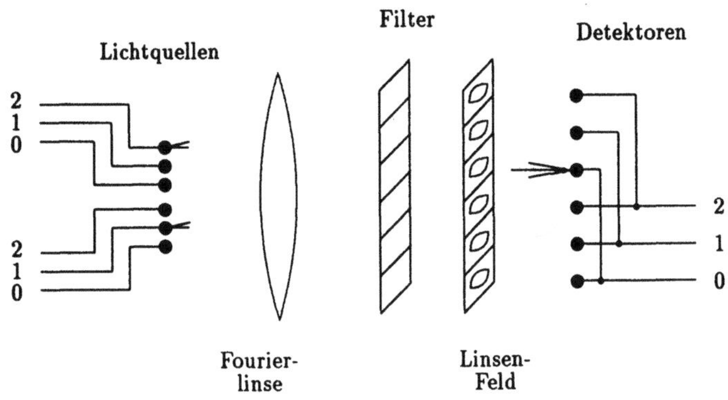

Abbildung 2: Addierer modulo 3

technik, wobei man in Restklassenringen rechnet.

Rechnen in Restklassenringen

Ein häufig angewandtes Verfahren der Algorithmentechnik besteht darin, eine Parallelisierung entsprechend des *Teile-und-Herrsche-Prinzips* vorzunehmen. Dabei wird ein umfangreiches Problem auf mehrere gleichartige, weniger aufwendige Probleme zurückgeführt, die dann simultan lösbar sind. Ein gutes Beispiel hierfür ist die chinesische Restetechnik.

Auf herkömmlichen digitalen Rechenanlagen wird immer modulo einer Zweierpotenz gerechnet, wobei der Exponent der Wortbreite entspricht. Durch geschicktere Wahl des Modulus, als Produkt relativ primer Zahlen $m_0, m_1, \ldots, m_s$, kann man eine gleichwertige, auf Wortebene parallele arithmetische Einheit entwerfen ([9]). Jede der Untereinheiten arbeitet hierbei modulo eines m_i. Die beiden wesentlichen Vorteile sind

- die in der Optik bestens ausnutzbare Parallelität des Verfahrens und
- die gegenüber digitalen Rechenanlagen kleinere Anzahl von Zuständen, die in einem Wort repräsentiert werden. Hiermit wird es möglich, eine wenig störanfällige Kodierung in der Amplitude, Phase oder Polarisationsrichtung des Lichts vorzunehmen.

Abbildung 2 zeigt eine mögliche Implementierung eines Addierers modulo 3. Die Distanz zweier Lichtquellen ist proportional zur Summe der zugeordneten Werte. In der Fourierebene entsteht ein Beugungsbild, das für jede Lichtquellendistanz eine bestimmte Charakteristik zeigt. Gleicherorts sind verschiedene Filter angebracht, die jeweils auf genau eines der charakteristischen Beugungsmuster ansprechen, d. h. Licht durchlassen, und sonst stark streuen. Jedem Filter ist ein Schwellwert-Detektor zugeordnet. So läßt sich das Ergebnis der Addition leicht bestimmen.

Optische Bauelemente eignen sich nicht nur dazu, eine digitale ALU aufzubauen. Für hybride

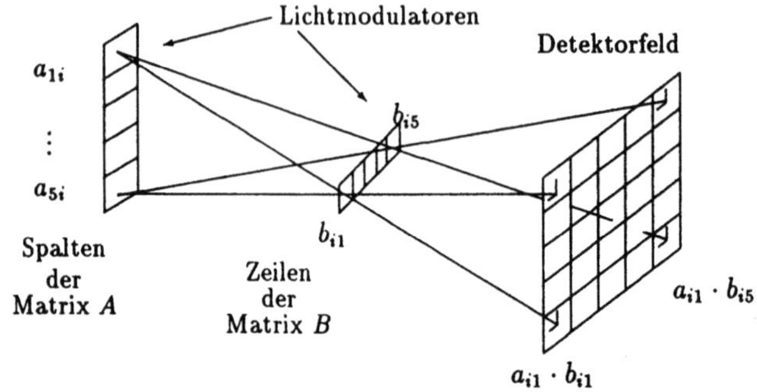

Abbildung 3: Aufbau für eine optische Matrix-Multiplikation

Rechensysteme sind auch Spezialprozessoren interessant, die umfangreiche Operationen, wie beispielsweise eine Matrix-Multiplikation, schnell bestimmen.

Matrix-Multiplikation

Als eher klassisches Beispiel für optische Implementierungen arithmetischer Funktionen ist die Matrix-Multiplikation bekannt. Der dafür vorgesehene optische Aufbau in Abbildung 3 zeigt zwei zeilenförmige optische Lichtmodulatoren und ein quadratisches Detektorfeld. Die beiden Lichtmodulatoren werden mit der Spalten- bzw. Zeileninformation der zu multiplizierenden Matrizen geladen. Auf Empfangsseite integrieren Detektoren das ankommende Licht in aufeinanderfolgenden Rechenstufen. Das Resultat in jedem der Detektoren ist zuletzt die Summe der eingegangenen Lichtsignale.

Wie auch für viele andere Anwendungen ist es aus Geschwindigkeitsgründen extrem wichtig, daß die Lichtmodulatoren optisch angesteuert werden können. Während sich die Schaltgeschwindigkeiten elektrooptischer Modulatoren derzeit im Bereich von Mikrosekunden bewegen, ist es durchaus zu erwarten, daß man auf der Basis optisch steuerbarer Fabry-Perot-Zellen oder gar mit neuen Werkstoffen (vgl. z. B. [12]) in den Nano- und Picosekundenbereich vorstößt.

Räumliche Lichtmodulatoren sind überhaupt sehr vielseitig einsetzbar, nicht nur zur Implementierung arithmetischer und logischer Operationen, sondern auch zur Verschaltung der Prozessoren eines Parallelrechners. Lichtwege, die gleichzeitig Kommunikationspfade darstellen, können mittels der Modulatoren geöffnet oder blockiert werden. Als aktive Elemente sind sie sogar in der Lage Verknüpfungsfunktionen wahrzunehmen — eine wichtige Eigenschaft für die Konstruktion von Parallelrechnern. Mindestens ebenso interessant in diesem Zusammenhang ist die Theorie der Kommunikationstrukturen, denn diese nehmen entscheidenden Einfluß auf den Durchsatz von Daten in parallelen Systemen.

Kommunikation in Parallelrechnern

Obwohl es Vorschläge gibt, serielle Rechner sowohl voll-optisch als auch in Hybrid-Technologie zu konstruieren, sollte man doch ein besonderes Augenmerk auf parallele Architekturen richten. Schließlich bietet die hohe räumliche Bandbreite des Lichts dafür exzellente Voraussetzungen.

Ein wichtiges bei Parallelrechnern zu lösendes Problem ist die Kommunikation der Prozessoren untereinander. Die Optik bietet dabei den besonderen Vorteil, daß sich Lichtstrahlen nahezu störungsfrei kreuzen. Somit können beliebige Kommunikationsstrukturen im freien Raum realisiert werden. Nach wie vor bleiben jedoch Probleme bei der Zuteilung von Übertragungskanälen und, falls nötig, auch beim Routing von Datenpaketen.

Aus technischen Gründen ist es bei großer Prozessorzahl sicherlich kaum sinnvoll, einen vollständigen Graphen als Verbindungsstruktur zu wählen. In herkömmlichen Rechnern wird deshalb als Schaltbaustein der Kreuzschienenverteiler eingesetzt. Dessen Implementierung ist auch in voll-optischen sowie in optisch gesteuerten Rechnern möglich.

Optische Kommunikation

Während Elektronen prinzipiell Leiterbahnen oder elektrische Kabel als Übertragungsmedium erfordern, genügt bei Photonen im Prinzip der freie Raum. Licht gestattet also vielseitigere Möglichkeiten, um Nachrichten zu übertagen. Müssen größere Entfernungen zwischen Sender und Empfänger zurückgelegt werden, so geschieht dies durch Glasfaserkabel. Hierbei verspricht die Kodierung von Daten auf Modulationsebene (Amplitude, Frequenz, Phase, Polarisation) weitere Fortschritte bzgl. Durchsatz und Sicherheit. Für kurze Übertragungsstrecken, on-chip oder chip-to-chip, eignet sich ganz besonders auch die Freiraumkommunikation.

Es gibt heute bereits marktreife hybride Multi-Chip-Module, deren Kommunikationsstruktur durch ein optisches Freiraumübertragungssystem umgesetzt wird [6]. Bei der Konstruktion dieser Module kommen die üblichen VLSI-Hilfsmittel zum Tragen, allerdings mit der Neuerung, elektronische Chips gemeinsam mit optischen Linsen und Gittern auf einem transparenten Substrat zu plazieren. In dieser Technologie übernehmen die Chips nach wie vor die Verarbeitung der Daten. Ein Nachrichtenaustausch zwischen zwei Chips wird durch Mikrolaser eingeleitet und setzt sich über planare, lithographisch hergestellte Gitter bzw. Linsen fort, die an der Substratoberfläche gemäß eines VLSI-Entwurfs positioniert sind. Als Lichtempfänger dienen einfache CMOS-Detektoren.

Diese Technologie kann man auch dazu verwenden, elektronische Schaltbausteine wie den Kreuzschienenverteiler optisch anzusteuern. Solche Schaltelemente spielen bei der Vernetzung von Prozessoren eine wichtige Rolle. Deshalb sollten sie möglichst schnell schalten und Signale weiterleiten können. Ein optischer Kreuzschienenverteiler wäre natürlich für eine effiziente Kommunikation optimal. Interessanterweise erfüllt ein ähnlicher optischer Aufbau, wie er zur Matrix-Multiplikation vorgestellt wurde, diesen Wunsch (Abbildung 4). Der zentral angeordnete zweidimensionale Lichtmodulator ist elektrooptisch beschreibbar und enthält eine binäre Permutationsmatrix, die hier beispielsweise Prozessor 5 mit Prozessor 3 verbindet.

Ein weiterhin schwieriges Problem bereiten die Ansteuerung und die Zuteilung von Schaltpunk-

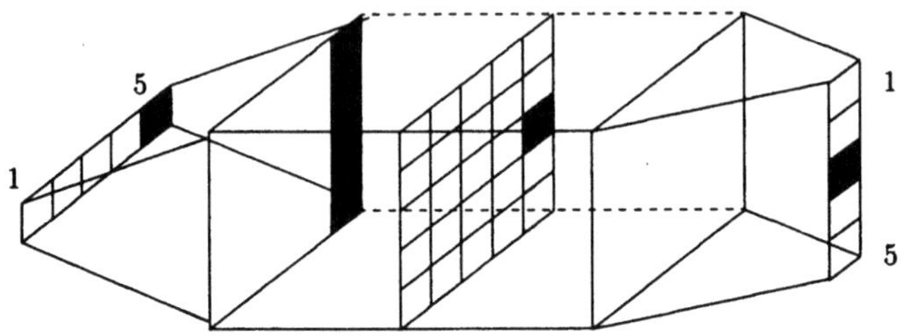

Abbildung 4: Aufbau eines optischen Kreuzschienenverteilers

ten (Arbitrierung) des Kreuzschienenverteilers. Dafür gibt es mittlerweile neue Lösungsvorschläge, so z. B. durch die Einführung eines Rotorbausteins, wie er in [2] beschrieben steht. Dieser Baustein, kurz Rotor genannt, realisiert im Zeitmultiplexing alle Kanten des vollständigen Graphen als Verbindungsstrecke, wobei kollisionsfrei eine maximale Bandbreite erreicht wird. In einem System von n Prozessoren können in $n-1$ Zeitschritten jeweils $\frac{n}{2}$ Prozessorpaare bidirektional miteinander kommunizieren. Dabei löst das feste Zeitscheibenverfahren automatisch die Probleme bei der Zuteilung. Die Steuerung geschieht durch ein einfaches Schieberegister, wie es heute auch schon optisch implementierbar ist.

Verbindungsnetzwerke

Ein Kreuzschienenverteiler kann wegen der aufwendigen Zuteilungslogik nicht beliebig groß gebaut werden. Ebenso ist ein Rotor mit vielen Anschlüssen nicht erstrebenswert, da dann zuviele Zeittakte vergehen, bis ein und dieselbe Verbindung wieder geschaltet wird. Andererseits besitzen herkömmliche Verbindungsnetzwerke einen zu großen Durchmesser, um in elektrooptischen Rechnern sinnvoll eingesetzt werden zu können. Durch die ständige Umsetzung von optischer auf elektronische Signalführung entstehen große Zeitverluste. Es ist daher sinnvoll, nach neuen Strukturen mit geringem Durchmesser zu suchen. Interessante Ansätze liefern hier die Algebra und die Kombinatorik auf dem Gebiet der Design-Theorie [3].

Ein Blockdesign ist eine regelmäßige Inzidenzstruktur, die aus Punkten und Blöcken besteht. Die Blöcke setzen sich aus einer, für unsere Zwecke festen, Anzahl von Punkten zusammen. Jedes Paar von Punkten tritt in mindestens einem Block gemeinsam auf. Identifiziert man die Punkte eines Designs mit Schaltelementen (Rotoren) und die Blöcke mit Prozessoren, so entsteht ein Parallelrechnerentwurf, bei dem zwei beliebige Prozessoren über maximal zwei Schaltelemente miteinander verbunden sind. Die Anzahl der Prozessoren spielt dabei keine Rolle. Es ist sogar möglich, die Anzahl der Prozessoranschlüsse konstant zu halten. Aufgrund der Struktur solcher Netzwerke ergeben sich jedoch mitunter recht lange Signalwege, die bei einer herkömmlichen Lösung nicht tragbar sind. Durch die Ausnutzung optischer Freiraumverbindungen werden diese Netzwerke erst richtig interessant.

Optische Speicher

Die compact disk (CD) hat optische Speichermedien populär gemacht. Sie bietet kostengünstig ein hohes Speicherpotential bei schnellem und sicherem Zugriff. Der wesentliche Nachteil von CD-ROMs (read only memory) und CD-WORMs (write once read many) besteht darin, daß sie weder lösch- noch wiederbeschreibbar sind. Mit magnetooptischen Materialien kann man dieses Manko jedoch beseitigen.

Durch fokussiertes Laserlicht lassen sich Daten auf einem magnetooptischen Speichermedium beliebig oft überschreiben. Kurzzeitige Lichtimpulse hoher Intensität beeinflussen lokal die Orientierung der magnetischen Dipole des Materials, so daß sie eine definierte Richtung einnehmen und sonst genau in die entgegengesetzte Richtung zeigen. Das Auslesen von Daten geschieht bei niedriger Lichtintensität, wobei ein wichtiger magnetoopticher Effekt, der Faraday-Effekt, ausgenutzt wird. Diesem Effekt zufolge wirkt die Orientierung der magnetischen Dipole auf die Polarisation der einfallenden Laserstrahlen. Da ein beschriebenes Speichermedium nur zwei Dipol-Orientierungen zuläßt, gibt es auch nur zwei ausgezeichnete Polarisationsrichtungen des reflektierten Laserlichts, wodurch binäre Daten repräsentiert werden. Immerhin lassen sich auf einer $5\frac{1}{4}$-Zoll magnetooptischen Platte ca. 600 MBytes speichern.

Die Algorithmentechnik befaßt sich in diesem Bereich vornehmlich mit verschiedenen Formen der Datenkodierung und -organisation. Dabei gilt es, einen schnellen und sicheren Zugriff auf das Speichermedium zu gewährleisten. Beim CD-ROM beispielsweise mußten durch algebraisch fundierte Kodiermethoden gleich mehrere Probleme gelöst werden.

- Eine Kodierung auf Modulationsebene garantiert, daß zwei 1-Nutzbits wenigstens durch zwei 0-Bits getrennt liegen (notfalls werden 0-Bits eingefügt). Der Laser wäre sonst nicht in der Lage, aufeinanderfolgende 1-Nutzbits zu trennen.

- Anhand der Kodierung muß die Rotationsgeschwindigkeit der Platte prüfbar sein, um evtl. Korrekturen einzuleiten.

- Ein geschachtelter Reed-Solomon-Code dient der effizienten Korrektur von fehlerhaften Daten, wobei eine zusätzlich vorgenommene Kodespreizung (gezieltes Permutieren der Datenbytes) Bündelfehlern massiv entgegenwirkt.

Eine andere, sehr vielversprechende Variante der optischen Speicherung bietet die Holographie. Liegt die zu speichernde Information in zweidimensionalen digitalen Mustern vor, so lassen sich davon Fourier-Hologramme erstellen. Diesen Vorgang können wir als eine natürliche Form einer fehlerkorrigierenden Kodierung auffassen, denn die Information jedes einzelnen Bits ist auf das gesamte Hologramm verteilt. Folglich produzieren beispielsweise einzelne Staubpartikel, die mit Bündelfehlern gleichzusetzen sind, kaum Fehler beim optischen Auslesen des Speichers.

Ein Nachteil besteht allerdings darin, daß es nicht ohne weiteres möglich ist, einzelne Bits in einem Hologrammspeicher zu verändern. Schließlich hat jedes Bit Einfluß auf das komplette Hologramm. Von daher eignen sich Speicher dieser Art wohl vornehmlich als ROMs, doch als solche fallen sie durch außergewöhnliche Leistungen auf. Neben einer enormen Speicherkapazität von ca. 6 MBit/cm^2 bieten Hologramm-Speicher einen optischen, massiv parallelen Datenzugriff.

In dieser Technologie sind große optische Datenbanken denkbar, wobei Informationen auf einer Anordnung vieler Hologramme organisiert und verteilt liegen. Der optische Zugriff geschieht über ein Licht-Ablenksystem, um die verschiedenen Hologramme einzeln anzusteuern. Dabei könnten optische Spezialprozessoren aufgrund der Datenfülle zeitaufwendige String- bzw. Pattern-Matching-Probleme erledigen. Eine weitere wichtige Anwendung parallel zugreifbarer optischer Speicher ist wiederum in der Parallelrechnertechnik zu finden. Die in der Komplexitätstheorie definierte *PRAM* (parallel random access machine) läßt sich wahrscheinlich nur in optischer Technologie verwirklichen, da in diesem Modell massiv parallele Speicherzugriffe vorgesehen sind, die sich in VLSI-Technologie bisher nicht effizient in den Griff bekommen lassen.

Zukünftige Bedürfnisse und Zusammenfassung

Anhand der drei wichtigen Betätigungsfelder Arithmetik, Kommunikation und Datenspeicherung haben wir aufgezeigt, daß die Algorithmentechnik auch im Bereich optisches Rechnen viele, zum Teil bewährte, aber auch neue Methoden zur Konstruktion von Hard- und Software anbieten kann. Ein sicherlich großer Vorteil der algorithmentechnischen Verfahren besteht darin, daß sie jeweils zu einem übergeordneten und damit technologieunabhängigen Modell führen.

Die bislang vorgenommenen elektronischen Implementierungen von Prozessoren, Kommunikationsnetzen und Speichern haben sich zweifellos bei diversen Anwendungen bewährt. Da jedoch zunehmend höhere Leistungen an Rechensysteme gestellt, nicht selten sogar Supercomputer verlangt werden, erscheint es naheliegend, moderne Entwicklungen der elektrooptischen Hybrid-Technologie beim Entwurf z. B. von Rechnern, Modulen oder Chips mit in Betracht zu ziehen. Um die Hard- und Software von Hybrid-Systemen erstellen zu können, sind neue Entwicklungswerkzeuge nötig, sowohl für den Entwurf als auch zum Testen. Sobald diese Werkzeuge in ausreichendem Maß zur Verfügung stehen, wird die neue Technologie erst richtig zum Tragen kommen.

Literatur

[1] P. B. Berra, A. Ghafoor, M. Guizani, S. J. Marcinkowski, P. Mitkas, „Optics and Supercomputing", Proceedings of the IEEE, Vol.77, No.12, December 1989, S.1797–1815

[2] T. Beth, V. Hatz, „Permutationsschaltung", Offenlegungsschrift, Deutsches Patentamt, Nr. DE 4002455A1, 1990

[3] T. Beth, D. Jungnickel, H. Lenz, „Design Theory", B.I.-Wissenschaftsverlag, 1985

[4] K. H. Brenner, „Programmable optical processor based on symbolic substitution", Applied Optics, Vol.27, No.9, May 1988, S.1687–1691

[5] C. D. Capps, R. A. Falk, T. L. Houk, „Optical arithmetic/logic unit based on residue arithmetic and symbolic substitution", Applied Optics, Vol.27, No.9, May 1988, S.1682–1686

[6] A. Dickinson, M. M. Downs, „An Optical Multichip Module", Proceedings IEEE International Conference on Computer Design 1991, Cambridge, S.490–493

[7] D. G. Feitelson, „Optical Computing: A Survey for Computer Scientists", MIT Press, 1988

[8] V. Hatz, T. Beth, „Transputersysteme für topologieunabhängige Programmierung", Tagungsband des Transputer-Anwender-Treffen '91, Aachen, erscheint in Springer Informatik Fachberichten

[9] J. D. Lipson, „Elements of Algebra and Algebraic Computing", Addison-Wesley, 1981

[10] A. Lohmann, W. Stork, G. Stucke, „Optical perfect shuffle", Applied Optics, Vol.25, No.10, May 1986, S.1530–1531

[11] M. J. Murdocca, A. Huang, J. Jahns, N. Streibl, „Optical design of programmable logic arrays", Applied Optics, Vol.27, No.9, May 1988, S.1651–1660

[12] R. Ruthen, „Clear Advantage: Opaque gases can be made transparent using laser light", Scientific American, September 1991, S.26

[13] J. Shibata, T. Kajiwara, „Optics and electronics are living together", IEEE Spectrum, February 1989, S.34–38

[14] O. N. Tufte, D. Chen, „Optical techniques for data storage", IEEE Spectrum, February 1973, S.26–32

Optik und Informatik

D.Schreier; F.Zimmer
TU Dresden - Fak. Informatik
Inst. Datenbanken/Künstliche Intelligenz
Mommsenstr. 13
0-8027 Dresden

Vom 18. bis 21. März 1991 fanden sich Optiker, Physiker und Informatiker in Oberbärenburg bei Dresden unter der Schirmherrschaft der Deutschen Physikalischen Gesellschaft und der Deutschen Gesellschaft für angewandte Optik zusammen. Die Vorträge und Diskussionen standen unter dem Thema "Optik und Informatik".

Enge Beziehungen zwischen der Optik und Informatik bestehen bei der Anwendung und technischen Umsetzung der Informatik auf den Gebieten der Informationsübertragung, -speicherung und Bildverarbeitung, wobei das optische Rechnen mit einbezogen wird. Andererseits fördert die Informatik, insbesondere mit Hilfe ihrer formalen Methodiken und rechentechnischen Mittel, die Optik. Geläufige Beispiele sind die synthetische Holographie, die optische Nachbildung neuronaler Netze und die optische Bildverarbeitung.

Bei der optischen Bildverarbeitung /1/ sollen Szenen oder Objekte im Sinne einer Vorverarbeitung für Computer einschließlich der visuellen Betrachtung so aufbereitet werden, daß sie aufgenommen und mit geringem Aufwand verarbeitet werden können bzw. den Anforderungen des Beobachters gerecht werden.

Die in diesem Zusammenhang angestellten Untersuchungen führten bisher in einzelnen Fällen zu technisch umsetzbaren Ergebnissen, z. B. der Thorwirthmethode /2/.

In der Praxis sind zahlreiche optische Eigenschaften zu berücksichtigen, um eine Bildverarbeitung zu verwirklichen:

* die Erfassung von dreidimensionalen Szenen durch zweidimensionale Aufnahmeeinrichtungen
* die Beleuchtung

* die optischen Eigenschaften
  wie Farbe, Brechungsindex, Reflexionsverhalten
* die Abbildungseigenschaften in der Optik
* die Parameter des optisch-elektronischen Wandlers.

In Zukunft wird die hohe Parallelität der in der Optik möglichen Verarbeitung von Daten für die Bildverarbeitung im engeren und die Informatik im weiteren Sinne, z. B. bei der Entwicklung von Rechnerarchitekturen und der Verwirklichung neuronaler Netzwerkmodelle, an Einfluß gewinnen.

Gerade die Entwicklung leistungsfähiger neuronaler Systme verlangt neuartige technische Lösungen, deren theoretische und experimentelle Grundlagen durch die Physik mit geschaffen werden. Als Beispiele seien optische Verbindungsnetzwerke oder optische räumliche Modulatoren (Schaltmatrizen) für die Wichtung der Verbindungen genannt.

Wünschenswert wäre, daß die Informatiker in stärkerem Maße den Wert der Optik und Physik für ihr Gebiet erkennen und ein Teil der Physiker sich Aufgaben der Informatik zuwendet.

Mit steigendem Entwicklungsstand der neuronalen Netze wird sich die Aufmerksamkeit der Informatiker mehr und mehr auf die Gewinnung und die Ausgabe der Informationen richten müssen. Wir denken dabei an den audio-visuellen Bereich und die motorische Steuerung. Eine enge Zusammenarbeit mit Physikern wird unumgänglich sein.

Dazu ist erforderlich, daß Forschungsthemen von Optikern, Physikern und Informatikern gemeinsam zur Lösung grundlegender interdisziplinärer Aufgaben bearbeitet werden. In diesem Zusammenhang sei die Vorverarbeitung optischer Bildsignale bis zur Schaffung einer künstlichen Retina /3/ angeführt.

In der Lehre halten wir Vorlesungen und Praktika für erstrebenswert, die die Verbindung zwischen Optikern, Physikern und Informatikern zum Ausdruck bringen. Entsprechende Lehrinhalte, z. B. zur optischen Bild- und Informationsverarbeitung, zur Physik der Bildverarbeitung und -erfassung, zu optischen Rechnerarchitekturen, zur Neuroinformatik sind unter dem Gesichtspunkt physikalischer Systeme eine Bereicherung für die Praktische und Theoretische Informatik und sollten in die Lehrpläne dieser Teilgebiete mit aufgenommen werden.

Literatur

/1/ Häusler, G.: Optical Preprocessing for Automated Inspection. - Optics and Lasers in Engineering 10 (1989) S. 251-263

/2/ Thorwirth, G.: Inkohärente strukturzonale Analyse zur optischen Bildverarbeitung.-Beiträge zur Optik und Quantenelektronik, Vorträge der 23. Frühjahrsschule Optik (1991)S.52, 53

/3/ Mead, C.A.; Mahowald, M.A.: A silicon model of early visual processing.- Neural Network 1 (1988)1, S.91-97

Möglichkeiten zur Parallelisierung des Error-Backpropagation Algorithmus

B. Klauer, K. Waldschmidt
FB 20 Lehrstuhl für Technische Informatik
J. W. Goethe Universität
Postfach 11 19 32
6000 Frankfurt am Main 11

In dieser Arbeit wird ein Verfahren vorgestellt, das auf der Error-Backpropagation Methode aufsetzt und sowohl die Simulation als auch die Lernphase neuronaler Netze hochgradig parallelisiert.

Ergebnisse:

1. Ein paralleles Simulationsverfahren

2. Ein parallelisierter Error-Backpropagation Algorithmus

3. Ein modifiziertes Berechnungsmodell für Neuronen, das den gleichzeitigen Ablauf von Simulation und Lernverfahren erlaubt.

Einführung

In der herkömmlichen Error-Backpropagation Methode folgt auf eine Berechnungsphase eine Lernphase, die anhand der Ausgabe des Netzwerkes und einem gewünschten Sollwert die Gewichte der Verbindungsstrukturen modifiziert. Gemäß einer bestimmten Reihenfolge werden dabei in den meisten Simulatoren für neuronale Netze die Ausgabewerte der einzelnen Neuronen sequentiell berechnet. Danach werden die Synapsen ebenfalls sequentiell modifiziert.

Die systemimmanente massive Parallelität neuronaler Netze wird in den meisten Simulations- und Lernverfahren nicht berücksichtigt. Aufgrund der Monoprozessor-Struktur vieler Rechenanlagen und der von Neumann-Sichtweise vieler Programmierer arbeiten viele Simulations- und Lernsysteme noch nach einem strikt sequentiellen Verfahren. Erste Ansätze zur Parallelisierung werden für Transputer angegeben. [1]

In dieser Arbeit wird die Parallelisierbarkeit der Error-Backpropagation Methode gezeigt. Dazu wird der Algorithmus in zwei Phasen zerlegt. Die Berechnung des Verhaltens des neuronalen Netzes auf ein bestimmtes Eingangsmuster wird im folgenden als Simulationsphase bezeichnet. Die Auswertung des Ergebnisses, die Fehlerberechnung für die einzelnen Neuronen und die Korrektur der Gewichte der Synapsen bezeichnen wir als Lernphase. Die aufeinander folgende Anwendung einer Simulationsphase und einer Lernphase mit einem konkreten Ein- und Ausgabemuster bezeichnen wir als Lektion. In dieser Arbeit wird gezeigt, wie sowohl die Simulationsphase als auch die Lernphase durch Parallelisierung beschleunigt werden können. Es wird zudem gezeigt, daß beide Phasen unter gewissen Voraussetzungen gleichzeitig ausführbar sind.

Voraussetzungen

1. Die Voraussetzungen zur Anwendung des Error-Backpropagation Verfahrens sind gegeben.

2. Das Neuronale Netz enthält keine Rückkopplungen.

3. Jedes Neuron läßt sich eindeutig einer Tiefe zuordnen.

Die Tiefe eines Neurons verstehen wir wie folgt:
Ein neuronales Netz gemäß 2. kann als zyklusfreier gerichteter Graph verstanden werden. Jedem Neuron entspricht ein Knoten des Graphen, jeder Synapse eine gerichtete Kante. Eingangsneuronen entsprechen Knoten, auf die keine Kante zeigt. Von Ausgangsknoten geht keine Kante aus. Unter der Tiefe eines solchen Knotens verstehen wir die Pfadlänge zu den Ausgangsknoten. Diese Pfadlänge ist eindeutig, wenn alle Pfade, die mit den ausgehenden Kanten beginnen, mit gleicher Pfadlänge zu einem Ausgangsknoten führen.

Parallelisierung

Nebenläufigkeit in der Simulationsphase

Neuronen, mit gleicher Tiefe können zu einer Schicht zusammengefaßt werden. Zwischen Neuron einer Schicht bestehen unter den oben angegebenen Voraussetzungen keine Datenabhängigkeiten. Sie können daher konfliktfrei parallel berechnet werden.

Pipelining in der Simulationsphase

Die Schichten neuronaler Netze können unter den oben genannten Voraussetzungen als Stufen einer Pipeline aufgefaßt werden. Wird ein Eingangsmuster an die Eingangsschicht angelegt und der Ausgabevektor der ersten verdeckten Schicht berechnet, so können nach der folgenden Eingabe sowohl die erste Schicht als auch die zweite Schicht eine Berechnung durchführen.

Nebenläufigkeit in der Lernphase

Wie in der Simulationsphase, so bestehen auch in der Lernphase keine Datenabhängigkeiten zwischen den Neuronen einer Schicht, daher können auch in der Lernphase die Synapsen der Neuronen, die in einer Schicht liegen, stets parallel modifiziert werden.

Pipelining in der Lernphase

Es liegt nahe, ähnlich wie in der Simulationsphase, auch in der Lernphase die Schichten eines neuronalen Netzwerkes als Pipeline zu organisieren und einen errechnenten Fehler pipelineartig zur Eingangsschicht zu propagieren. Aus zwei Gründen ist dies jedoch nicht ohne weiteres möglich.

I. Bei der Bewertung des Fehlers eines Neurons wird der Ausgabewert o_{pj} des Neurons, der während der Simulationsphase berechnet wurde, benötigt. Bei der Modifikation der Synapsen, die den eigentlichen Lernprozeß darstellt, wird dieser Ausgabewert ebenfalls benötigt. Das geht aus den Gleichungen für die Berechnung des Fehlers eines Neurons und für die Berechnung der Korrekturwerte für die Synapsen hervor [2]:

$$\delta_{pj} = o_{pj}(1-o_{pj}) \sum_k \delta_{pk} w_{kj}$$

$$\Delta w_{ij}(n+1) = \alpha(\delta_{pj} o_{pj}) + \beta \Delta w_{ij}(n)$$

Abb. 1

Durch die Anwendung von Pipelining während der Simulationsphase würden diese Werte in jeder Berechnungsphase der Pipeline überschrieben. Die Ausgabewerte, die während der Simulationsphase erzeugt werden, werden im folgenden als Spur bezeichnet.

II. Nach jedem Pipelinetakt in der Lernphase muß ein neuer Fehlerwert zur Verfügung stehen, damit das Verfahren effizient einsetzbar ist.

Lösungen

Die Lösung für **Problem II**. besteht darin, daß man die pipelineartig organisierten Phasen Simulation und Lernphase parallel synchronisiert arbeiten läßt. Die Simulationsphase liefert nach dem Einschwingvorgang Ausgabewerte in genau dem Takt, in dem sie von der Lernphase benötigt werden. Ein spezielles Rechenwerk in den Ausgangsneuronen berechnet dabei aus den aktuellen Ausgaben und den gewünschten Werten den für die Lernphase relevanten Fehler. Wird ein Eingangsmuster an die Eingangsneuronen angelegt, muß der zugehörige Ausgabewert zur Berechnung des Fehlers bis zum Erscheinen des entsprechenden Ausgabewertes an der Ausgangsschicht des Netzes verzögert werden.

Problem I. besteht jedoch weiter, da durch das Pipelining in der Simulationsphase die Simulationsspur bei jedem Takt überschrieben wird. Die Zwischenergebnisse stehen daher für den

Lernprozeß nicht mehr zur Verfügung. Durch eine Modifikation im Aufbau der Neuronen kann dieses Problem jedoch gelöst werden. Ein Modellneuron sei entsprechend Abb. 2 aufgebaut. Die gewichteten Ausgaben der vorhergehenden Schicht werden aufsummiert und an die Ausgabefunktion, die den neuen Ausgabewert des Neurons berechnet, weitergegeben. In der Lernphase wird aus der gewichteten Summe der Fehler der nachfolgenden Neuronen und dem Ausgabewert ein neuer Fehlerwert und ein Korrekturwert für die eingehenden Synapsen berechnet. Erweitert man dieses Modell, indem man die Verbindung zwischen der Ausgabefunktion und der Fehlerfunktion durch eine Warteschlange gemäß Abb. 3 ersetzt, so können die in der Simulationsphase erzeugten Zwischenergebnisse bis zur Lernphase gepuffert werden. In der Simulationsphase werden die Zwischenergebnisse nicht nur den Synapsen der folgenden Schicht zur Verfügung gestellt, sondern gleichzeitig in die Warteschlange eingereiht. Dort werden sie in der Lernphase wieder angefordert.

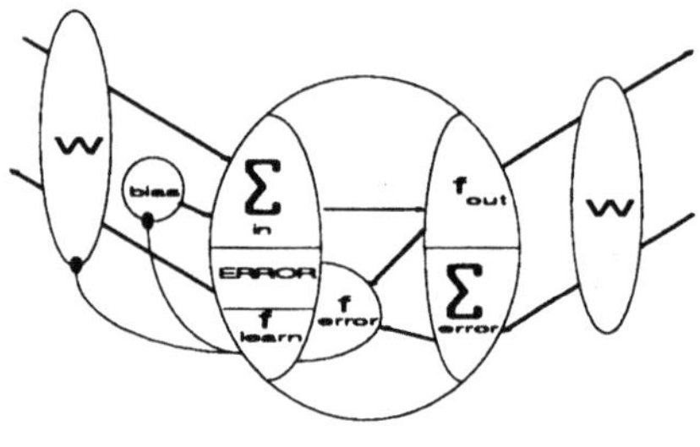

Abb. 2

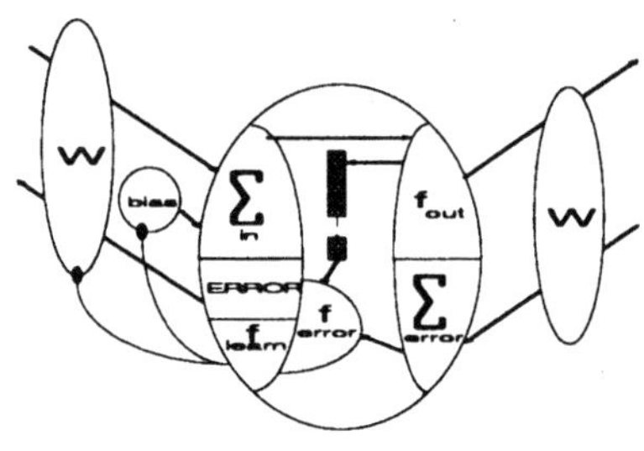

Abb. 3

Die Kapazität der Warteschlange

Die minimale Kapazität k der Warteschlange eines Neurons in einer verdeckten Schicht läßt sich mit Hilfe der Tiefe der Schicht berechnen. Als Tiefe τ einer Ebene bezeichnen wir die Entfernung der Ebene zur Ausgangsschicht. Die Ausgangsschicht hat dem entsprechend die Tiefe $\tau=0$. Die Eingangsschicht hat die Tiefe τ_{max}, die wir gleichzeitig als Tiefe T des neuronalen Netzes definieren. Für die Kapazität κ der Warteschlangen der Neuronen in den verdeckten Ebenen gilt dann:

$$2 * \tau \leq \kappa$$

Dabei wird der Ausgabewert eines Neurons dann in die Warteschlange eingereiht, wenn ein neuer berechnet wird. Die oben angegebene Relation für die Tiefe der Warteschlange ergibt sich, wenn man die Berechnungen zählt, die in dem Neuron durchgeführt werden, bis der jüngste berechnete Wert wieder benötigt wird. Berechnet ein Neuron zum Zeitpunkt t_0 in Tiefe τ eine Ausgabe so wird diese Ausgabe des Neurons in τ Schritten zur Ausgabeschicht propagiert. Nach weiteren τ Schritten hat das Lernverfahren die Schicht mit der Tiefe τ wieder erreicht. Zu diesem Zeitpunkt muß das Ergebnis der Berechnung zum Zeitpunkt t_0 dem Lernverfahren zur Verfügung stehen. Die Neuronen der Ausgangsschicht benötigen diese Warteschlange nicht. Die Ausgabewerte der Ausgangschicht gehen unverzögert in die Berechnung des Fehlers der Ausgangsneuronen ein. Der Simulations- und Lernvorgang läßt sich gemäß Abb. 4 darstellen. In Abb. 4 wird die Berechnung von 10 Lektionen unter Ausnutzung des oben beschriebenen Parallelismus in einem neuronalen Netz mit zwei verdeckten Schichten dargestellt. Am linken Rand der Grafik sind die Berechnungszeitpunkte als diskrete Zeitschritte dargestellt. Zum Zeitpunkt t_1 wird an der Eingangsschicht das zur Lektion a gehörige Eingangsmuster zur Verfügung gestellt. Zum Zeitpunkt t_2 werden sämtliche Ausgaben der Schicht S_2 berechnet. Gleichzeitig wird durch die Eingangsschicht das zur Lektion b gehörige Eingangsmuster bereitgestellt. Zum Zeitpunkt t_3 werden die Ausgaben der Schicht S_3 berechnet, die zur Lektion a gehören. In S_2 werden die Ausgaben, die zur Lektion b gehören, berechnet,

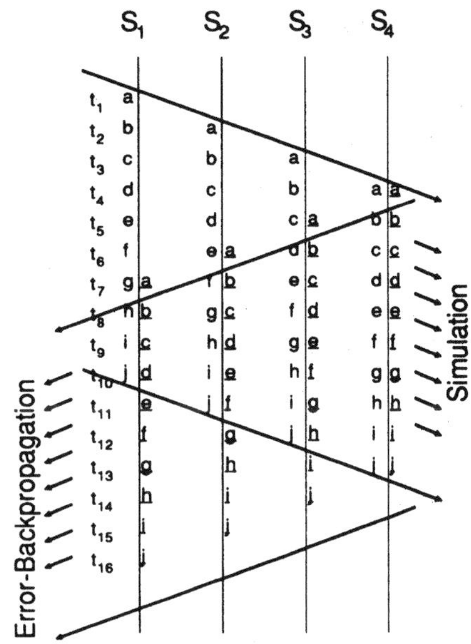

Abb. 4

während von der Eingangsschicht das zur Lektion c gehörende Eingangsmuster bereitgestellt wird. Zum Zeitpunkt t_4 werden die in der Grafik ungestrichen dargestellten Lektionen berechnet. Danach setzt in S_4 die Lernphase ein. Das Ergebnis, daß bei der Berechnung der zur Lektion a gehörenden Ausgabe entsteht, muß an dieser Stelle nicht gepuffert werden, da die Fehlerberechnung unmittelbar nach der Ausgabeberechnung erfolgt. An dieser Stelle wird jedoch zum ersten mal auf das gepufferte Ergebnis der Lektion a in S_3 zurückgegriffen. Die Lernphase einer Lektion ist durch die gestrichenen Zeichen dargestellt. Zu den Zeitpunkten t_1 bis t_6 wird die Pipeline gefüllt. Zu den Zeitpunkten t_7, t_8, t_9 und t_{10} arbeitet das System mit maximaler Parallelität. Zu den Zeitpunkten t_{11} bis t_{16} wird die Pipeline geleert. Zu einem Zeitschritt werden folgende Operationen durchgeführt:

1. Ergebnis der Simulationsphase aus der Warteschlange holen (ENQ).

2. Simulationsphase, Lernphase

3. Ergebnis der Simulationsphase in die Warteschlange einreihen (DEQ).

Während des Füllens und des Leerens der Pipeline werden nicht alle der oben angegebene Operationen ausgeführt. Die Lernphase und die Operation DEQ werden in der Grafik durch die gestrichenen Symbole markiert. Die ungestrichenen Symbole markieren die Operationen Simulation und ENQ. Bei den Eingangsneuronen entfällt die Simulationsphase, statt dessen wird ein neuer Eingangswert eingestellt. Bei den Ausgangsneuronen entfällt die ENQ Operation, statt dessen wird ein Soll-Wert für die Ausgabe eingestellt. Eine modifizierte Fehlerfunktion berechnet aus den aktuellen Ausgabewerten und den Soll-Ausgabewerten den Fehlerwert der Ausgangsneuronen.

Effizienzbetrachtung

Das in dieser Arbeit vorgestellte Verfahren vergleichen wir mit der sequentiellen Berechnungsmethode. Die Komplexität C_{seq} der rein sequentiellen Berechnungsmethode für eine Ausgabe bei gegebener Eingangsbelegung schätzen wir wie folgt ab:

$$C_{seq} = n*c_s+m*c_n$$

c_S: Zeitkonstante für die Multiplikation des Ausgabewertes eines Neurons mit dem Gewicht einer Synapse.

c_N: Zeitkonstante für die Berechnung des Ausgabewertes eines Neurons.

n: Anzahl der Synapsen

m: Anzahl der Neuronen

Die Komplexität l_{seq} des sequentiell berechneten Lernvorganges kann ähnlich abgeschätzt werden.

$$l_{seq} = n*l_S+m*l_N$$

l_S: Zeitkonstante für die Multiplikation des Fehlerwertes eines Neurons mit dem Gewicht einer Synapse.

l_N: Zeitkonstante für die Berechnung des Fehlers eines Neurons.

Die Rechenzeit L_{seq} für die sequentielle Berechnung einer Lektion kann als Summe

$$\begin{aligned} L_{seq} &= c_{seq} + l_{seq} \\ &= n*c_S+m*c_N + n*l_S+m*l_N \\ &= n*(c_S+ l_S) + m*(c_N+l_N) \end{aligned}$$

abgeschätzt werden.

Gehen wir bei der Berechnung der Rechenzeit L_{parl} davon aus, daß die Synapsen zwischen zwei Ebenen sowie die Neuronen einer Ebene parallel berechenbar sind, so können wir in beiden Phasen n und m durch T mit $T = \tau_{max}$ ersetzen. Für die Rechenzeit L_{parl} gilt dann:

$$L_{parl} = T*(c_S+ l_S+ c_N+l_N)$$

Für k Lektionen werden dann L'_{parl} Berechnungsschritte benötigt.

$$L'_{parl} = k*T*(c_S+ l_S+ c_N+l_N)$$

$L_{par1} < L_{seq}$ gilt genau dann, wenn m>T. In diesem Fall existieren mehr Neuronen als Schichten. Es muß daher genau eine Schicht mit mindestens zwei Neuronen existieren. Der Grenzfall m=T stellt den worst-case dar, in dem die Parallelisierung die Rechenzeit nicht veringert. Das zugehörige Netz (s. Abb. 5) kann jedoch als sehr unwahrscheinlich angesehen werden.In der Regel kann m>>τ angenommen werden.

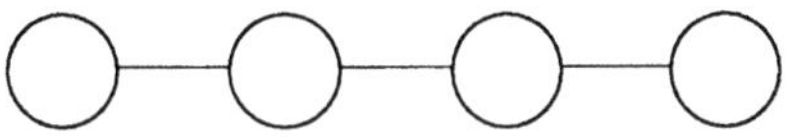

Abb. 5

In den bisherigen Berechnungen blieb das Pipelining-Prinzip und die gleichzeitige Anwendung von Simulations- und Lernphase unberücksichtigt. Das Füllen der pipelineartig organisierten Schichten kostet in beiden Phasen jeweils t Berechnungen. Nach T Berechnungsschritten erscheint das erste Ergebnis der Simulationsphase. Nach weiteren T Schritten ist die erste Lernphase beendet. Nach 2 T Schritten ist die gesamte Pipeline gefüllt und die erste Lektion beendet. Danach wird mit jedem Schritt eine weitere Lektion abgeschlossen. Bei k Lektionen werden also

$$L_{par2}=(c_S+ l_S + c_N+l_N)*(2T+k)$$

Berechnungsschritte benötigt. Das Verfahren arbeitet für k>2T effizient. In den meisten Fällen gilt k>>2T. Der Term 2T kann daher vernachlässigt und die Anzahl der Rechenschritte mit $L_{par2}=(c_S+ l_S + c_N+l_N)*k$ angenähert werden. Die Rechenzeit im Vergleich zu L'_{par1} verringert sich so um den Faktor T.

Korrektheit

Die Korrektheit des Verfahrens ist nicht gesondert zu beweisen, da das Error-Backpropagation-Verfahren selbst nicht verändert wurde. Die Parallelisierbarkeit der einzelnen Komponenten wurde oben gezeigt.

Simulation

Zur Simulation des Verfahrens wurden das modifizierte Neuron als Klasse in C++ implementiert. Der quasiparallele Ablauf von Simulation und Modifikation auf einer Maschine konnte bereits erfolgreich getestet werden.

Literatur:

[1] Eckmiller, Hartmann, Hauske, Parallel processing in neural systems and computers, North-Holl. 1990

[2] Rummelhart, McClelland and the PDP res. group, Parallel distributed processing V1, MIT 1988

[3] Cohonen, Content adressable memories, 2.ed. Springer 1987

[4] Brause, Neuronale Netze, Teubner 1991

[5] Hwang, Briggs, Computer architecture and parallel processing, McGraw-Hill 1984

[6] Rückert, Integrationsgerechte Umsetzung von assoziativen Netzwerken mit verteilter Speicherung VDI 1990

[7] B.Soucek, M.Soucek, Neural and massively parallel computers, Wiley & Sons, 1988

[8] Ritter, Martinez, Schulten, Neuronale Netze, Addison-Wesley 1991

GIBT ES EIN ZUR STARKEN WECHSELWIRKUNG ANALOGES PRINZIP BEI DER NEURONALEN INFORMATIONSVERARBEITUNG?

Alfred NISCHWITZ, Helmut GLÜNDER*‡

Lehrstuhl für Nachrichtentechnik, Technische Universität, Arcisstr. 21, 8000 München 2
email: alfred@lnt.e-technik.tu-muenchen.de
*Institut für Med..Psychologie, Ludwig-Maximilians-Universität, Gœthestr. 31, 8000 München 2

Neuere Ergebnisse elektrophysiologischer Untersuchungen am primären visuellen Cortex der Katze lassen vermuten, daß zeitlichen Korrelationen einzelner Nervenimpulse bzw. Impulsgruppen eine wichtige Rolle bei der Verarbeitung visueller Reize zukommt. Man nimmt gegenwärtig an, daß auf der Sensorfläche zusammengehörige, z.B. benachbarte (Objekt-)Merkmale gleicher Qualität, wie Bewegungsrichtung, Kontrast, Orientierung der Konturtangente, Kontur-Krümmung, Textur, Farbe, etc., zentralnervös durch synchronisierte Aktivität derjenigen Neuronen markiert sind, die diese Merkmale sowie deren Lage im Cortex repräsentieren [1,2]. Fehlt reizbedingt diese Zusammengehörigkeit, z.B. bei unterschiedlich bewegten Objekten, so erzeugen die entsprechenden Neuronen oder Neuronen-Populationen ihre Impulse nicht mehr synchron, jedoch durchaus mit gleicher Rate. Eine in diesem Kontext wichtige Frage ist daher die nach den Mechanismen dieser Synchronisation.

Abb.1 Lateral gekoppeltes eindimensionales Netzwerk mit zeitlich konstant anliegenden externen Gesamterregungen E und den Ausgängen A

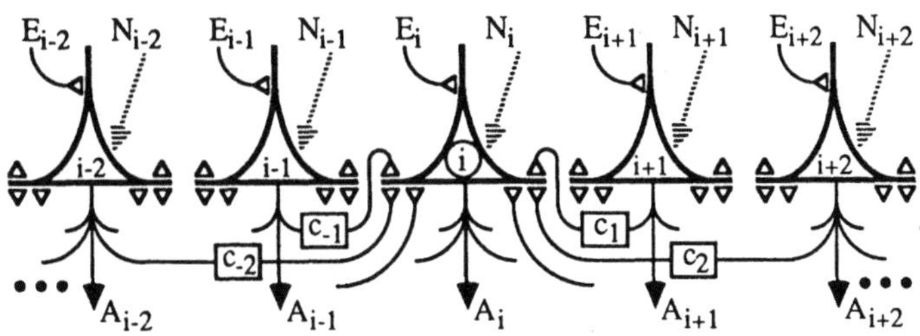

Die bekannten Erklärungskonzepte basieren meistens auf erregenden, sehr selten hemmenden, in jedem Fall aber merkmals-spezifischen Kopplungen von Neuronen [3,4,5,6,7]. Einen Sonderfall stellt die hier nicht weiter betrachtete divergente Vorwärtskopplung dar [8]. Wir haben Netzstrukturen der in Abb.1 skizzierten Art untersucht. Das Verhalten eines einzelnen Modellneurons geht aus Abb.2 hervor: Die an allen Neuronen anliegende, zeitlich konstante externe Erregung E wird gemäß einer Exponentialfunktion (RC-Glied) aufintegriert und beim Erreichen der Schwelle θ ein Impuls erzeugt, worauf der Integrationprozeß vom Startwert D_{re} aus neu beginnt. Von Nachbarzellen eintreffende Impulse verändern das momentane Integrationsergebnis zusätzlich um einen konstanten Betrag, wie dies in idealisierter Form für zu verschiedenen Zeiten auftretende und jeweils gleichstarke hemmende oder erregende Einflüsse in Abb.2 eingezeichnet ist.

Abb.2 Zeitliche Integrations-Charakteristik und Impulsgenerierung beim verwendeten Neuronenmodell. Abhängig vom Zeitpunkt des Eintreffens einer ideal wirkenden impulsförmigen Hemmung oder Erregung wird die Impulserzeugung offensichtlich unterschiedlich stark verzögert, bzw. beschleunigt

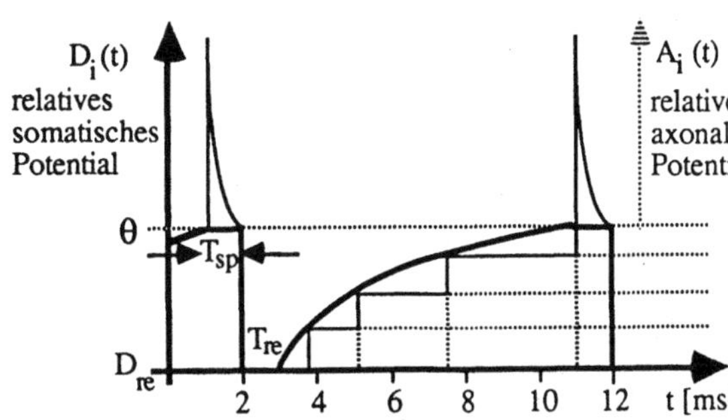

Es zeigt sich, daß sowohl über erregende als auch hemmende laterale Kopplungen die Synchronisation der Ausgangsimpulse von Modellneuronen möglich ist, wobei sich allerdings das Synchronisationsverhalten in beiden Fällen stark unterscheidet.

‡ gefördert durch die "Volkswagen-Stiftung" im Rahmen des 'Synergetik'-Programms, Projekt I/65 914

Bei erregender Übertragung ohne nennenswerte Verzögerung nimmt die synchronisierende Wirkung lateral übermittelter Impulse mit dem Synchronisationsgrad zu. Demzufolge stellt sich hierbei nach einer Einlaufphase sehr gute Synchronizität ein [7]. Ursache für dieses Verhalten ist die exponentielle Integrationscharakteristik der Neurone: Weichen die Integrationszustände D_i nur wenig voneinander ab, so werden die Impulse der zuerst überschwellig gewordenen Neurone ausreichen um ihre schon nahe der Schwelle befindlichen Nachbarn sofort überschwellig zu machen. Offensichtlich sinkt die Wahrscheinlichkeit für diese Impulstriggerung mit abnehmender Synchronizität.

Im Gegensatz zur erregenden wirkt unverzögerte hemmende Übertragung desynchronisierend. Da biologische Netzwerke aber Interneurone zur Umsetzung von Erregung auf Hemmung erfordern und damit Laufzeiten enstehen, ist es sinnvoll das Synchronisationsverhalten für verzögerte hemmende Kopplungen zu untersuchen. Für Verzögerungen τ, die etwa die Hälfte des mittleren Impulsabstandes eines Neurons betragen, nimmt die synchronisierende Wirkung lateral übermittelter hemmender Impulse mit dem Synchronisationsgrad ab. Wie aus den Beispielen in Abb.3 hervorgeht, ist die individuelle Impulsperioden-Verlängerung durch laterale Hemmung bei starker Asynchronizität sehr effektiv. Folglich wird rasch moderate Synchronizität erreicht und Ausreißer werden schnell korrigiert, wogegen bei zunehmender Synchronisation die Bindung des Zeitpunktes der individuellen Impulsauslösung an den des Kollektivs abnimmt.

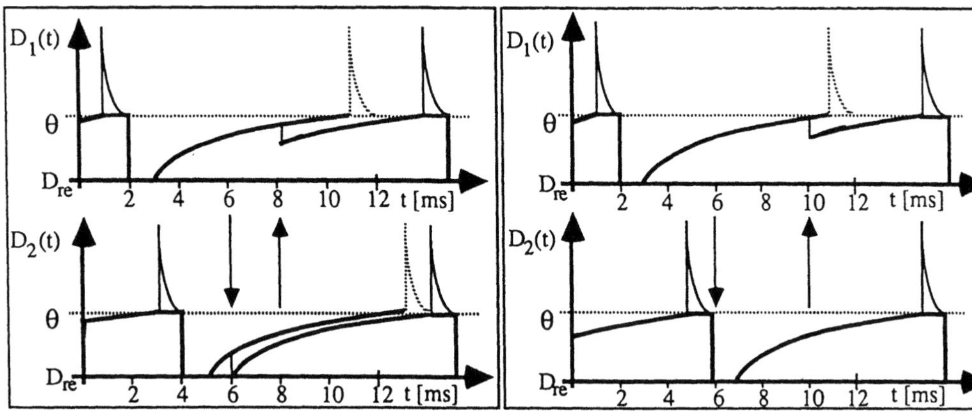

Abb.3 Signalverläufe zweier hemmend und mit τ = 5 ms verzögert gekoppelter Neurone (links: moderate Anfangssynchronisation; rechts: schlechte Anfangssynchronisation)

Das geschilderte Verhalten der Wechselwirkungen zwischen unterschiedlich gekoppelten Modellneuronen legt Vergleiche mit physikalischen Grundkräften nahe, nämlich mit der elektromagnetischen Wechselwirkung einerseits und der starken andererseits. Bei ersterer wächst bekanntlich die Kraft zwischen zwei Ladungen mit abnehmendem Abstand, bei letzterer sinkt sie dagegen mit der Annäherung zweier 'Farbladungen', d.h. Quarks. Folglich kommt Quarks innerhalb eines Hadrons sogenannte 'asymptotische Freiheit' zu, d.h. sie können sich letztlich nicht beliebig voneinander entfernen, da hierfür mehr Energie nötig wäre, als zur Erzeugung neuer Teilchen/Antiteilchen-Paare.

[1] Eckhorn R., Bauer R., Jordan W., Brosch M., Kruse W., Munk M. und Reitboeck H. J. (1988) Coherent oscillations: a mechanism of feature linking in the visual cortex? *Biol. Cybern.* **60**: 121-130.

[2] Gray C. M., König P., Engel A. K. und Singer W. (1989) Oscillatory responses in cat visual cortex exhibit intercolumnar synchronization which reflects global stimulus properties. *Nature* **338**: 334-337.

[3] Atiya A. und Baldi P. (1989) Oscillations and synchronizations in neural networks: an exploration of the labeling hypothesis. *Int. J. Neural Systems* **1**: 103-124.

[4] Eckhorn R., Reitboeck H. J., Arndt M. und Dicke P. (1990) Feature linking via synchronization among distributed assemblies: simulations of results from cat visual cortex. *Neural Computation* 2: 293-307.

[5] Schillen T. B. (1990) Simulation of delayed oscillators with the *MENS* general purpose modelling environment for network systems. In: Eckmiller R., Hartmann G. und Hauske G. (Hrsg.) *Parallel Processing in Neural Systems and Computers*. Elsevier, Amsterdam, pp. 135-138.

[6] Bush P. C. und Douglas R. J. (1991) Synchronization of bursting action potential discharge in a model network of neocortical neurons. *Neural Computation* 3: 19-30.

[7] Nischwitz A., Glünder H. und Klausner P. (1991) Synchronization of spikes in populations of laterally coupled model neurons. In: Kohonen T., Mäkisara K., Simula O. und Kangas J. (Hrsg.). Band 2. Elsevier, Amsterdam, pp. 1771-1774.

[8] Abeles M. (1982) *Local Cortical Circuits. An Electrophysiological Study*. Springer, Berlin.

Anwendungsbezogene Architekturen und Systeme

Topologieerhaltung und Dimensionsbestimmung: Methoden der nichtlinearen Dynamik in Neuroinformatik und Spracherkennung

Klaus Pawelzik und Hans-Ulrich Bauer
Institut für Theoretische Physik, Universität Frankfurt
und
SFB 185 "Nichtlineare Dynamik"
Robert-Mayer-Str. 8-10
W-6000 Frankfurt/Main 11

1. Einleitung

In unserem Beitrag stellen wir zwei Methoden vor, die im Rahmen der Neuroinformatik bzw. in technischen Anwendungsproblemen eingesetzt werden können und die ihren Ursprung in einem Teilgebiet der Physik haben, in der nichtlinearen Dynamik. Mit diesen beiden Beispielen wollen wir den inneren Zusammenhang zwischen Themen der Physik und der Informatik deutlich machen. In unserem ersten Beispiel setzen wir uns mit der Messung der Topologieerhaltung in neuronalen Karten auseinander. Neuronale Karten stellen sensorische Information in einem Ortscode dar. Diese Form der Repräsentation ist für technische Anwendungen ebenso wichtig wie für reale Hirne. Ein bekanntes Beispiel für die Repräsentation somatosensorischer Information durch eine neuronale Karte ist der Homunculus welcher die Zuständigkeit des sensorischen Cortex für die Hautoberfläche illustriert. Neuronale Karten können sich unter dem Einfluß sensorischer Information selbst organisieren [1]. Ein viel verwandter Algorithmus zur Selbstorganisation von Karten ist das Kohonen-Modell [2,3].

Das Kohonen-Modell ist sowohl in technischen Anwendungen wie Roboterarmsteuerung [4] und Spracherkennung [5], als auch für die Erklärung vieler Aspekte kortikaler Organisation erfolgreich eingesetzt worden [6,7]. Die Grundidee dieses Modells ist einfach: Die Kohonen-Karte approximiert eine Abbildung, welche die Topologie einer durch Datenpunkte gegebenen Verteilung im Eingangsraum bestmöglich erhält. Die Dimension des Ausgangsraums der Karte muß jedoch vor der Anwendung des Algorithmus festgelegt werden, die Nachbarschaftserhaltung wird durch den Algorithmus nur im Rahmen der bereits vorher festgelegten Dimensionsverhältnisse geleistet. Eine Kohonenkarte kann auch als eine nichtlineare Transformation auf Haupt(hyper)flächen der Eingangsdatenverteilung aufgefaßt werden [Martinez, Ritter, Schulten], wobei die Dimension der Haupt(hyper)flächen von vorneherein festgelegt ist. Im Gegensatz zu linearen Verfahren wie der Karhunen-Loeve Transformation fehlte bisher für Kohonenkarten ein Kriterium, welches die "Güte der Abbildung", d.h. das Ausmaß der Nachbarschaftserhaltung, mißt.

Diese Lücke kann mit einem Maß, dem Topographischen Produkt, gefüllt werden, welches vor wenigen Jahren in der nichtlinearen Dynamik entwickelt worden ist [8]. Dort ist man mit dem Problem konfrontiert, seltsame Attraktoren aus experimentellen Zeitreihen mittels Verzögerungskoordinaten zu rekonstruieren [9]. Das Topographische Produkt, auch "Waberprodukt" genannt, hilft dabei, sicherzustellen, daß es eine stetige und stetig invertierbare Abbildung zwischen dem Attraktor

eingeführte Topographische Produkt vor und zeigen, wie es auch verwandt werden kann, um Homeomorphieeigenschaften selbstorganisierender Karten zu testen. Dieses Kriterium hilft dem Anwender, seine Kartenmodelle im topologischen Sinne zu optimieren.
Die zweite in der Physik entwickelte Methode, deren Anwendung auf technische Problemstellungen wir in diesem Beitrag diskutieren wollen, behandelt die direkte Messung der Informations-Dimension eines Datensatzes. Die Messung der Informations-Dimension berührt die Frage nach der Struktur multivariater Datensätze und ist von allgemeinem Interesse in vielen experimentellen und angewandten Problemen. Beispielsweise ist die Frage nach der Existenz und der Dimension von Hauptflächen in vergleichsweise hochdimensionalen (D=10-20), durch Spektralkoeffizienten gegebenen Sprachdaten von großer Bedeutung für die Spracherkennung. In der nichtlinearen Dynamik gibt es einen Kanon an Methoden, Dimensionen von Verteilungen zu bestimmen, welche seltsame Attraktoren approximieren; einen derartigen Algorithmus stellen wir im 3. Abschnitt vor. Die direkte Messung der Dimension eines Datensatzes, und die Bestimmung der optimalen Dimension eines Ausgangsraums, in den die Daten möglichst nachbarschaftserhaltend abgebildet werden, sind zwei Methoden, die sich gut ergänzen, obwohl sie verschiedene Fragen beantworten. Eine naheliegende Möglichkeit, die Methoden zu kontrollieren, ist, die jeweilig ermittelten Dimensionen zu vergleichen. Diesen Test führen wir im 4. Abschnitt am Beispiel eines Sprachdatensatzes durch. Dabei finden wir, daß optimale Karten für die Repräsentation dieser Daten dreidimensionale Ausgangsräume haben, ein Ergebnis, welches nicht nur durch die direkte Messung der Informations-Dimension des Datensatzes bestätigt wird, sondern auch mit Ergebnissen einer anderen Arbeitsgruppe zur Erkennungsrate übereinstimmt.

2. Topologieerhaltung in Kohonen-Karten

Beim Kohonen-Modell handelt es sich um einen adaptiven Algorithmus, welcher eine Abbildung von einem Eingangsraum **A** in einen Ausgangsraum **V** entsprechend einer Verteilung ρ der Eingangsdaten erzeugt. Der Algorithmus kann hier nur in aller Kürze dargestellt werden, eine ausführlichere Beschreibung der Kohonen-Karte findet der interessierte Leser u.a. bei Ritter, Martinetz und Schulten [3], eine Beschreibung des Topographischen Produkts in [10]. Die Kohonen-Karte besteht aus Knoten ("Neuronen") j in **A**, denen Zeiger $\mathbf{w}_j$ in **V** zugeordnet sind. Ein Datenpunkt ("Stimulus") **v** im Eingangsraum **A** wird auf denjenigen Knoten j im Ausgangsraum **V** abgebildet, welcher den Abstand $d^{\mathbf{V}}(\mathbf{w}_j, \mathbf{v})$ im Eingangsraum minimiert.
Die Karte wird an einen Datensatz angepaßt, indem eine Vielzahl von Lernschritten durchgeführt wird. Während der Lernschritte werden die Zeiger $\mathbf{w}_j$ verschoben werden. Für jeden Lernschritt wird ein Stimulus **v** entsprechend der Verteilung ρ zufällig ausgewählt und auf j abgebildet. Dann wird der Zeiger $\mathbf{w}_j$ zusammen mit weiteren Zeigern $\mathbf{w}_{j'}$ in Richtung auf **v** verschoben. Die j' stammen aus einer kleinen Umgebung von j in **V**, d.h. es sind die Ausgangsraumnachbarn von j. An dieser Stelle des Algorithmus wird die Nachbarschaftserhaltung erzwungen, und die Topologie des Ausgangsraumes bekommt entscheidende Bedeutung.

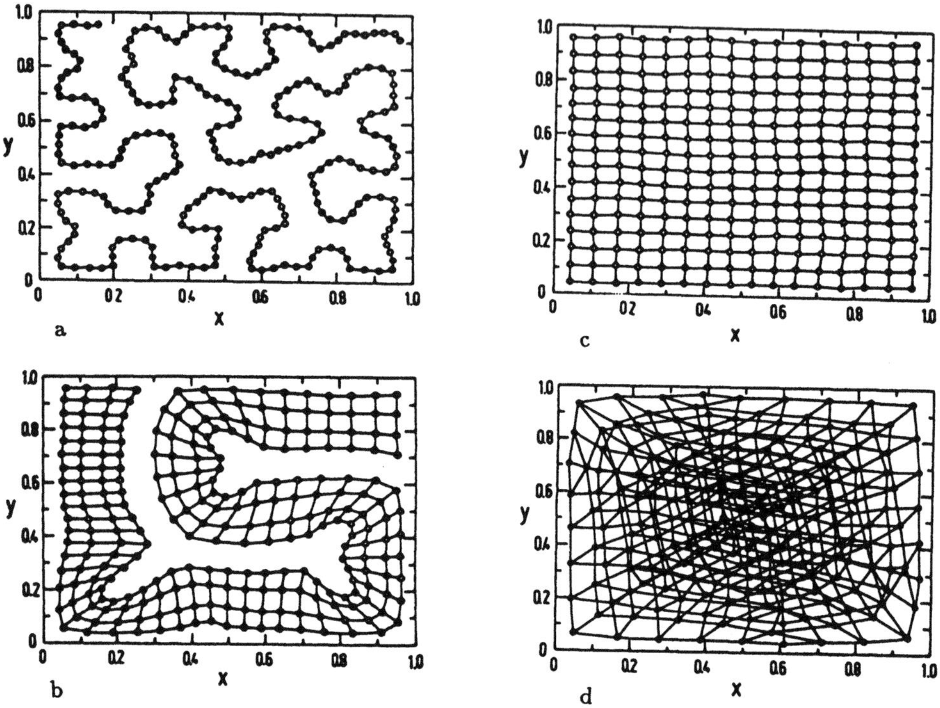

Abb. 1 Kohonen-Karten, welche Stimuli aus einem 2-dimensionalen, quadratischen Eingangsraum auf einen 1-dimensionalen (a), 2- dimensional langgestreckten (b), 2-dimensional quadratischen (c), und 3-dimensionalen (d) Ausgangsraum abbilden. Die Karten werden visualisiert, indem die Positionen der Zeiger $\mathbf{w}_j$ im Eingangsraum gezeigt werden (offene Kreise), und die Zeiger von im Ausgangsraum benachbarten Knoten durch Linien verbunden werden. Man sieht, daß die Abbildung des Quadrats auf eine Linie eine Einstülpung der Linie im Eingangsraum nach der Art einer Peano-Kurve erfordert, daß auch ein 2-dimensionaler Ausgangsraum Einstülpungen der Karte erfordern kann, wenn er zu langgestreckt ist, daß die Abbildung eines Quadrats auf ein Quadrat sehr geordnet aussieht, und daß ein 3-dimensionaler Ausgangsraum zu unübersichtlichen Nachbarschaftsrelationen führt.

In vielen praktischen Anwendungen wird dann davon ausgegangen, daß die so erzeugte Abbildung die Nachbarschaftsverhältnisse der Inputverteilung erhält, d.h. daß benachbarte Punkte der Verteilung im Eingangsraum immer auf benachbarte Knoten im Ausgangsraum abgebildet werden und weiterhin, daß benachbarte Knoten im Ausgangsraum benachbarte Urbilder im Eingangsraum haben. Mit anderen Worten wird durch den Kohonen-Algorithmus eine homeomorphe Abbildung approximiert. Eine Homeomorphie der Abbildung ist jedoch nur dann möglich, wenn der Träger der Inputverteilung dieselbe Dimension wie der Ausgangsraum hat. Darüber hinaus sollte der Ausgangsraum Ausdehnungen haben, welche denjenigen der Inputverteilung grob entsprechen, da sonst eine Faltung des Ausgangsraums in den Eingangsraum erfolgen kann (Abb. 1b). In einfachen Fällen, wie den in Abb. 1 gezeigten, lassen sich Verletzungen der Homeomorphie durch visuelle Inspektion beurteilen. Abgesehen von der Willkür einer solchen Beurteilung durch "Draufgucken" ist es in Eingangsräumen, die eine Dimension $D^V > 2$ haben, unmöglich, die Karte überhaupt darstellen, mithin kann auch ihre Nachbarschaftserhaltung nicht beurteilt werden.

Zur Beurteilung der Topologieerhaltung im allgemeinen Fall punktweise gegebener Abbildungen zwischen Räumen beliebiger (endlicher) Dimension stellen wir nun ein quantitatives Verfahren vor, welches ursprünglich im Zusammenhang mit dem Problem der Einbettung der Dynamik experimenteller Zeitreihen entwickelt wurde. Die Grundidee hierbei ist es, die Transformation von Nachbarschaften unter der Abbildung eines Punktes mit der Transformation durch die Inversen zu vergleichen. Punkte welche die Nachbarschaften weit verlassen, tragen im sogenannten Topographischen Produkt stark bei.

Im einzelnen betrachten wir die Transformation naher Nachbarn eines Knotens j im Ausgangsraum **A** und naher Nachbarn der Zeiger $\mathbf{w}_j$ im Eingangsraum **V**. $n_k^{\mathbf{A}}(j)$ sei dann der k-nächste Nachbarpunkt von j in **A** und $n_k^{\mathbf{V}}(j)$ der k-nächste Nachbarpunkt des Zeigers $\mathbf{w}_j$ von j in **V**.

Mit dieser Notation definieren wir zunächst die Quotienten

$$Q_1(j,k) = \frac{d^V(\mathbf{w}_j, \mathbf{w}_{n_k^A(j)})}{d^V(\mathbf{w}_j, \mathbf{w}_{n_k^V(j)})}, \tag{.1}$$

$$Q_2(j,k) = \frac{d^A(j, n_k^A(j))}{d^A(j, n_k^V(j))}. \tag{.2}$$

Bildet man nun die Produkte

$$P_1(j,k) = \left(\Pi_{l=1}^k Q_1(j,l)\right)^{\frac{1}{k}}, \tag{.3}$$

$$P_2(j,k) = \left(\Pi_{l=1}^k Q_2(j,l)\right)^{\frac{1}{k}}, \tag{.4}$$

so kann man sich anhand einfacher Beispiele klarmachen daß generell

$$P_1(j,k) \geq 1,$$
$$P_2(j,k) \leq 1$$

gilt. Große Abweichungen von P_1 und P_2 von 1 sind ein Hinweis auf mögliche Abweichungen von der Homeomorphie der zugrundeliegenden Abbildung. Eine tatsächliche Verletzung der Homeomorphie liegt jedoch erst dann vor, wenn das Produkt

$$P_3(j,k) = \left(\Pi_{l=1}^k Q_1(j,l) Q_2(j,l)\right)^{\frac{1}{2k}}, \tag{.5}$$

von 1 abweicht. Dies erkennt man, betrachtet man die Beiträge zu P_1 und P_2 genauer. Es wird dann klar, daß neben Verletzungen der Nächst-Nachbar-Ordnung Krümmungen der Abbildung (und ihrer Inversen) wesentlich zu $P_1 \neq 1$ (und $P_2 \neq 1$) beitragen. Dieser Effekt wird durch die Produktbildung (.5) unterdrückt.

Ein weiterer wichtiger Grund für die Wahl von (.5) ist jedoch die Tatsache, daß $P_1 > 1/P_2$, wenn sich der Eingangsraum in den Ausgangsraum faltet, und $P_1 < 1/P_2$, falls sich der Ausgangsraum in den Eingangsraum faltet. Mit anderen Worten weist ein Wert von P_3 größer bzw. kleiner 1 auf eine zu große bzw. zu kleine Ausgangsraumdimension $D^{\mathbf{A}}$ hin.

Während die bisher betrachtete Formel nur ein lokales Maß für Homeomorphie darstellt, führen wir nun im nächsten Schritt noch eine Mittelung über alle N Punkte des Kohonen-Gitters und alle Nachbarschaftsordnungen durch und gelangen zum globalen topographischen Produkt

$$P = \frac{1}{N(N-1)} \sum_{j=1}^{N} \sum_{k=1}^{N-1} \log(P_3(j,k)). \tag{.6}$$

In Abb. 2 zeigen wir Auswertungen des topographischen Produktes für die in Abb. 1 gezeigten Beispiele. Man erkennt, daß das topographische Produkt im Falle gelungener Einbettung den Wert $P = 1$ einnimmt, während ein Falten des Ausgangsraumes in den Eingangsraum ebenso detektiert wird wie der umgekehrte Fall zu hoher Ausgangsraumdimensionalität.

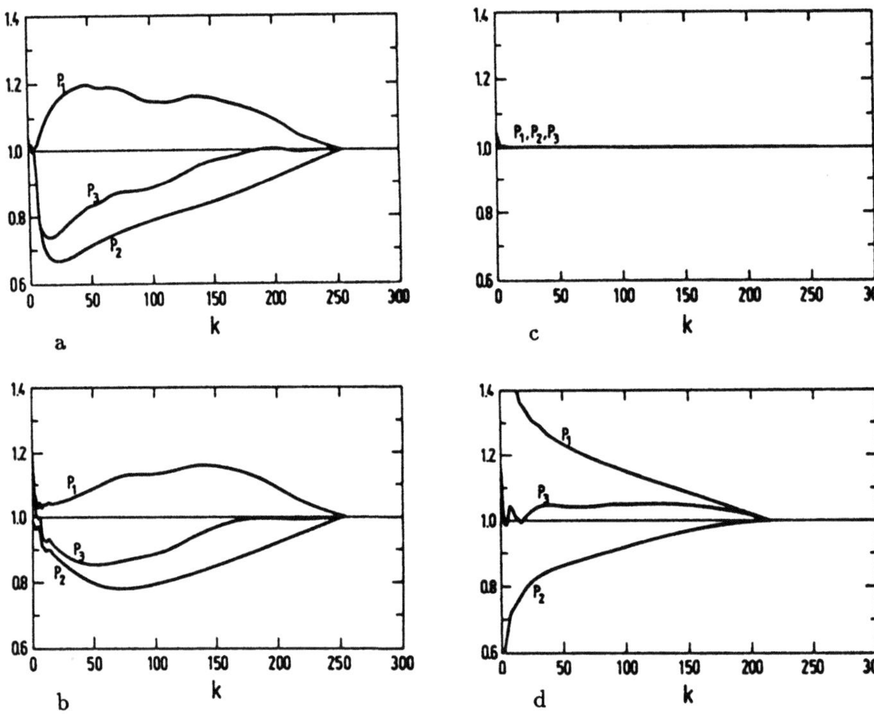

Abb. 2 Auswertung der Teilprodukte $P_1(k)$ (untere Kurven), $P_2(k)$ (obere Kurven) und $P_3(k)$ (mittlere Kurven). Die Teilprodukte sind jeweils über alle Knoten gemittelt, und in Abhängigkeit von der Nächstnachbarordnung aufgetragen. Man sieht, daß in den Fällen (a) und (b) —Einbettung zu "niedrigdimensional"— das Produkt $P_3(k)$ im Mittel einen Wert < 1 annimmt, im Fall (d) —Einbettung zu hochdimensional— einen Wert > 1. Im Fall (c) wird die Korrektheit der Einbettung durch $P_3 = 1$ angezeigt. Die gemittelten Werte P betragen -0.090 (a), -0.067 (b), 0.001 (c), und 0.0382 (d).

3. Dimension von Sprachdaten

Wir wenden uns nun der zweiten Methode zu, mit der wir den Zusammenhang zwischen Physik und Informatik verdeutlichen wollen. Hier geht es um die direkte Messung von Informationsdimensionen aus Datensätzen. In vielen Ansätzen der technischen Mustererkennung wird davon ausgegangen, daß komplexe Muster sich geeignet transformieren lassen, so daß zu ihrer Darstellung einige wenige Parameter ausreichen. Beispiele hierfür sind Karhunen-Loeve-Transformationen und Wavelets. Dies scheint analog zur Erkenntnis der nichtlinearen Dynamik, daß Systeme mit wenigen Freiheitsgraden komplexes Verhalten hervorbringen können. Aktivitätsmuster chaotischer Systeme lassen sich durch Punkte auf einer niedrigdimensionalen Mannigfaltigkeit im Phasenraum

traktoren erfolgt üblicherweise durch Dimensionsalgorithmen, bei denen Skalierungseigenschaften der durch die Daten gegebenen Dichte $p(\varepsilon)$ ausgewertet werden. Findet man $p(\varepsilon) \propto \varepsilon^D$ so ist D eine (untere) Abschätzung der Dimension der Mannigfaltigkeit. In einer doppeltlogarithmischen Darstellung $p(\varepsilon)$ versus ε hat man dann Geraden der Steigung D.
Die zum Nachweis niedrigdimensionaler Attraktoren entwickelten Methoden lassen sich jedoch auch unabhängig von Untersuchungen der Dynamik auf multivariate Datensätze anwenden. Ein Beispiel hierfür sind Sprachdaten welche typischerweise als Abfolge 10-20-dimensionaler Vektoren vorliegen. Abb. 3 zeigt eine Anwendung der Methode zur Dimensionsbestimmung auf 19-dimensionale Spektralvektoren, welche aus Aufnahmen der Worte "null" bis "neun" von einem und von mehreren Sprechern gewonnen wurden [11]. Es ergeben sich Abschätzungen für die Dimensionen der Hauptfläche von $D = 2.34$ (sprecherabhängig) und $D = 2.66$ (sprecherunabhängig). Dieses Ergebnis deutet darauf hin, daß zur Darstellung dieser Daten mindestens 3 Parameter erforderlich sind.

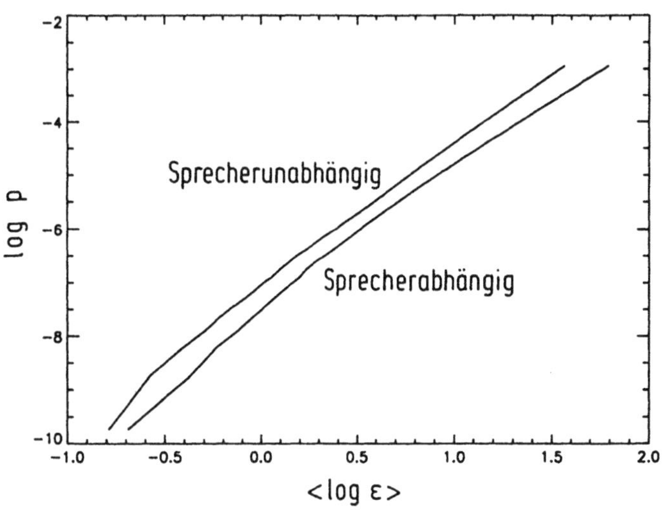

Abb. 3 Lokale Dichte $\log(p(\epsilon))$ über der gemittelten Längenskala $< \log(\epsilon) >$. Die Steigungen der Linien geben die Dimensionen der zu Grunde liegenden Datensätze an.

4. Optimale Karten für die Spracherkennung

Für die im 3. Abschnitt vorgestellten Daten wurden Kohonen-Karten unterschiedlicher Ausgangsraumdimension trainiert ($D^V = 19, D^A = 1,2,3,4$). Die folgende Tabelle fasst die Auswertung des Topographischen Produktes für diese Karten zusammen:

	d=1	d=2	d=3	d=4
sprecherabhängig	-0.158	-0.0282	0.0191	0.0367
sprecherunabhängig	-0.229	-0.0358	0.0066	0.0344

Die gefundenen Werte zeigen, daß eine Einbettung der Sprachdaten mittels Kohonen-Karten mit 3-dimensionalen Ausgangsräumen am besten gelingt. Dieses Ergebnis stimmt mit der Dimensionsanalyse in Abschnitt 3 überein.
Die naheliegende Vermutung, daß Kohonen-Karten mit 3-dimensionalen Ausgangsräumen auch die Spracherkennungleistung optimieren, wird durch ein kürzlich durchgeführtes Experiment bestätigt [12]. Hierbei wurden die Sprachdaten, die auch unseren Experimenten zu Grunde lagen, durch Kohonen-Karten unterschiedlicher Ausgangsraumdimension abgebildet. Zur Worterkennung wurden dann die Trajektorien im Ausgangsraum mit einem dynamischen time-warping Algorithmus klassifiziert. Es zeigte sich, daß die Erkennungsleistung diese Hybridsystems für $D^{\mathbf{A}} = 3$ um 7% besser war als bei einer Vorverarbeitung mit einer zweidimensionalen Kohonen-Karte (Erkennungsraten 0.795 im Vergleich mit 0.725).

5. Literatur

[1] C. von der Malsburg, "Self-Organization of Orientation Sensitive Cells in the Striate Cortex", Kybernetik **14**, 85-100 (1973).
[2] T. Kohonen, "Self-Organization and Associative Memory", 3rd Ed., Springer (1989).
[3] H. Ritter, T. Martinetz, K. Schulten, "Neuronale Netze", Addison Wesley (1990).
[4] T. Martinetz, H. Ritter, K. Schulten, "Three-Dimensional Neural Net for Learning Visuomototr-Coordination of a Robot Arm", IEEE Trans. Neur. Net. **1**(1), 131-136 (1990).
[5] J. Kangas, "Time-Dependent Self-Organizing Maps for Speech Recognition", in Artificial Neural Networks (Proc. ICANN 91 Helsinki), ed. T. Kohonen et al., North-Holland, II-1591 (1991).
[6] K. Obermayer, H. Ritter, K. Schulten, "A Model for the Development of the Spatial Structure of Retinotopic Maps and Orientation Columns", in: Principles of Organization in Organisms, Proc. Vol. XII, Addison-Wesley, in print (1991).
[7] F. Wolf, H.-U. Bauer, T. Geisel, "Origin of Field-Discontinuities and Islands in Visual Cortical Maps", submitted to Science
[8] W. Liebert, K. Pawelzik, H.G. Schuster, "Optimal Embeddings of Chaotic Attractors from Topological Considerations", Europhys. Lett. **14**, 521-526 (1991).
[9] F. Takens, "Detecting Strange Attractors in Turbulence", in: Dynamical Systems and Turbulence, ed. D. Rand, L.-S. Young, Springer Lecture Notes in Mathematics **898**, 366 (1981).
[10] H.-U. Bauer, K. Pawelzik, "Quantifying the Neighborhood Preservation of Self-Organizing Feature Maps", submitted to IEEE Trans. Neur. Net. (1991).
[11] Daten aus der DPI-Datenbasis des III. Physikalischen Instituts, Uni. Göttingen (Strube, Gramß), (1990).
[12] W.D. Brandt, H. Behme, H.W. Strube, "Bildung von Merkmalen zur Spracherkennung mittels phonotopischer Karten", in: Fortschritte der Akustik-DAGA 91, DPG GmbH, Bad Honnef, Germany, 1057 (1991).

Automated Detection and Interpretation of Earthquake Seismograms by Adaptive Pattern Recognition

Manfred Joswig

Institut für Geophysik, Ruhr-Universität Bochum

Postfach 10 21 48 D-4630 Bochum 1 Germany

Introduction

The measurement and interpretation of ground motion in seismograms is the primar source for our knowledge of the earth's interior. As in other scientific areas, seismology has seen great improvements of its experimental facilities in the last decades. State-of-the-art observatories monitor earthquakes magnitudes below they are recognized by humans. Additionally, these observatories record explosions in quarries up to 1000 km apart and nuclear explosions all over the world. Over a year, these events sum up to some 10.000 data sets that must be found and processed in TeraBytes of raw data which are contaminated by local noise bursts of car traffic, sonic bangs and other sources. This task of routine preselection and processing is about to exceed the possibilities of human data analysis; any further progess will depend on automated methods. However, these methods must yield human-like performance not to degrade the achieved sensitivity and reliability.

In seismology, all processes of automated evaluation base on seismograms, i.e., they are - at least in their initial steps - approaches of time series analysis. In this paper, we want to focus on the initial tasks of (a) detection which will perform an effective data reduction to the significant seismic events and (b) parameter extraction for some further reasoning that may be performed by subsequent rule-based systems. We introduce two approaches that are based on images as their internal knowledge representation and utilize any fragmentary theory to adapt a simple set of few patterns to a wide variety of environmental conditions when the actual pattern recognition is performed.

Event Detection by Sonograms

Earthquake detection started with digital filters as the most simple approach some 30 years ago. Selecting the most promissing frequency bands, the environmental noise should be constant and any significant deviation would flag an event. While this idea is sufficient to detect earthquakes, its obvious weakness is the high false alarm rate caused by any non-stationary noise burst. So the next step was to apply pattern recognition techniques which search for earthquakes by positive agreement. The difficulty here, however, is that we can not predict the expected seismograms; their variety represents the differences in source mechanism, distances to the receiver station, different wave types of body and surface waves and their conversions, scattering and anisotropic modifications during different propagation paths. Therefore pattern recogniton can succeed at this stage only, if we find a description of seismograms which restricts their variaty to a reasonable simplification. While descriptions by sentences or grammar structures were difficult and of less success (Anderson, 1982; Liu and Fu, 1983), the characterization of seismograms by coarse grain spectral images yielded good results (Joswig, 1990a).

In the last years, we broadened this approach and tried to copy more closely, how humans perform this recognition task once they gain some working knowledge as the competent scientist. We found many hints to cognitive plausibility, they in general conform to the theory of mental images (Kosslyn, 1980; Joswig, 1990b). For the scientific task, we start the recognition by viewing to a suitable image. Most important, this image is not just a screendump from any data visualization program or - in former times - an ink-recording of the time series, but the subjective weigthing and selection of information. The common characteristics of mental images are the drastic reductions in dynamic range and number of pixels as well as some noise blanking to focus just on deviations. In this way, they correspond to the initial data like the pen-sketch of an artist to the appropriate photographic picture. Our goal in creating these images is to achieve data descriptions that remain constant over a wide range of changes in its environmental conditions like the noise background and thus allow for recognition by a common pattern.

In case of seismogram detection, the transformation to best suited images in Fig. 1 is based on the same calculus as the optimum filter detector; it just enhances the detectable energy spots relative to background noise (Joswig, 1991a). However, this improvement is quite significant as shown in Fig. 2, where we compare the seismograms of adjacent seismic stations while recording the same earthquake. Because of the higher ambient noise, the signal gets invisibale at the lower traces. This is true also for the common way to transfrom these data to short term spectral images which in seismology are called sonograms or spectrograms. With the optimum transformation, in contrast, the similarity is seen on all traces and the success of pattern recogniton appears obvious.

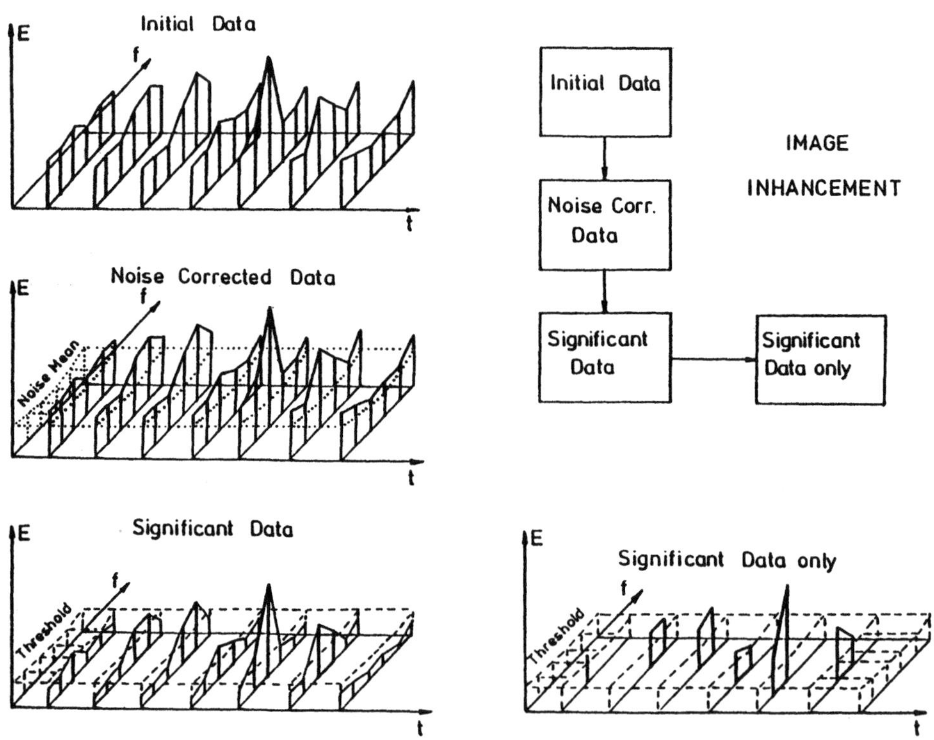

Fig. 1 Image enhancement by noise blanking

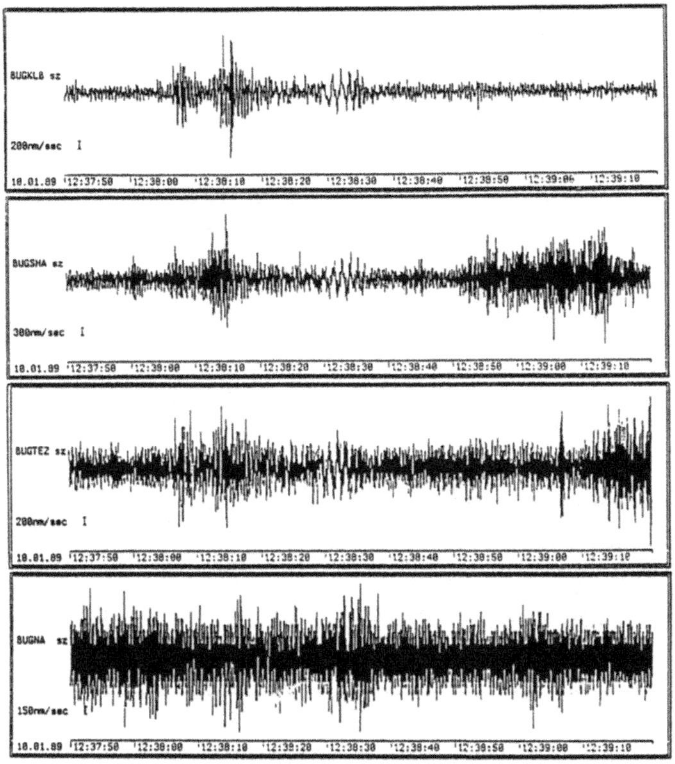

Seismograms of a 'Hamm' rockburst at the BUG array in 40 km distance

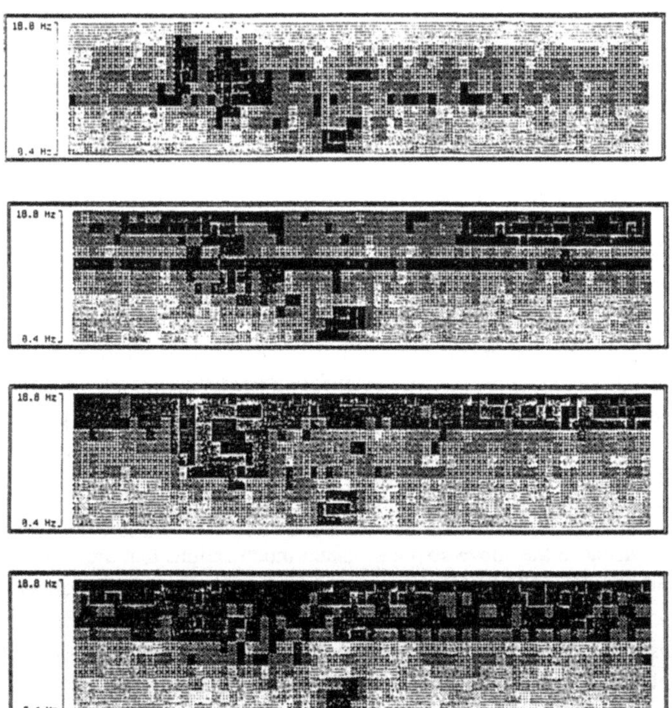

Power spectral density plots (SONOGRAMs) with log f / log P - scaling

Fig. 2a+b Seismograms and sonograms of one earthquake at four stations

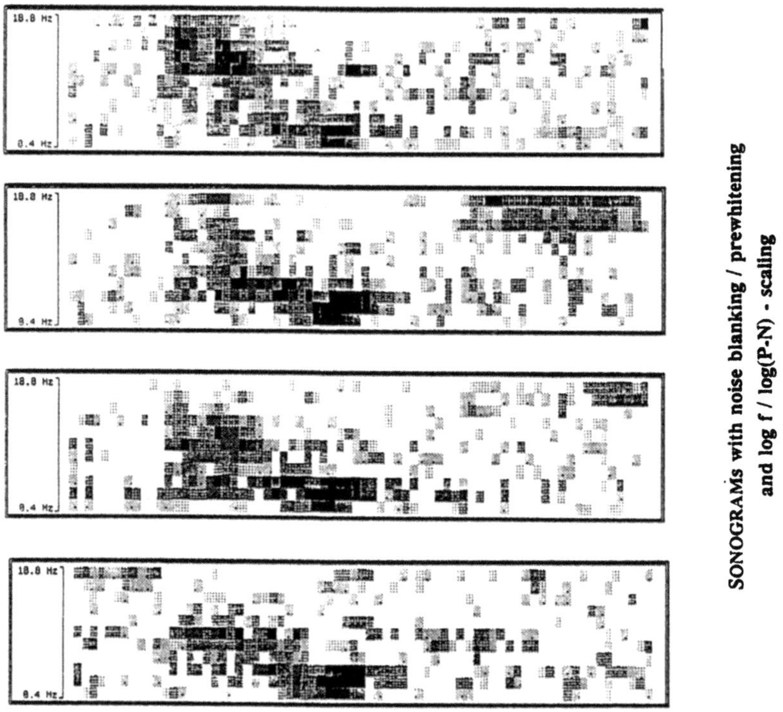

Fig. 2c Optimum sonograms by the image enhancement of fig. 1

Adaptive Pattern Recognition

The second step in human-like event recognition is the pattern adaptation. The seismologist knows by experience how the permanent noise floor and any expected event signature will interact in dependence of their relative amplitudes: the weak parts of the signal will be masked once they fall below the noise level. This knowledge allows him to predict how much of an initial pattern remains valid in the actual situation. By modelling this effect in the image modification of Fig. 3, the final approach of pattern recognition is able to detect events below the threshold of human routine performance which is always biased by some inattentiveness and the restriction to standard procedures (Joswig, 1991b). For example, Fig. 4 shows the detection and classification of a weak earthquake from the Recklinghausen cluster at 4 adjacent stations with different ambient ground noise. The recognition is performed independently on each trace and takes into account some even higher noise bursts at SHA and TEZ also. In the routine processing, the seismologist would reliably identify the event on the KLB trace only.

The cognitive equivalent to the above step are mental modifications that are confirmed for every day vision (Shepard and Metzler, 1971). In the contents of our application, it is important to notice that these modifications are based on some physical modells matured in the scientist over the time of data analysis. However, these modells are in no way complete theoretical frameworks. Instead they allow us to cope with some half-knowledge and fractional modells when analyzing the data. For the task of

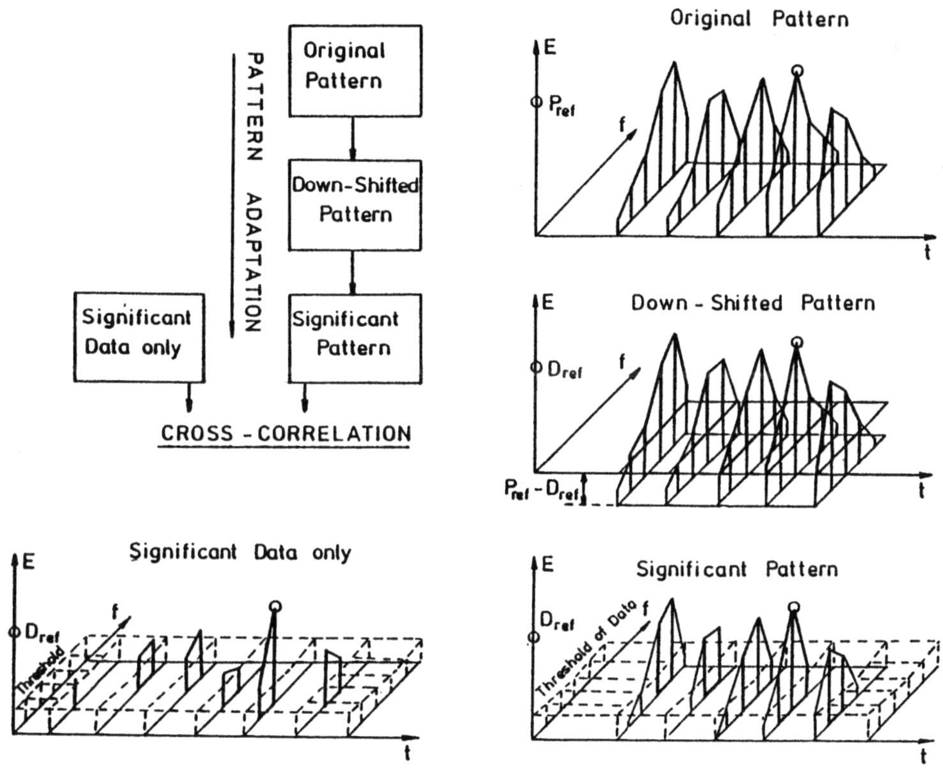

Fig. 3 Pattern adaptation relative to the actual amplitude ratios

earthquake detection, we have even with the approach of pattern adaptation no chance to predict, how our signals will look like in detail. Any calculation of synthetic seismograms, today still restricted by very simplifying assumptions about earth layers, lateral homogeneity or at least isotropic behaviour, may guide our imagination but is not needed for the purpose of earthquake detection. What is sufficient is just the description of events by one representative per cluster - not its understanding. In this way adaptive pattern recognition on mental images is like other knowledge-based approaches as well ideally suited to deal with uncomplete knowledge that can not be derived by theory.

Instead of adaptive pattern recognition it is, of course, possible to apply ANN to the optimally enhanced images in Fig. 2. This would require some long-running training cycles to adjust the internal weights to a set of given answers. The result, however, would just reflect the average situation of all events in question; it must be worse than the individual tuning by adaptation once a mathematical model is known and can be utilized.

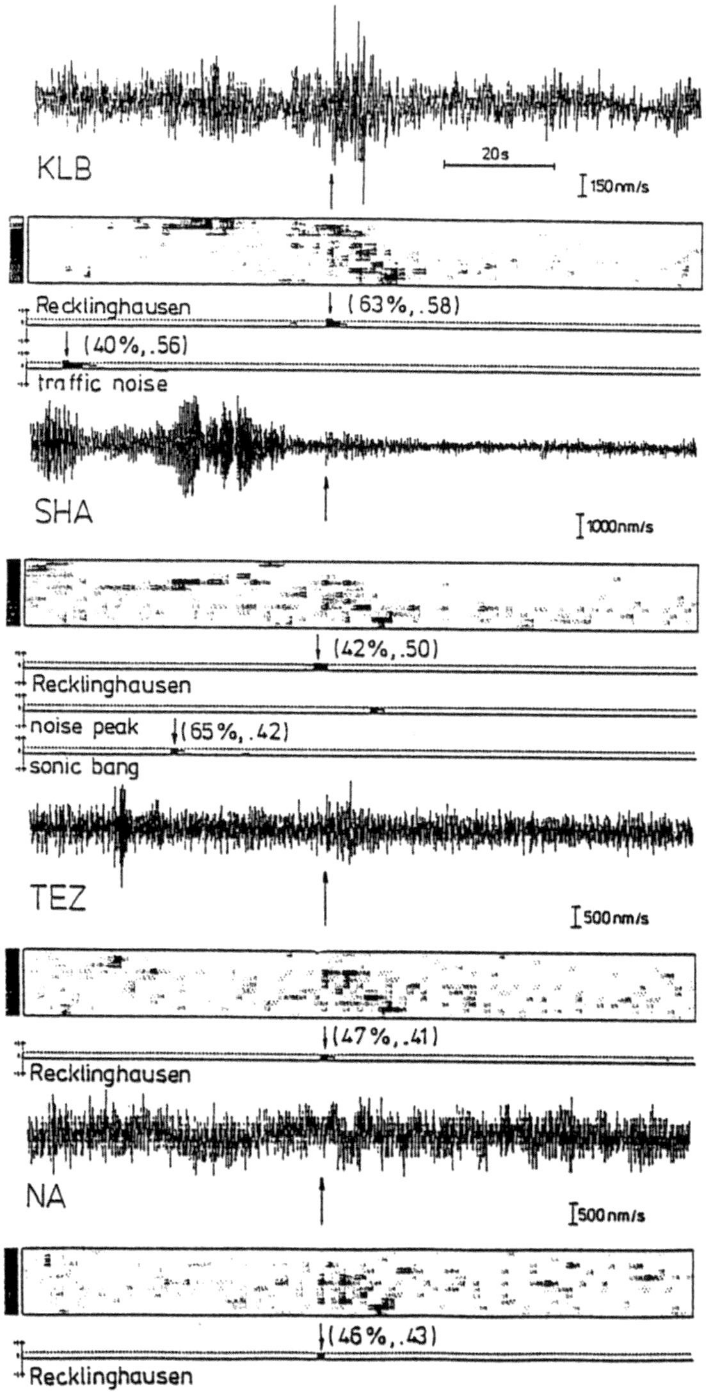

Fig. 4 Independent recognition of a weak 'Recklinghausen' event at four stations

Phase Picking on Polarization Images

When the knowledge about the principle properties of mental images and their modification were derived for the detection of earthquakes, it was straight foreward to test this approach for another application. We chose the picking of phase onsets for the different body wave types - compressional versus shear waves, they lay within a seismogram and demand higher timing resolution than the above detection task. The knowledge of onset times is crucial for epicenter determination and thus is carried out routinely on all the measured seismograms. It is, however, one of the most delicate tasks in seismogram evaluation because one has to determine onset times within the event and not relative to the stationary noise background. The decisions are based on changes of the three-dimensional particle motion over time, i.e., a four-dimensional problem. While for the detection task some image representation already existed in form of the sonograms and we just introduced the idea of pattern recognition and the maximum image enhancement, for the phase picking no suited images were available.

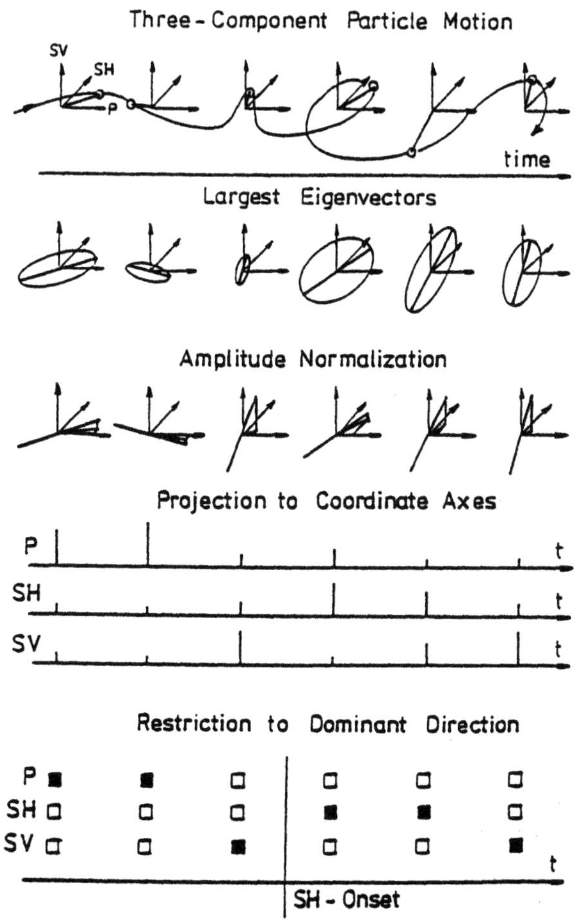

Fig. 5 Composition of polarization images (only one freq.-band is shown)

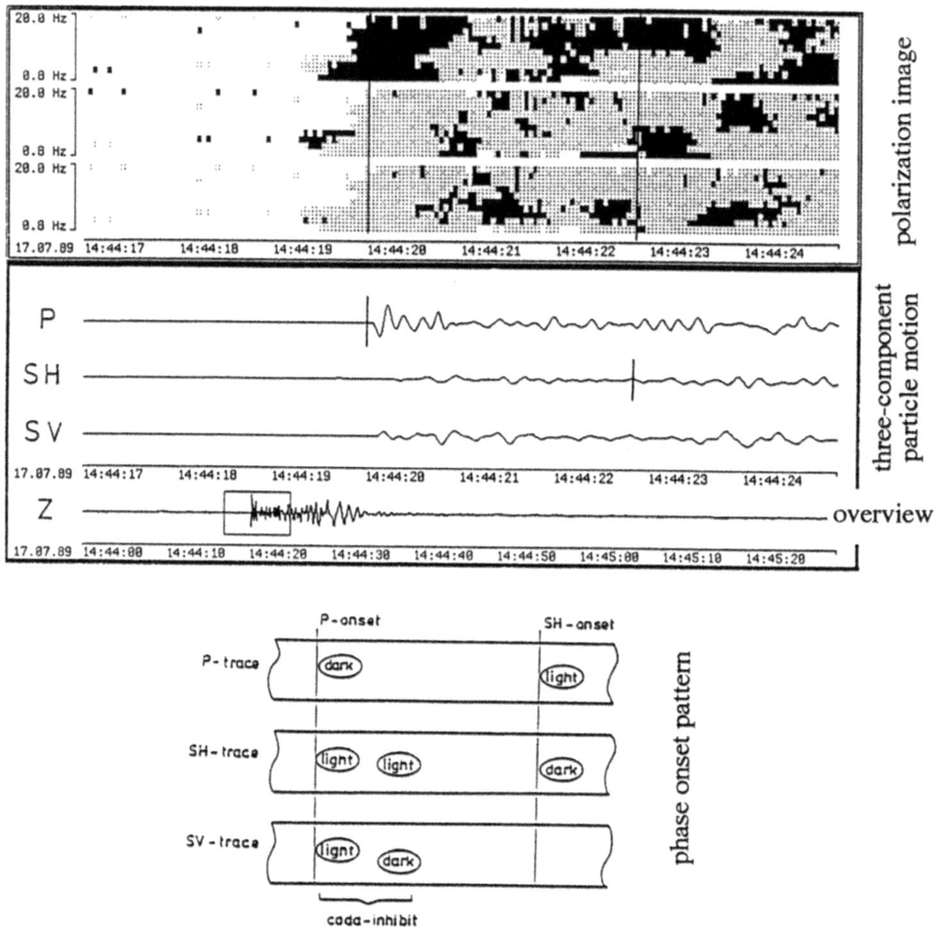

Fig. 6 Phase picking of the SH shear wave onset on polarization images as of fig. 5

Knowing the neccssary features from the sonograms, it took us one year to find the proposed images that are stable for shear wave phase picking (Klumpen and Joswig, 1991). Initially we perform spectral analysis, a restriction to fewer pixels and noise blanking as the known steps of image enhancement. Then we restrict the particle motion to one dominant direction at any pixel, see Fig. 5. This modification follows a simple model of body waves that travel to the receiving station. Once we know the angles of azimuth and incidence, the coordinate system can be aligned appropriately so that compressional energy should show up in line with the incidence (the P trace) while shear waves cause perpendicular particle motion (the SH and SV traces). This model, however, is a oversimplification for real world data that show additional effects like frequency-dependent changes in the angle of incidence, shear wave splitting and mode conversion in the codas of each wave type. These effects are not fully understood at date and yielded failure for all those approaches that tried to identify phase onsets by model conformance (Montalbetti and Kanasewich, 1970; Christoffersson et al., 1988). By adaptive pattern recognition, however, we once again can cope with these data just by describing the effects by the pattern in Fig. 6 - which is far away from any understanding. So the phase picking by PR on polarization images is the first approach that prooved success in case of high-frequency local earthquake data.

Conclusions

Pictorial knowledge representation was introduced as an appropriate approach in scientific recognition tasks performed on mass data for further symbolic postprocessing. These images differ from the inital data by lower resolution and limited dynamic range, they focus primarily on changes by background blacking and they adjust to different environmental conditions by model-based modifications. The optimum transformation is always problem dependent and must be changed for every new task that should be performed on the initial data.

The effects of scaling can be seen as the difference between photographic pictures and an abstracted pen-sketch. The latter is far more robust and thus more suited for the automated pattern recognition. The pen-sketches are designed to correspond best to the subjective weighting of recognized impressions by the scientist, they follow the ideas of mental images and mental modifications in human cognition.

Recognition by mental image patterns allows for symbolic processing, but the images themselves stay pictorial and are not part of the image interpretation. Although some modification procedures are based on exact physical models, the approach of mental images is especially suited to automate the handling of real-world data. They can be recognized by the scientist, but they are often not fully understood or predictable by the actual theories of natural sciences. In the pattern recognition part, some similarities exist with ANN; however, they hide the fundamental difference that here the input of few patterns is sufficient to process a wide variety of situations. ANN in its training cycles just requires the opposite relation. In this way mental images are a top-down approach like expert systems are, but with an analogue instead of the propositional knowledge representation.

Acknowledgment

This work was supported by Deutsche Forschungsgemeinschaft under grant Ha 842/8-1.

References

Anderson, K. R., "Syntactic analysis of seismic waveforms using augmented transition network grammers", *Geoexploration*, vol. 20, pp. 161-182, 1982.

Christoffersson, A., E. S. Husebye, and S. F. Ingate, "Wavefield decomposition using ML-probabilities in modelling single-site-3-component records", *J. Geophys.*, vol. 93, pp. 197-214, May 1988.

Joswig, M., "Pattern recognition for earthquake detection", *Bull. Seism. Soc. Am.*, vol. 90, no. 1, pp. 170-186, 1990a.

Joswig, M., "Wissensbasierte Erdbebenerkennung mit Sonogrammen als mentalen Bilder", in *GWAI-90*, ed. H. Marburger, pp. 144-149, Springer-Verlag, IFB 251, Berlin, 1990b.

Joswig, M., "Noise level adaptation for seismic pattern recognition", in *Proc. XXII. General Assembly of ESC 1990 in Barcelona*, Zürich, (in press), 1991a.

Joswig, M., "Earthquake recognition by the Sonogram-detector with master event adaptation", *Bull. Seism. Soc. Am.*, (submitted), 1991b.

Klumpen, E. and M. Joswig, "High-frequency S-phase discrimination for local earthquakes", in *Proc. XXII. General Assembly of ESC 1990 in Barcelona*, Zürich, (in press), 1991.

Kosslyn, S. M., *Image and mind*, Harvard Univ. Press, Cambridge, MA., 1980.

Liu, Hsi-Ho and King-Sun Fu, "An application of syntactic pattern recognition to seismic discrimination", *IEEE Transa. Geosc. Rem. Sens.*, vol. GE-21, no. 2, pp. 125-132, 1983.

Montalbetti, J. and E. R. Kanasewich, "Enhancement of teleseismic body phases with a polarization filter", *Geophys. J. Royal astr. Soc.*, vol. 21, pp. 119-129, 1970.

Shepard, R. N. and J. Metzler, "Mental rotation of three-dimensional objects", *Science*, vol. 171, pp. 701-703, 1971.

Anwendung der Renormierungsgruppentheorie in der Signal- und Bildverarbeitung

S.Hamann † und G. Sobotta
Messerschmitt–Bölkow–Blohm GmbH Ottobrunn *
und
† Physik Department der Technischen
Universität München in Garching

Zusammenfasssung:

Die Renormierungsgruppentheorie wird auf das Problem der Signal- und Bildentrauschung angewendet. Ergebnisse für Binärdaten und erstmals auch für diskrete Signale werden präsentiert.

Der Vergleich mit bisherigen Monte–Carlo–Verfahren zeigt, daß der Renormierungsgruppenansatz erhebliche Vorteile in der Restaurationsgüte und der Rechenzeit bringt.

1 Einleitung

Ein zentrales Problem der Signal- und Bildverarbeitung ist die Restauration von Signalfolgen und Bildern. Dieses Problem wurde in früheren Jahren mit Hilfe der klassischen Filtertheorie bearbeitet. Dabei ging es sowohl um die Eliminierung der systematischen Fehler als auch um die Entrauschung. Diese Verfahren findet man in Standardwerken beschrieben (z.B. Lücke 1985 [13], Gonzales und Wintz 1987 [8]).

Seit einigen Jahren werden Monte–Carlo–Verfahren zur Bildrestauration angewendet. Die Monte–Carlo–Methode nahm ihren Anfang mit der Einführung elektronischer Rechner. Sie wurde zur Behandlung von Zustandsgleichungen entwickelt (Metropolis, Rosenbluth, Rosenbluth, Teller und Teller 1953 [14]). Diese Methode wurde zum Standardwerkzeug der theoretischen Physik. Es wurden insbesondere komplexe nichtlineare Probleme wie Phasenübergänge untersucht, wobei Gittermodelle mit Spins als Freiheitsgraden im Vordergrund standen (z.B. K.Binder 1979 [1]). Die Suche nach Grundzuständen komplexer Gittermodelle für Spingläser und das damit verbundene nichtlineare Optimierungsproblem wurde mit Hilfe der Monte–Carlo–Methode studiert (Kirckpatrik, Gellatt und Vecchi 1983 [12]), womit sich diese Methode als ein sehr universelles Werkzeug etablierte (vgl. z.B. Kinzel 1988 [11]). In Anlehnung an das Ausglühen einer Legierung nennt man dieses Verfahren simuliertes Ausglühen.

Vor einigen Jahren wurde dieses Optimierungsverfahren auf die Probleme der Signal- und Bildverarbeitung übertragen. Die Brücke zwischen dieser Datenverarbeitung und der Physik stellen Modelle mit Gibb'schen Zufallsfeldern dar, die zunächst für die Beschreibung magnetischer Legierungen eingesetzt wurden (vgl. z.B. Derin und Elliott 1987 [3], Geman und Geman 1984 [5], Geman und Graffigne 1987 [6]). Für die Bildentrauschung lag die Vorstellung zugrunde, daß ein verrauschtes Bild einem Gittermodell entspricht, das in einem Hochtemperaturzustand ist. Zur Entrauschung wurde die Relaxation in den Grundzustand simuliert, der mit dem entrauschten Bild identifiziert

*MBB-Z-0411-91-Pub

wurde. Dabei wurden die Analogien zwischen Binärbildern und Gittermodellen mit zweiwertigen Spins, also Isingmodellen, genutzt, die auf diejenige zwischen Grauwertbildern und Gittermodellen mit mehrwertigen Spins, also Pottsmodellen erweitert wurden.

Diese Entwicklung ließ es ahnen, daß andere erfolgreiche Methoden zur Behandlung komplexer nichtlinearer Probleme aus der theoretischen Physik ihren Eingang in die Signal- und Bildverarbeitung finden würden. Den ersten Schritt in diese Richtung tat B.Gidas (1989) [7], dem es gelang, die Renormierungsgruppentheorie auf das Problem der Signal- und Bildrestauration anzuwenden. Diese Theorie wurde nach einer Vielzahl orgineller Anläufe auf dem Gebiet der Quantenfeldtheorie für statistische Gittermodelle von L.P. Kadanoff (1966) [9] strukturell formuliert. Die erste mathematische Ausarbeitung gelang K.G. Wilson (1971) [17], womit ein Durchbruch in der Theorie komplexer nichtlinearer Probleme gelang. Zusammenfassende Darstellungen dieser recht anspruchsvollen Theorie sind sehr zahlreich (z.B. in Domb und Green (1976) [4]). Als Einführung sind populäre Darstellungen sehr anregend (K.G.Wilson [18]).

In dieser Arbeit soll anhand der Signalrestauration der Formalismus des Verfahrens (Abschnitt 2) erläutert werden. In Abschnitt 3 werden Ergebnisse präsentiert und diskutiert. Einige Rechneraspekte werden in Abschnitt 4 angesprochen. In der in Abschnitt 5 folgenden Zusammenfassung werden die Probleme der Industrieforschung angeschnitten, die die Anwendung der Renormierungsgruppentheorie als nützlich erscheinen lassen.

2 Verfahren

Das Verfahren zur Restauration gliedert sich in folgende Teile:

i) Berechnung der Energiefunktion $\mathcal{H}$ aus den gegebenen verrauschten Daten S^R.

ii) Berechnung der entrauschten Daten S^0 als Zustand mit minimaler Energie.

Dieses vollständige Programm wurde bisher nicht durchgeführt, wobei das Defizit noch im Punkt i) liegt. Um $\mathcal{H}$ aus den Daten S^R zu berechnen, kann man die Maximum-Likelihood-Methode anwenden. Dazu betrachtet man die Wahrscheinlichkeit P mit der Temperatur T im Parameter $\beta = \frac{1}{k_B T}$

$$P_{\beta\mathcal{H}}(S) = \frac{1}{Z} e^{-\beta\mathcal{H}(S)} \tag{1}$$

wobei die Zustandssumme Z durch

$$Z = \sum_{S'} e^{-\beta\mathcal{H}(S')} \tag{2}$$

gegeben ist. Das optimale $\mathcal{H}$ folgt nun aus der Maximierung der Verteilungsfunktion für die vorgegebenen verrauschten Daten $P_{\beta\mathcal{H}}(S^R)$. Das eigentliche Problem sind die Nebenbedingungen:

i) Für $S = S^R$ soll die Temperatur T hoch sein.

ii) Für $T \to 0$ soll $S \to S^0$ gelten, d.h. der Grundzustand am wahrscheinlichsten sein.

iii) Es soll nur ein Grundzustand existieren, der zunächst unbekannt ist.

Ein Ansatz zur mathematischen Formulierung dieser Nebenbedingung läßt sich unter Verwendung der Energie $U_{\mathcal{H}}$ und der Entropie $S_{\mathcal{H}}$ formulieren:

$$\begin{aligned} \text{zu i)} \quad U_{\mathcal{H}}(T) &= U_1 > U_0 && (3) \\ \text{zu ii)} \quad U_{\mathcal{H}}(T \to 0) &= U_o && (4) \\ \text{zu iii)} \quad S_{\mathcal{H}}(T \to 0) &= S_0 = 0 && (5) \end{aligned}$$

Diese Nebenbedingungen lassen sich in dem Log-Likelihood-Ansatz mit den Lagrangemultiplikatoren γ_0, γ_1 und δ_0 integrieren, womit man das zu optimierende Funktional $W[\mathcal{H}]$ mit

$$U_{\mathcal{H}}(T) = \sum_S P_{\beta\mathcal{H}}(S)\mathcal{H}(S) \tag{6}$$

$$S_{\mathcal{H}}(T) = -k_B \sum_S P_{\beta\mathcal{H}}(S) \ln P_{\beta\mathcal{H}}(S) \tag{7}$$

in der Form

$$\begin{aligned} W[\mathcal{H}] &= \ln P_{\beta\mathcal{H}}\left(S^R\right) \\ &\quad +\gamma_1 (U_{\mathcal{H}}(T) - U_1) \\ &\quad +\gamma_0 (U_{\mathcal{H}}(T \to 0) - U_0) \\ &\quad +\delta_0 (S_{\mathcal{H}}(T) - 0) \end{aligned} \tag{8}$$

erhält.

Die Optimierung von $W[\mathcal{H}]$ ist ziemlich aufwendig. Eine Lösung ist uns nicht bekannt. Es ist auch nicht bekannt, ob die Nebenbedingungen geeignet und vollständig sind. Daher wurden bisher Ansätze mit suboptimalem $\mathcal{H}$ untersucht. Dabei hat sich der von B.Gidas [7] verwendete Ansatz bewährt. Für das Entrauschen binärer Signalfolgen wird dabei von Isingspins

$$S_i = \pm 1 \quad \text{für} \quad i = 1, 2, \ldots, 2^N + 1 \tag{9}$$

ausgegangen, wobei man $S = \{s_i\}$ als Signalfolge auffassen kann. Die Energiefunktion $\mathcal{H}$ wird möglichst einfach angesetzt, nämlich

$$\mathcal{H} = J^0 + \sum_{i=1}^{2^N+1} h^i s_i + \sum_{i=1}^{2^N} J^{i,i+1} s_i s_{i+1} \tag{10}$$

wobei sogar angenommen wird, daß

$$J^{i,i+1} = J \quad \text{für} \quad i = 1, 2, \ldots, 2^N \tag{11}$$

gilt. Es wird dann der Ansatz der zufälligen Magnetfelder gemacht, d.h.

$$h^i = s_i^R \cdot \frac{1}{\sigma^2} \tag{12}$$

wobei σ^2 die Varianz der angenommenen Verteilung eines additiven Gaußschen Rauschens mit verschwindendem Mittelwert ist. Die hohe Temperatur, für die S^R der wahrscheinlichste Zustand ist, wird bei $T = 1$ angesetzt und $J = 1$ gewählt, womit man $\frac{J}{k_B T}$ für $k_B = 1$ erhält.

Diese Vereinfachungen verdeutlichen, daß nur Signalfolgen ohne Texturen hier behandelt werden können, also letztlich einfach strukturierte Grundzustände S^0 gesucht werden. Das zugehörige Isingmodell hat lokal eine ferromagnetische Ordnung im Grundzustand.

Dieser Ansatz für die Energiefunktion hat sich für eine beschränkte Klasse von Signalfolgen sowohl für die Monte-Carlo-Optimierung als auch für die Renormierungsgruppenansätze bewährt, die hier erläutert werden. Die Idee der Renormierungsgruppentheorie besteht vereinfacht darin,

i) eine Reduktion der Freiheitsgrade durchzuführen und

ii) zur Kompensation die restlichen Freiheitsgrade mit "renormierten" Wechselwirkungen zu verkoppeln.

Dieses Kunststück gelingt nur für wenige Modelle exakt (vgl. Kadanoff 1977 [10]). Für das erläuterte Modell erhält man aus (10), d.h. der unrenormierten Energiefunktion,

$$\mathcal{H}^{(0)} = J^{(0)0} + \sum_{i=1}^{2^N+1} h^{(0)i} s_i^{(0)} + \sum_{i=1}^{2^N} J^{(0)i,i+1} s_i^{(0)} s_{i+1}^{(0)} \tag{13}$$

die renormierte Energiefunktion

$$\mathcal{H}^{(1)} = J^{(1)0} + \sum_{i=1}^{2^{N-1}+1} h^{(1)i} s_i^{(1)} + \sum_{i=1}^{2^{N-1}} J^{(1)i,i+1} s_i^{(1)} s_{i+1}^{(1)} \tag{14}$$

wobei von den

$$2^N + 1 \quad \text{Spins} \quad \{s_1^{(0)}, s_2^{(0)}, \ldots, s_{2^N+1}^{(0)}\}$$

nach Eliminierung der

$$2^N \quad \text{Spins} \quad \{s_2^{(0)}, s_4^{(0)}, \ldots, s_{2^N}^{(0)}\}$$

die

$$2^{N-1} + 1 \text{ Spins } \{s_1^{(0)}, s_3^{(0)}, \ldots, s_{2^N+1}^{(0)}\}$$

übrigbleiben, die als "renormierte" Spins

$$2^{N-} + 1 \text{ Spins } \{s_1^{(1)}, s_2^{(1)}, \ldots, s_{2^N+1}^{(1)}\}$$

in der "renormierten" Energiefunktion $\mathcal{H}^{(1)}$ auftauchen.

Durch mehrfache Wiederholung dieses Dezimierungsverfahrens läßt sich das $2^N + 1$–Spinproblem auf ein 2–Spinproblem mit der Energiefunktion

$$\mathcal{H}^{(N)} = J^{(N)0} + h^{(N)1} s_1^{(N)} + h^{(N)2} s_2^{(N)} + J^{(N)1,2} s_1^{(N)} s_2^{(N)} \tag{15}$$

reduzieren. Die konkrete Berechnung der renormierten Wechselwirkungen kann hier nur skizziert werden (vgl. [7]). Dazu genügt es, die unrenormierten Verteilungsfunktionen für drei Spins zu betrachten:

$$\rho_{abc}(s_a, s_b, s_c) = e^{-\beta(\mathcal{H}_{ab}(s_a, s_b) + \mathcal{H}_{bc}(s_b, s_b))} \tag{16}$$

Die renormierte, unnormierte Verteilungsfunktion:

$$\rho_{ab}^{(1)}(s_a^{(1)}, s_b^{(1)}) = e^{-\beta^{(1)} \mathcal{H}_{ab}^{(1)}(s_a^{(1)}, s_b^{(1)})} \tag{17}$$

ergibt sich aus

$$\rho_{ab}^{(1)}(s_a, s_c) = \sum_{s_b = \pm 1} \rho_{abc}(s_a, s_b, s_c) \tag{18}$$

und

$$s_a^{(1)} := s_a \tag{19}$$

$$s_b^{(1)} := s_c \tag{20}$$

Ähnlich läßt sich die Dezimierung der Spins $\left\{s_2^{(0)}, s_4^{(0)}, \ldots, s_{2^N}^{(0)}\right\}$ durchführen.

Die Optimierungsaufgabe zur Berechnung des Grundzustandes S^0 läßt sich nun vom 2–Spin–System aus beginnen. Dazu ist

$$\mathcal{H}^{(N)}\left(s_1^{(N)0}, s_2^{(N)0}\right) = \min_{s_1^{(N)}, s_2^{(N)}} \mathcal{H}^{(N)}\left(s_1^{(N)}, s_2^{(N)}\right) \tag{21}$$

zu berechnen. Damit erhält man bereits

$$s_1^0 = s_1^{(0)0} = s_1^{(N)0} \tag{22}$$
$$s_{2^N+1}^0 = s_{2^N+1}^{(0)0} = s_2^{(N)0} \tag{23}$$
$$\tag{24}$$

Reduziert man nun die Renormierungsstufe um 1 auf $N-1$, so erhält man mit den bekannten Spins

$$s_1^{(N-1)0} = s_1^{(N)0} \tag{25}$$
$$s_3^{(N-1)0} = s_2^{(N)0} \tag{26}$$
$$\tag{27}$$

das Optimierungsproblem zur Bestimmung von $s_2^{(N-1)0}$:

$$\mathcal{H}^{(N-1)}\left(s_1^{(N-1)0}, s_2^{(N-1)0}, s_3^{(N-1)0}\right) = \min_{s_2^{(N-1)}} \mathcal{H}^{(N-1)}\left(s_1^{(N-1)0}, s_2^{(N-1)}, s_3^{(N-1)0}\right) \tag{28}$$

In der $(N-2)$–ten Renormierungsstufe erhält man nun zwei äquivalente Teilaufgaben, nämlich

$$\mathcal{H}_{123}^{(N-2)}\left(s_1^{(N-2)0}, s_2^{(N-2)0}, s_3^{(N-2)0}\right) = \min_{s_2^{(N-2)}} \mathcal{H}_{123}^{(N-2)}\left(s_1^{(N-2)0}, s_2^{(N-2)}, s_3^{(N-2)0}\right) \tag{29}$$

und

$$\mathcal{H}_{345}^{(N-2)}\left(s_3^{(N-2)0}, s_4^{(N-2)0}, s_5^{(N-2)0}\right) = \min_{s_4^{(N-2)}} \mathcal{H}_{345}^{(N-2)}\left(s_3^{(N-2)0}, s_4^{(N-2)}, s_5^{(N-2)0}\right) \tag{30}$$

Es ist genau diese Aufspaltung in elementare Teilaufgaben, die dafür sorgt, daß das Optimierungsproblem in jeder Renormierungsstufe $n = N, N-1, \ldots, 1, 0$ immer einfach lösbar bleibt.

Dieses Verfahren ist nun für alle Temperaturen T durchführbar, denn abgesehen von der 0–ten Renormierungsstufe, in der T nur in $\beta = \frac{1}{k_B T}$ enthalten ist, aber nicht in $\mathcal{H}^{(0)}$, wird $\mathcal{H}^{(n)}$ für $n = 1, 2, \ldots, N$ von T abhängig. Die maximale Entrauschung erfolgt im Limes $T \to 0$, d.h., die renormierten Kopplungskonstanten in $\mathcal{H}^{(n)}$ müssen für diese Optimierung vorher berechnet werden. Sie enthalten die Information über den Grundzustand S^0.

Das Optimierungsproblem läßt sich auch in der Sprache der unnormierten Verteilungsfunktionen oder Transfermatrizen ρ formulieren. Für das 2–Spin–System für $T \to 0$ erhält man $s_1^{(N)0}, s_2^{(N)0}$ aus

$$\rho_{12}^{(N)}\left(s_1^{(N)0}, s_2^{(N)0}\right) = \max_{s_1^{(N)}, s_2^{(N)}} \rho_{12}^{(N)}\left(s_1^{(N)}, s_2^{(N)}\right) \tag{31}$$

Dann folgt $s_2^{(N-1)0}$ aus

$$\rho_{123}^{(N-1)}\left(s_1^{(N-1)0}, s_2^{(N-1)0}, s_3^{(N-1)0}\right) = \max_{s_2^{(N-1)}} \rho_{123}^{(N-1)}\left(s_1^{(N-1)0}, s_2^{(N-1)}, s_3^{(N-1)0}\right) \tag{32}$$

wobei

$$\rho_{abc}^{(N-1)}(s_a, s_b, s_c) = \rho_{ab}^{(N-1)}(s_a, s_b)\rho_{bc}^{(N-1)}(s_b, s_c) \tag{33}$$

exakt gilt, d.h., nur Transfermatrizen, die das System vollständig beschreiben, eine Rolle spielen.

In unseren aktuellen Arbeiten haben wir uns fast vollständig auf Transfermatrizen bei Signalfolgen und Transfertensoren (vgl. Sobotta 1984,1985 [15] und [16]) bei Bildern gestützt, da für die Datenrestauration die Berechnung der renormierten Kopplungsparameter aufwendiger als die direkte Verwendung der Transfermatrizen und –tensoren ist. Dieser Vorteil ist insbesondere bei der Restauration diskreter Signalfolgen erkennbar. Für Q mögliche Signalwerte erhält man als Freiheitsgrade im Gittermodell Pottsspins $s_i = 1, 2, \ldots, Q$. Damit bleibt die Struktur der Renormierungsgruppentransformation (18) hier erhalten:

$$\rho_{ac}^{(1)}(s_a^{(1)}, s_c^{(1)}) = \sum_{s_b=1}^{Q} \rho_{abc}(s_a, s_b, s_c) \tag{34}$$

die man nun als Matrixgleichung in der Form

$$\rho_{ac}^{(n)} = \rho_{ab}^{(n-1)} \rho_{bc}^{(n-1)}, \quad n = 1, 2, \ldots, N \tag{35}$$

angeben kann. Ebenso bleibt die Struktur der Optimierungsgleichungen

$$\rho_{ab}^{(n)}(s_a^{(n)0}, s_b^{(n)0}) \rho_{bc}^{(n)}(s_b^{(n)0}, s_c^{(n)0}) = \max_{s_b^{(n)}} \rho_{ab}^{(n)}(s_a^{(n)0}, s_b^{(n)}) \rho_{bc}^{(n)}(s_b^{(n)}, s_c^{(n)0}) \tag{36}$$

unabhängig von Q. Dadurch konnte die Restauration diskreter Signalfolgen durch eine einfache Verallgemeinerung des Ansatzes von B.Gidas erstmals im Rahmen einer Renormierungsgruppentheorie durchgeführt werden. Die Ergebnisse werden im Abschnitt 3 vorgestellt.

Für die Restauration von Bildern ist die Renormierungsgruppentheorie für zweidimensionale Gittermodellle anzuwenden, die erheblich anspruchsvoller ist und hier nicht besprochen werden kann. Die von B.Gidas diskutierten Ansätze wurden für Binärbilder implementiert. Die Ergebnisse werden in Abschnitt 3 zusammengefaßt.

Durch die Verwendung von Transfermatrizen und Transfertensoren wurde eine Grundlage geschaffen, nun auf übersichtliche Art und Weise Grauwertbilder und auch Farbbilder zu restaurieren. Wesentlich für den eingeschlagenen Weg ist die Struktur des Formalismus, der keine Einschränkungen hinsichtlich der Spindimensionen (=3 z.B. für RGB–Bilder) und der Anzahl der Spinwerte aufweist.

3 Ergebnisse

Es wurden binäre und diskrete Signalfolgen und binäre Bilder restauriert. Die Daten wurden durch Computersimulation generiert. Dazu wurden

a) ungestörte Daten s^G generiert, also die realen Grundzustände,

b) gestörte Daten s^R generiert, indem das von B.Gidas vorgeschlagene Verfahren implementiert wurde, nämlich die Simulation eines additiven Gaußschen Rauschens mit anschließender Diskretisierung zur Erhaltung des Datentyps von s^G,

c) die Restauration mit der Monte–Carlo–Methode (MCS) des simulierten Ausglühens durchgeführt,

d) die Restauration mit der Renormierungsgruppentheorie (RNG) durchgeführt.

Folgende Fälle werden hier gezeigt:

1) Binäre Signalfolge mit $2^8 + 1 = 257$ Zeitpunkten (Abb.3.1)

2) Diskrete Signalfolge mit $2^8 + 1 = 257$ Zeitpunkten (Abb.3.2)

3) Binäres Bild mit $(2^6 \times 2^6) = 64 \times 64 = 4096$ Pixeln (und periodischen Randbedingungen) (Abb.3.3)

Zwei Kriterien sind zu vergleichen, nämlich

i) die Restaurationsgüte und

Monte- Carlo-Simulation

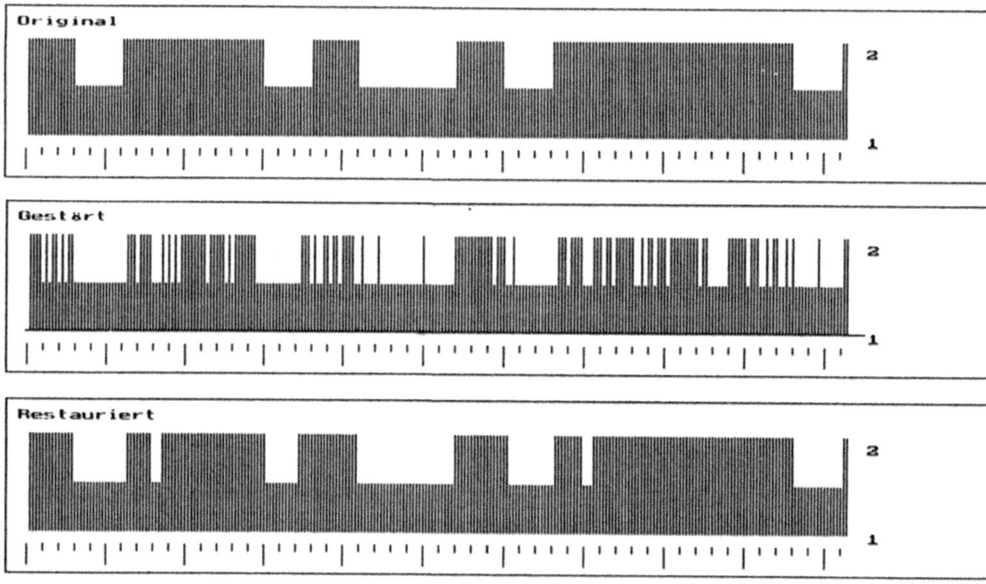

Renormierungsgruppenoptimierung

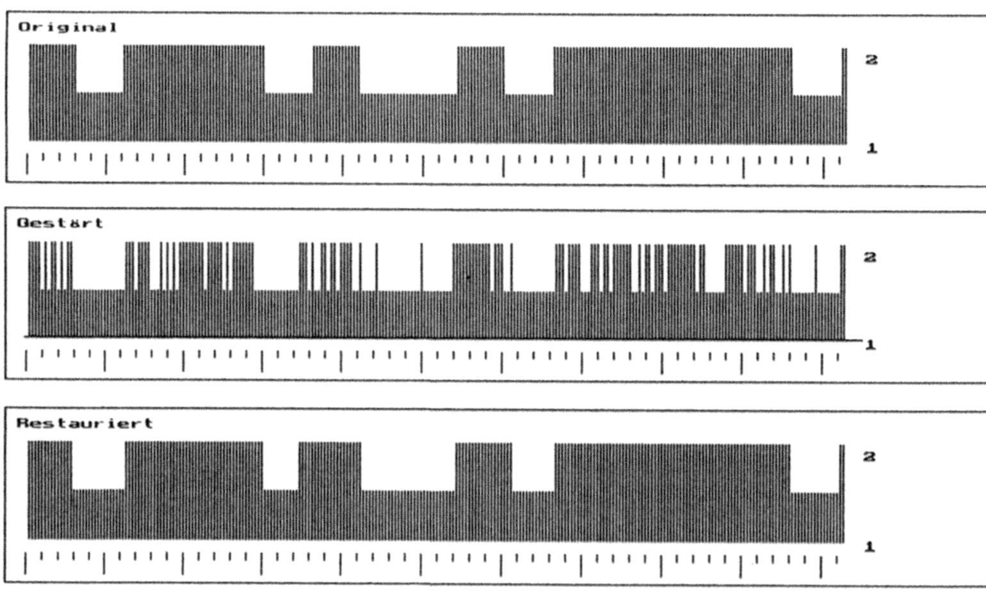

Abb.3.1 Restauration einer binären Signalfolge

Monte- Carlo-Simulation

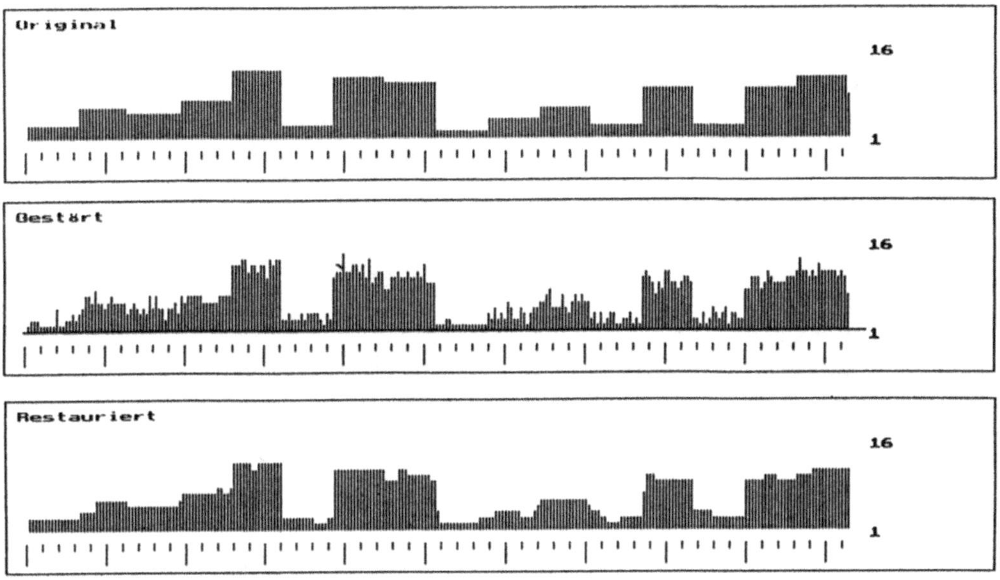

Renormierungsgruppenoptimierung

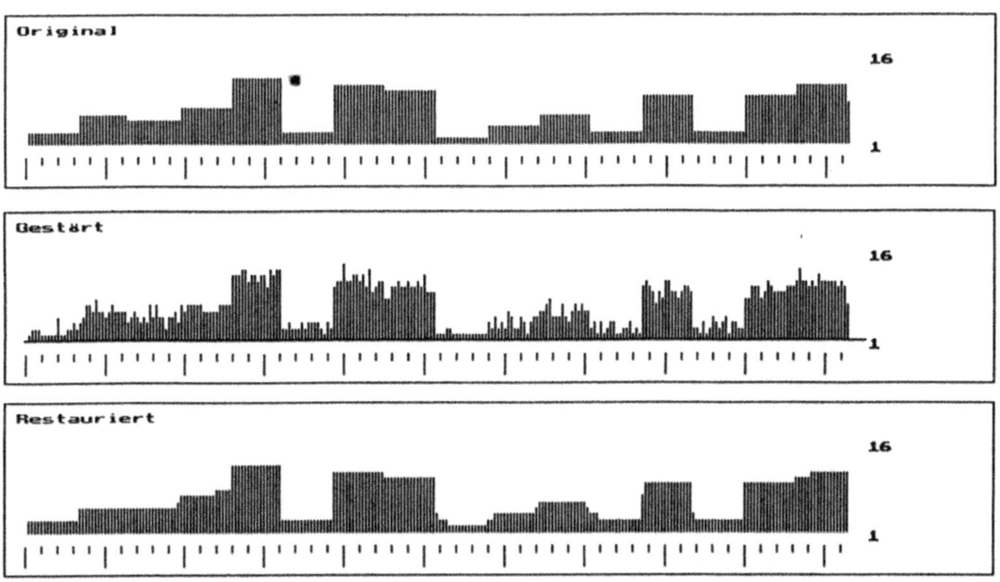

Abb.3.2 Restauration einer diskreten Signalfolge

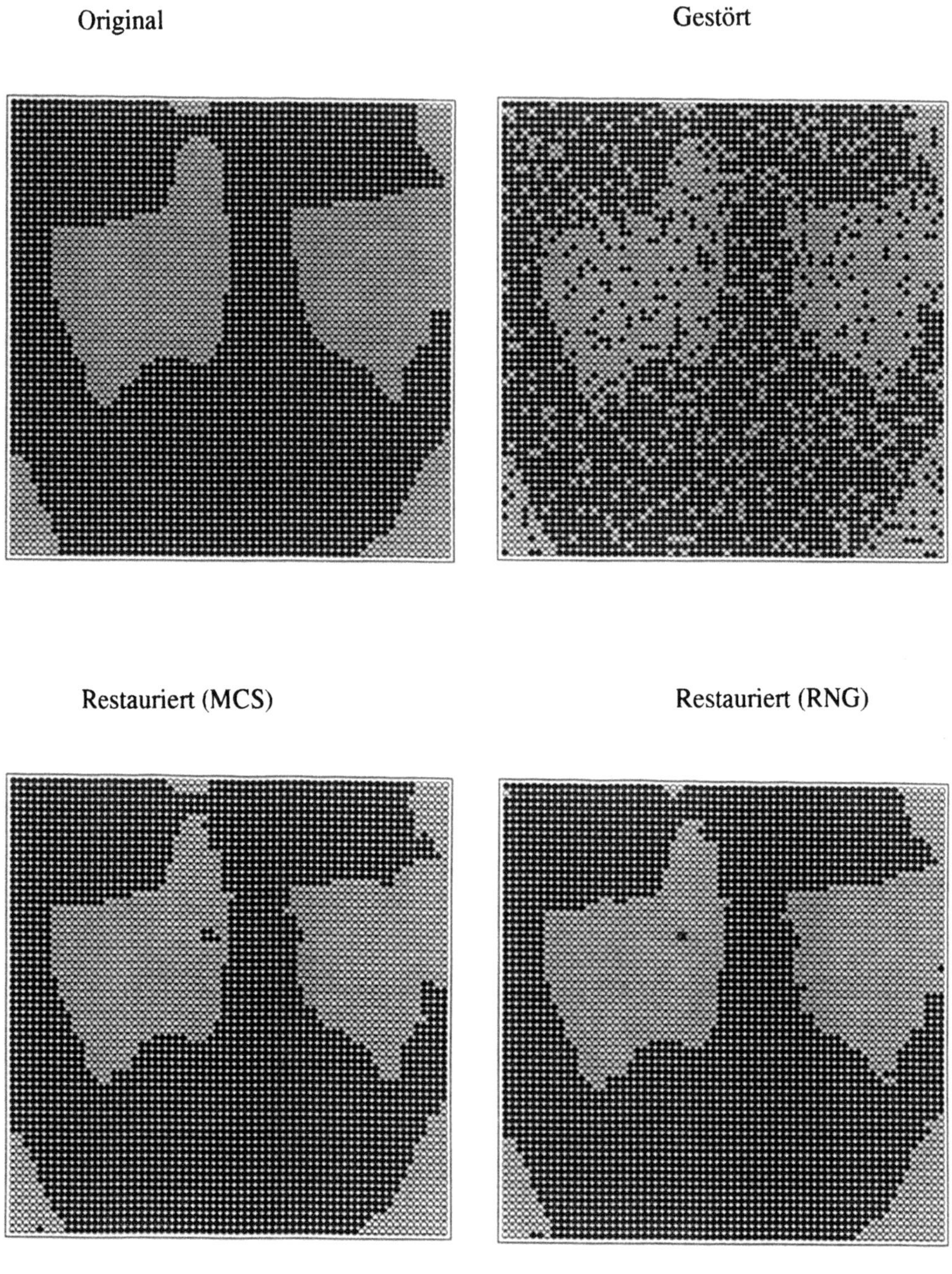

Abb.3.3 Restauration eines binären Bildes

ii) die Rechenzeit für die Restauration

Als Schätzer oder Abstandsmaß für die Restaurationsgüte wurde

$$V_0 = \sum_{i=1}^{I} \left| s_i^0 - s_i^G \right| \tag{37}$$

gewählt, wobei der Vergleichsschätzer

$$V_1 = \sum_{i=1}^{I} \left| s_i^R - s_i^G \right| \tag{38}$$

die Störqualität mißt. Die Ergebnisse sind in Tabelle 3.1 zusammengefaßt. Hierbei ist grundsätzlich die Annahme gemacht worden, daß der Wert von σ in der Simulation für die Bildstörung σ_s gleich dem Wert in der Modellenergiefunktion σ_M ist. Diese Annahme verwendet B.Gidas zur Konstruktion seiner Modellenergiefunktion.

Wir haben untersucht, welchen Einfluß die Wahl von σ_M auf die Ergebnisse hat. Für binäre Signalfolgen ist $V_0(\sigma_s = 1, \sigma_M)$ als Funktion von σ_M in Abb.3.4 dargestellt. Dieser erste Ansatz zeigt, daß B.Gidas bei seinem Modellansatz für eine suboptimale Energiefuntion $\mathcal{H}$ eine durchaus brauchbare Wahl getroffen hat.

Für die diskreten Signalfolgen wurde dieser Ansatz daher von uns für Pottsspins erweitert. Es wurde der Ansatz:

$$\mathcal{H} = J^0 + \sum_{i=1}^{2^N+1} h^i f(s_i - s_i^R) + \sum_{i=1}^{2^N} J^{i,i+1} g(s_i - s_{i+1}) \tag{39}$$

mit geraden Funktionen f und g gewählt. Im Falle der hier gezeigten Beispiele (Abb.3.2) wurde

$$f(x) = x^2 \tag{40}$$

$$g(x) = |x| \tag{41}$$

gewählt, womit der Spezialfall für Pottsspins mit zwei Zuständen tatsächlich auf den Fall der Isingspins reduziert wird.

Die Ergebnisse haben grundsätzlich gezeigt, daß der von B.Gidas eingeschlagene Weg zur Datenrestauration sehr nützlich ist. In der Restaurationsgüte und in der Rechenzeit erzielt man mit der Renormierungsgruppentheorie bessere Ergebnisse als mit der Monte-Carlo-Methode. Bei dieser Bewertung steht aus industrieller Sicht die Rechenzeit im Vordergrund, die sicherlich aus Sicht der Forschung anders gewichtet werden kann. Insofern kommt man zu dem Schluß , daß diese modernen Ansätze zur Bildrestauration erst durch die Anwendung der Renormierungsgruppentheorie für die Industrie interessant werden.

Tab.3.1 Restaurationsergebnisse

Datentyp	Datenmenge	Verfahren	σ^*	V_1^{**}	V_0^{**}	t [in s]***
Binäre Signalfolge	257	MCS	$1/\sqrt{2}$	46	19	11
		RNG	$1/\sqrt{2}$	46	4	2
Diskrete Signalfolge	257	MCS	1	197	72	12
		RNG	1	197	44	7
Binäres Bild	64 x 64 = 4096	MCS	$1/\sqrt{2}$	631	49	56
		RNG	$1/\sqrt{2}$	631	52	13

*) Es gilt $\sigma = \sigma_S = \sigma_M$; die Faktoren $1/\sqrt{2}$ ergeben sich durch den Übergang von Ising- auf Potts-Spins

**) Die Schätzer für binäre Daten wurden statt mit Isingspins mit Spins des 2-Zustands-Pottsmodell berechnet

***) Die Rechenzeiten wurden auf einem PC mit einer Intel i386/33MHz CPU ermittelt.

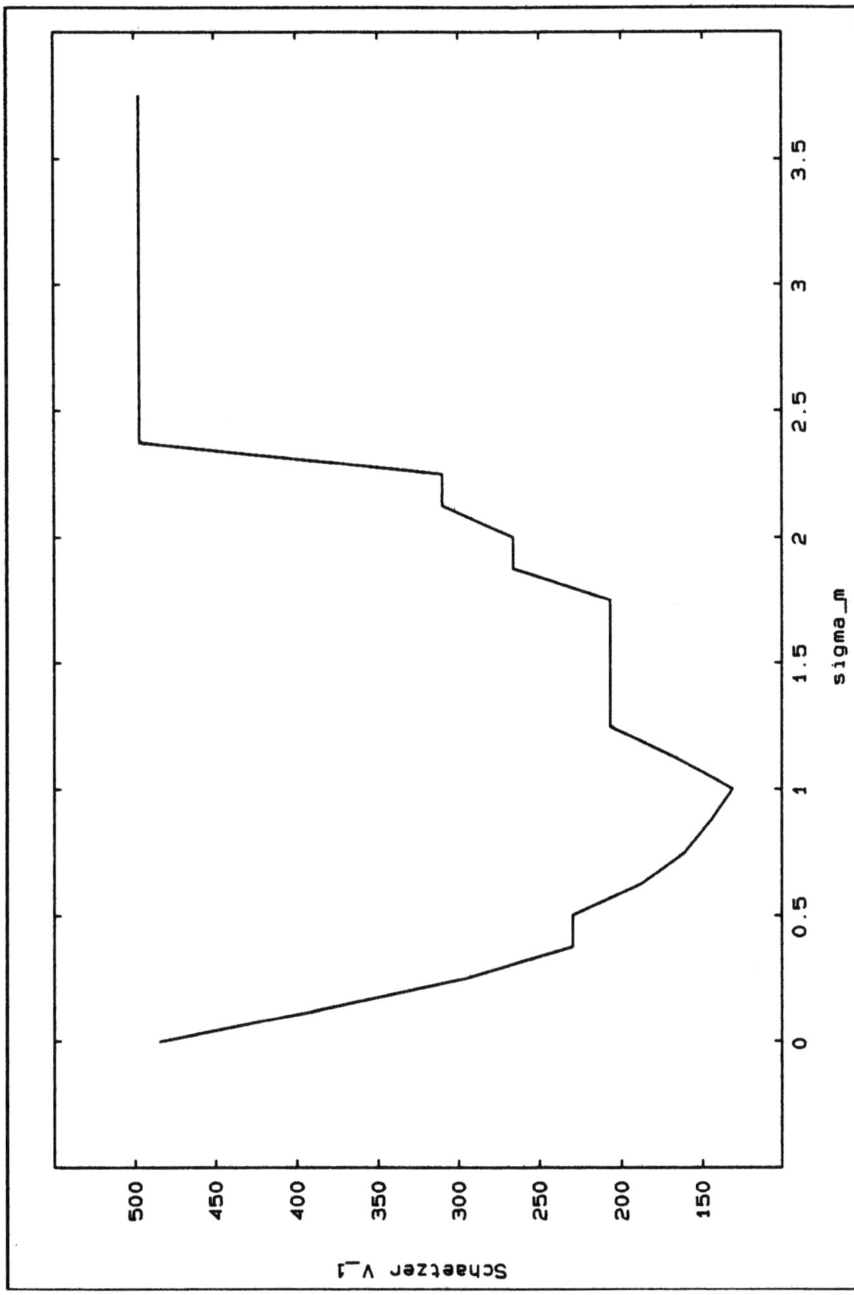
sigma_m
0
0.5
1
1.5
2
2.5
3
3.5
Schaetzer V_1
500
450
400
350
300
250
200
150

4 Rechneraspekte

Die Umsetzung der modernen Verfahren zur Bildrestauration werfen die Frage nach der Rechenzeitoptimierung auf, damit aber auch die Frage nach geeigneten Prozessorarchitekturen. Es soll daher in diesem Abschnitt diskutiert werden, welche Möglichkeiten der Parallelisierbarkeit gegeben sind.

In Abb.4.1 ist gezeigt, wie sich für die unterschiedlichen Renormierungsstufen n die Anzahl der parallel ausführbaren Matrizenmultiplikationen verändert. Allgemein erhält man in der n–ten Renormierungsstufe für $2^N + 1$ Spins:

- $2^{N-n} + 1$ Signale
- 2^{N-n} Transfermatrizen

und im $(n + 1)$–ten Renormierungsschritt

- $2^{N-(n+1)}$ Matrizenmultiplikationen,

die parallel auszuführen sind. Dabei erfolgt der Ergebnistransfer in die unmittelbar folgende Renormierungsstufe (vgl. Abb. 4.1).

Das Optimierungsverfahren hat die inverse Struktur (Abb.4.2). Damit liegen in der n–ten Renormierungsstufe für $2^N + 1$ Spins

- $2^{N-n} + 1$ Signale,

von denen

- $2^{N-(n+1)} + 1$ Signale optimiert sind

und von denen

- $2^{N-(n+1)}$ Signale parallel optimiert werden müssen,

wobei $n < N$ gilt. Der Fall $n = N$ ist aus Abb.4.2 zu entnehmen. Dabei sind die optimierten Signale aus der $(n + 1)$–ten Renormierungsstufe als bekanntes Ergebnis transferiert worden.

Damit zeigt sich, daß für die Renormierungsgruppenoptimierung hierarchisch aufgebaute Parallelrechner mit Pipeline–Struktur eine geeignete Architektur haben. Die Einzelprozessoren müssen dabei einerseits Matrizenmultiplikationen und andererseits Optimierungen durchführen, also ebenfalls parallelisierbare Operationen.

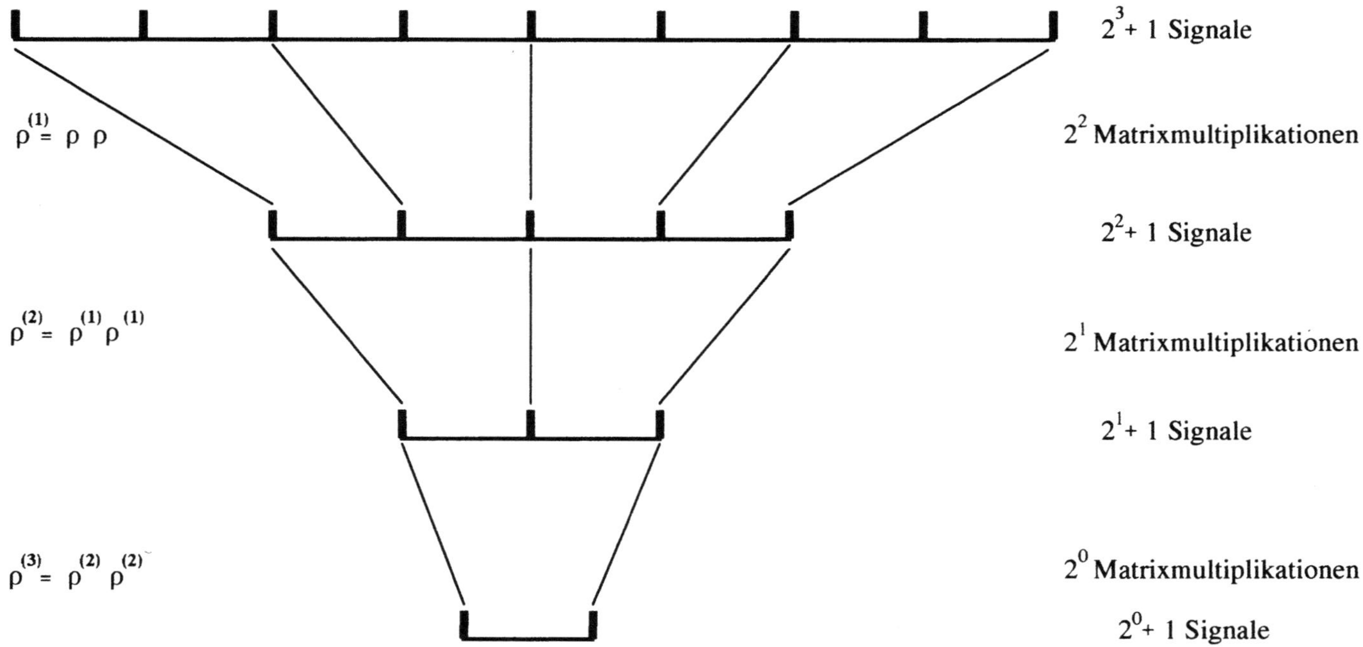

Abb. 4.1 Beispiel einer Renormierungsgruppenrechnung für diskrete Signale

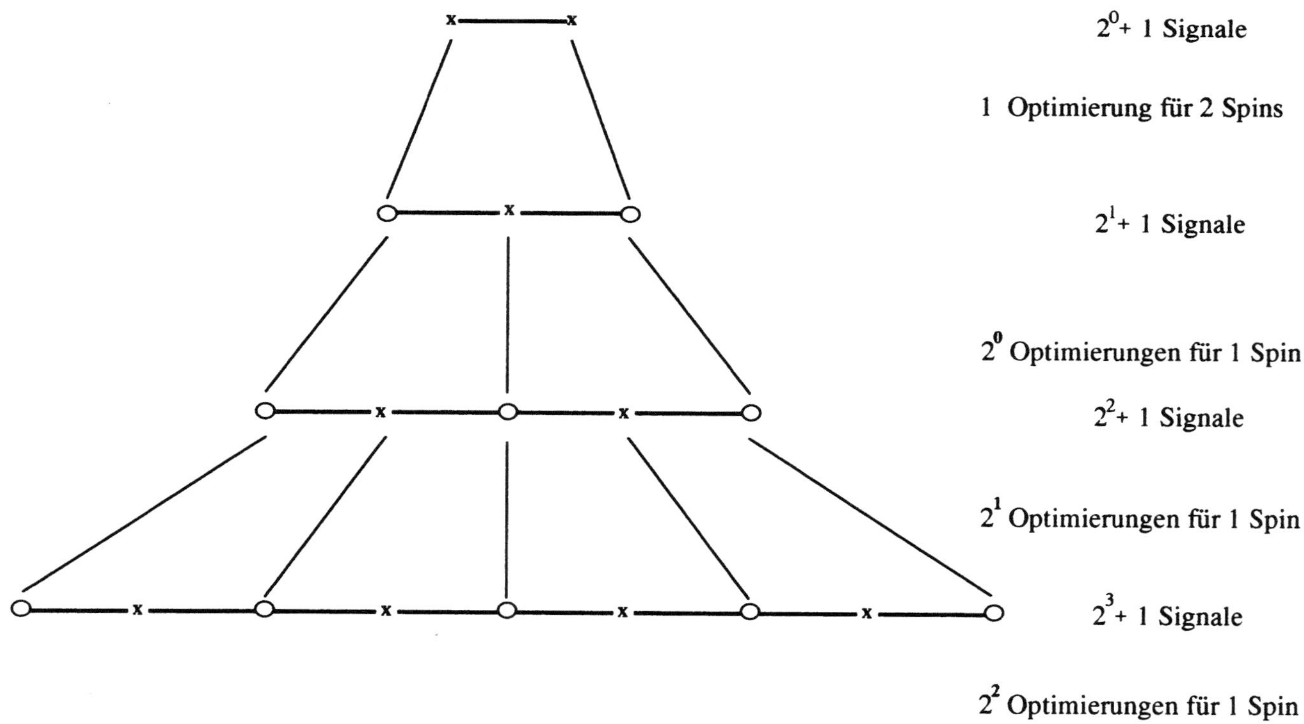

Abb. 4.2 Beispiel einer Renormierungsgruppenoptimierung für diskrete Signalfolgen

5 Zusammenfassung

An Beispielen der Entrauschung diskreter Signalfolgen und binärer Bilder konnte gezeigt werden, daß die Anwendung der Renormierungsgruppentheorie die Ergebnisse von Monte–Carlo–Verfahren in der Restaurationsgüte auf jeden Fall erreicht, meist übertrifft, und hinsichtlich der Rechenzeit entscheidend verbessert. Die Diskussion der Parallelisierbarkeit hat gezeigt, daß im Rahmen der Renormierungsgruppentheorie eine hochgradige Entkopplung der Rechnung in parallel durchzuführende elementare Teilschritte möglich ist.

Die bisherige Untersuchungen betreffen völlig unkorrelierte Störungen der Datenwerte, d.h. mehr Auflösungsprobleme der Datenwerte als Probleme bezüglich der Trennschärfe.

In der industriellen Anwendung müssen zwei Arten der Restauration unterschieden werden, nämlich für Fälle

i) Datenstörung vor der Digitalisierung

ii) Datenstörung nach der Digitalisierung

Im ersten Fall ist die Restauration oft die einzige Methode zur Datenentrauschung, während im zweiten Fall die Datencodierung so gewählt werden kann, daß mit der Decodierung auch eine Entrauschung unterstützt wird. Datenstörungen vor der Digitalisierung treten in unterschiedlichen Anwendungen auf, beispielsweise bei

i) Videobildern bei geringer Lichtausbeute

ii) Infrarotbildern

iii) Röntgenbildern bei geringen Strahlungsdosen

iv) Ultraschallbildern mit geringen Kontrastverhältnissen

Da beispielsweise in der medizinischen Diagnostik die Bildqualität nicht durch beliebige Steigerung der Strahlendosen oder anderer Intensitäten erreicht werden kann, scheint hier ein Anwendungsfeld für Verfahren zur Datenrestauration vorzuliegen.

Am Beispiel einer extremen Entrauschung eines Binärbildes (Abb.5.1) wird deutlich, daß hier ein Verfahren vorliegt, dessen Leistungsfähigkeit mit dem menschlichen Auge vergleichbar ist. Damit wird deutlich, daß das hier vorgestellte Verfahren in modernen technischen Sehsystemen, deren Leistungen nicht mehr an herkömmlichen Kameras orientiert sind, sondern eher an biologischen Vorbildern wie Wirbeltieraugen, eine Schlüsselrolle spielen könnte. Im konkreten Fall erkennt man das Potential auf dem Gebiet der Bildsegmentierung und Objekterkennung.

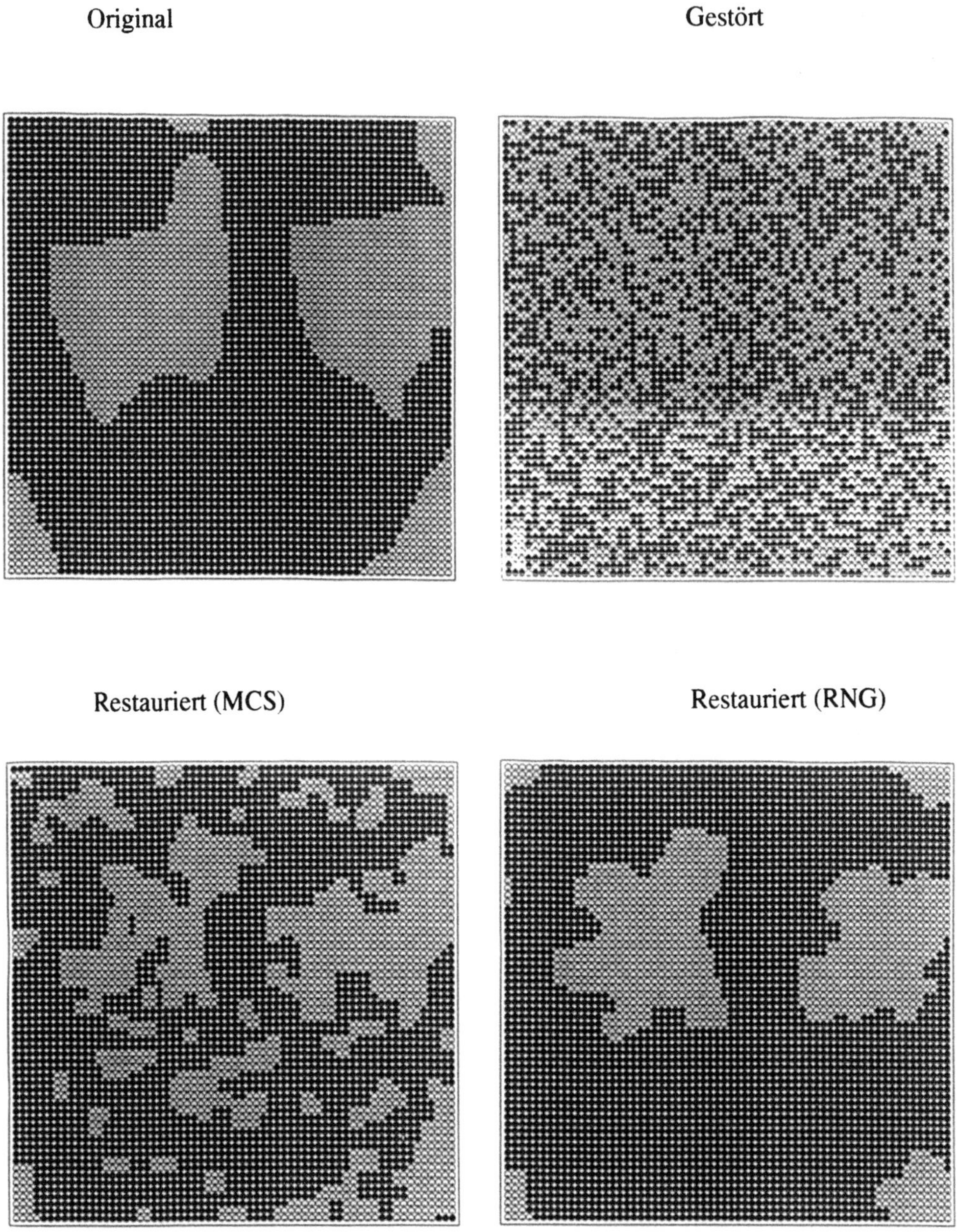

Abb.5.1 Restauration einer binären Bildes mit $\sigma = 5$

Literatur

[1] K.Binder (Hrg.), Monte Carlo Methods in statistical physics, Springer Verlag Berlin 1979

[2] K.Binder, Phys.Blätter Vol.40 No. 3, 53 (1984)

[3] H.Derin und H.Elliott, IEEE PAMI 9, 39, (1987)

[4] C.Domb und M.S.Green (Hrg.) , Phase Transitions Vol.6, Academic Press, London (1976)

[5] D.Geman und S.Geman, IEEE PAMI 6, 721 (1984)

[6] S.Geman und C.Graffinge, in A.M.Gleason (Hrg.), Proc. 1986 Int.Congr. Mathematicians, AMS New York (1987)

[7] B.Gidas, IEEE PAMI 11 No.2, 164 (1989)

[8] R.C.Gonzales und P.Wintz, Digital Image Processing, Addison Wesley Publishing Company Reading, Massachusetts (1987)

[9] L.P.Kadanoff, Physics (N.Y.) 2, 263 (1966)

[10] L.P.Kadanoff, Rev. Mod. Phys. 49, 267 (1977)

[11] W.Kinzel, Spektrum der Wissenschaft, 19 April (1988)

[12] S.Kirckpatrik, S.D.Gelatt und M.P.Vecchi, Science 220, 621 (1983)

[13] H.D.Lücke, Signalübertragung, Springer–Verlag (1985)

[14] N.Metropolis,A.W.Rosenbluth,M.N.Rosenbluth,A.H.Teller und E.Teller J.Chem.Phys. 21, 1087 (1953)

[15] G.Sobotta, in A.Pękalski und J.Sznajd (Hrg.), Lecture Notes in Physics 206, 137 Springer-Verlag, Berlin (1984)

[16] G.Sobotta, Physica 129 A , 343 (1985)

[17] K.G.Wilson, Phys.Rev. B4, 3174,3184 (1971)

[18] K.G.Wilson, Spektrum der Wissenschaft, 66 Oktober (1979)

Der Studiengang Naturwissenschaftliche Informatik an der Universität Bielefeld

R. Giegerich*
Technische Fakultät

W. Schepper†
Fakultät für Physik

Universität Bielefeld
Postfach 86 40
W-4800 Bielefeld 1

Zusammenfassung

An der Technischen Fakultät der Universität Bielefeld wird seit dem Wintersemester 1989/90 ein Studiengang „Naturwissenschaftliche Informatik“ aufgebaut. Dieser trägt der wachsenden Bedeutung der Informatik in den naturwissenschaftlichen Disziplinen Biologie, Chemie und Physik Rechnung. Motivation dieses neuartigen Studiengangs, sein inhaltlicher Aufbau und erste Erfahrungen bei der Durchführung werden nachfolgend beschrieben.

*robert@TechFak.Uni-Bielefeld.DE
†schepper@Physik.Uni-Bielefeld.dbp.DE

1 Informatik in den Naturwissenschaften

Wenn ein Biologe versucht, Strukturen in DNA-Sequenzen zu erkennen, wenn eine Chemikerin die räumlichen Strukturen von Makromolekülen bestimmt oder wenn ein Physiker Atome in einem Kristallgitter oszillieren läßt, dann findet das untersuchte Phänomen heute nicht mehr in natura, sondern im Inneren eines Rechners statt. Dabei kommen Resultate aus allen Teilgebieten der Informatik zum Einsatz: spezielle Rechnerarchitekturen, Prozeßsteuerung, Computer-Graphik, Datenbanken, Mustererkennung, um nur einige zu nennen. Die Fruchtbarkeit naturwissenschaftlicher Untersuchungen hängt von der geeigneten Wahl der verwendeten Hard- und Software ab. Wer neue Erkenntnisse produzieren oder neue Forschungsmethoden entwickeln will, muß sich auch an der Weiterentwicklung der informatischen Methoden beteiligen. Naturwissenschaftler kommen heute mit autodidaktisch erworbenen Kenntnissen in der Programmierung nicht mehr aus.

Daß der Einsatz der Informatik in den Naturwissenschaften weit hinter dem State-of-the-Art zurückgeblieben ist, kann man oft beobachten: Programme voller wertvoller naturwissenschaftlicher Ideen aus dem akademischen Bereich, beispielsweise in der theoretischen Chemie, können wegen softwaretechnischer Mängel nur selten zwischen verschiedenen Gruppen ausgetauscht werden (während kommerzielle Software nahezu unerschwinglich ist). Aus leicht eingeworbenen Sachmitteln beschaffte Spezialrechner werden nicht optimal genutzt, weil Know-How und Systemunterstützung fehlen. Oft werden die eigentlichen naturwissenschaftlichen Fragestellungen in den Hintergrund gedrängt durch einen überproportionalen Programmieraufwand für scheinbar einfache Detailprobleme, sobald der Rückgriff auf vorhandene Programmbibliotheken nicht ausreicht.

Phänomene wie diese, jedem Naturwissenschaftler aus seinem Umfeld vertraut, haben die naturwissenschaftlichen Fakultäten an der Universität Bielefeld bewogen, die Technische Fakultät beim Aufbau eines neuartigen Studiengangs zu unterstützen.

2 Aufbau des Studiengangs Naturwissenschaftliche Informatik

Der Studiengang Naturwissenschaftliche Informatik (NWI) ist ein Studiengang der Angewandten Informatik mit dem Anwendungsschwerpunkt Naturwissenschaften. Das Studium gliedert sich in zwei Hauptfächer: Informatik und eines der Fächer Biologie, Chemie oder Physik. Das Grundstudium gleicht dem eines klassischen Studiengangs der Diplom-Informatik, mit der Modifikation, daß anstelle eines Nebenfachs erste Grundkenntnisse in den drei Naturwissenschaften vermittelt werden. Das Naturwissenschaftliche Hauptfach wird zu Beginn des Hauptstudiums festgelegt und macht dort einen deutlich stärkeren Anteil als ein herkömmliches Nebenfach aus. Innerhalb des naturwissenschaftlichen Faches werden weitere Grundlagen gelegt und es erfolgt eine Orientierung auf eines der Teilgebiete, in denen die Informatik eine tragende Rolle spielt. Veranstaltungen zum Problemkreis „Technik und Gesellschaft" sind ein weiterer Bestandteil des Grund- und Hauptstudiums.

3 Berufsbild und Arbeitsfelder

Das Berufsbild des Studiengangs ist das des Informatikers in interdisziplinär besetzten Arbeitsgruppen auf naturwissenschaftlichem Gebiet. Er soll nicht z.B. den Diplom-Physiker ersetzen, sondern mit diesem den effektiven Einsatz von Methoden der Informatik zur Bearbeitung physikalischer Probleme gewährleisten. Als Arbeitsgebiete seien hier genannt: Simulation physikalischer oder evolutionärer Prozesse, Prozeßsteuerung z.B. in der Biotechnologie, Reaktions- und Stoffdatenbanken, computergestützte Synthese, Visualisierung chemischer Strukturen, Strukturanalyse von Gen- und Proteinsequenzen, Auswertung von Sensor- oder optischen Daten mit Methoden der Mustererkennung und der künstlichen Intelligenz. Besonders erwähnt sei hier auch das Gebiet der Robotik, auf dem stets Gruppen von Naturwissenschaftlern, Ingenieuren und Informatikern eng zusammenarbeiten, auch wenn dieses Gebiet nicht direkt den Naturwissenschaften zuzuordnen ist.

Arbeitsplätze werden dementsprechend in der Automatisierungstechnik, in der Luft- und Raumfahrt-Industrie, in der Mikroelektronik-Industrie, in der pharmazeutischen und chemischen Industrie und in Großforschungseinrichtungen gesehen.

Als Studiengang der Angewandten Informatik ist dieser Studiengang am besten mit der an der Universität Dortmund angebotenen Ingenieur-Informatik vergleichbar, die in mancher Hinsicht als Vorbild diente. Es fehlt allerdings die wirtschaftswissenschaftliche Komponente der Ausbildung zugunsten einer Vertiefung im naturwissenschaftlichen Fach.

4 Weder Informatiker noch Naturwissenschaftler? Von der Problematik eines interdisziplinären Studiengangs

Von vielen und auch von den maßgeblichen Seiten wird nach einer Verkürzung der Studienzeiten gerufen. Eine über 9 Semester hinausgehende Regelstudienzeit ist derzeit nicht durchsetzbar. Wer in solchen Zeiten einen interdisziplinären Studiengang aus der Taufe hebt, muß sich die Frage gefallen lassen, ob dies nicht zwangsläufig auf eine mangelhafte Qualifikation in beiden Fächern hinauslaufen muß. Tatsächlich ist ein solcher Effekt vorprogrammiert, wenn man einfach kanonische Lehrinhalte aus Informatik und beispielsweise Physik im Verhältnis 60 : 40 kombiniert. Im Studiengang NWI wird dieser Gefahr begegnet durch die folgenden Leitsätze:

- Im NWI-Grundstudium wird streng auf Vergleichbarkeit mit einem klassischen Studium der Diplom-Informatik geachtet. Ein Hochschulwechsel nach dem Vordiplom ist damit leicht möglich.

- Der Informatik-Anteil im Hauptstudium ist spezifisch auf die im Naturwissenschaftlichen Hauptfach wesentlichen Informatik-Methoden ausgerichtet.

- Die Ausbildung im Naturwissenschaftlichen Hauptfach vermittelt weitere allgemeine Grundlagen, konzentriert sich aber danach auf solche Teilgebiete, in denen der Bezug zur Informatik besonders wichtig ist, um diese in der nötigen Tiefe zu behandeln.

Dies heißt natürlich einerseits, daß alte Stoffpakete aufgetrennt und neue geschnürt werden müssen. In der Tat haben bereits die Fakultäten für Biologie, Chemie und Physik neue Lehrveranstaltungen für das Grundstudium der Naturwissenschaftlichen Informatiker entwickelt.

Es heißt aber auch, daß versucht werden muß, den wechselseitigen Bezug verstärkt innerhalb des Lehrstoffes auszudrücken. Beispiele:

- Klassische Syntaxanalyseverfahren aus dem Übersetzerbau werden zusammen mit ihrer Anwendung auf die Analyse von genetischen Sequenzen behandelt.

- Das Software-Praktikum behandelt naturwissenschaftlich motivierte Aufgabestellungen, wie die Auswertung von Spektralanalyse-Daten.

- Die physikalischen Grundlagen der Informatik (Kristalle, Halbleiter, Feldeffekt-Transistor usw.) werden in den Physik-Lehrstoff integriert.

Eine solche Integration kann nur im Laufe einiger Jahre entstehen. Für den Studiengang ist es jedoch ganz entscheidend, daß es gelingt, die Verflechtung von Naturwissenschaften und Informatik innerhalb aller Lehrveranstaltungen beider Disziplinen zum Ausdruck zu bringen.

5 Vielfältige Bezüge zwischen Informatik und Physik

Erst mit Computerhilfe ist in physikalischen Experimenten die Datenerfassung möglich in der oft kurzen Meßzeit, die von den zahlreichen, oft auch reparaturanfälligen Komponenten einer physikalischen Apparatur gesetzt wird. Es gibt aber auch physikalische Apparaturen, die über Monate stabil arbeiten, in denen wegen der geringen Zahl der relevanten Ereignisse wochenlange Meßzeiten typisch sind, die auch ohne Computerhilfe nicht möglich sind.

Gleich die ersten verfügbaren Prozeßrechner wurden bei Steuerungsaufgaben großer physikalischer Apparaturen und in der Datenerfassung der Experimente in physikalischen Instituten eingesetzt. Der relativ hohe Preis der anfangs verfügbaren Computer hat die Entwicklung von Bussystemen für die Meßdaten-Erfassung sehr vorangetrieben — vor allem von den Großforschungsinstituten der Hochenergiephysik initiiert. Erst über solche Cratesysteme wie z.B. das Camac-System war die Aufgabe, viele Meßdaten von vielen Meßstellen in das Nadelöhr des Computers aufzunehmen, lösbar. Über einen die Bussignale führenden Steckrahmen wurden viele Module der Prozeßperipherie (AD, DA, Zähler,...) mit dem Computer verbunden. In ähnlicher Weise wurden auch Aufgaben der Regelungs- und Steuerungs-Technik in der Industrie gelöst. Auch die über Spezialrechner (Vielkanalanalysator) mögliche, im Vergleich zu den Befehlsablaufzeiten schnellere Meßdatenerfassung wurde in solche Systeme integriert. Mit solchen Systemen mit einem Zentralrechner ist nicht die heute erreichbare Ausfallsicherheit zu erreichen. Durch den Aufbau verteilter Systeme ist die Datenerfassung schneller, sicherer und preiswerter geworden. Heute verleihen Computerchips den traditionellen Modulen der Prozeßperipherie zusätzliche Intelligenz. Transputer als Knotenrechner, die einen schnellen Datenaustausch untereinander ermöglichen, werden für Spezialaufgaben eingesetzt.

Nach der Datenerfassung bietet die Auswertung vielfältige Einsatzmöglichkeiten für den Computer. In umfangreichen Entfaltungsprogrammen (zum Teil nichtlineare Least-Square-Fits) sind die Messungen auf die physikalisch relevanten Größen zu reduzieren. Die unvermeidbaren apparaturtypischen Abweichungen von dem idealen Experiment (Winkel- und Energie-Divergenz der Strahlen) sind zunächst herauszurechnen. Es schließt sich die Aufgabe des Vergleichs mit der Theorie an.

Ein weiterer Berührungspunkt liegt in der Simulation physikalischer Prozesse auf dem Computer. Erst die Simulation zeigt dem Experimentalphysiker, in welchen der vielen möglichen Meßverteilungen und mit welchen Parametern interessante Messungen zu erwarten sind. Diese Simulation kann darauf ausgerichtet sein, die Experimente in den für diese Apparatur typischen Bereichen hoher Auflösung zu plazieren. Ein anderer Ansatzpunkt ist, Voraussagen über die zu erwartende Verteilung auf Grund publizierter Ergebnisse aus Experimenten anderer Gruppen zu gewinnen.

Die für Steuerungs-, Datenakquisitions- Auswertungs- und Simulations-Aufgaben entwickelten Programme sind oft sehr aufwendig. Der für die Experimente verantwortliche Physiker wird deshalb enorm entlastet, die Experimente liefern sehr viel schneller das erwartete Ergebnis, wenn Studenten mit fundierten Informatik-Kenntnissen sich diesem Teil des physikalischen Experiments widmen. Physikern und Informatikern, die schon bisher stets den Kontakt mit der Nachbardisziplin gesucht haben, boten sich immer vielfältige Einsatzmöglichkeiten in Forschungs-Instituten und der Industrie.

Das Ausbildungsziel des neuen Studiengangs „Naturwissenschaftliche Informatik“ ist der Versuch, bewährte Kenntnisse aus zwei Nachbardisziplinen auf eine Person zu vereinigen. Wenn es gelingt, den Studenten sowohl in Physik und Informatik fundierte und aufeinander abgestimmte Kenntnisse zu vermitteln, wird dem neuen Studiengang die Bewährung in der Praxis nicht versagt bleiben.

6 Erste Erfahrungen

Abschließend sollen noch einige der problematischen Punkte beim Aufbau des Studiengangs NWI angesprochen werden:

- Das Studiengang NWI wird von der Technischen Fakultät durchgeführt, die die Abteilungen Informationstechnik und Biotechnologie umfaßt. Dabei leisten die Fakultäten für Biologie, Chemie, Physik, Mathematik und Soziologie Beiträge. Eine so vielseitige Kooperation bedeutet natürlich enormen organisatorischen Aufwand. Die Vorstellungen zur Ausgestaltung der Lehrveranstaltungen konkretisieren sich erfahrungsgemäß frühestens dann, wenn die ersten NWI-Studenten den entsprechenden Studienabschnitt erreicht haben. Zum Glück bringt die erste Studentengeneration, die im Herbst 1991 das Vordiplom absolviert hat, eine gehörige Portion Pioniergeist mit.

- Bei der Formulierung der Studien- und Prüfungsordnung wurde darauf geachtet, daß die naturwissenschaftlichen Fächer sich in den quantitativen Anforderungen nicht unterscheiden. So bleibt es den verschiedenen Disziplinen überlassen, durch entsprechende Beiträge zum Grundstudium für sich als Naturwissenschaftliches Hauptfach zu werben. Aber auch äußere Rahmenbedingungen wie Berufschancen für Physiker, Studienzeiten in Chemie oder Numerus Clausus in Biologie beeinflussen die Wahl des Naturwissenschaftlichen Hauptfachs. Die Unsicherheit über die Aufteilung der Studierenden auf die naturwissenschaftlichen Fächer im Hauptstudium bringt erhebliche Schwierigkeiten bei der Planung des Lehrangebots, insbesondere bei den Praktikumsplätzen, mit sich.

- Die differenzierte Oberstufe der Gymnasien stellt für einen interdisziplinären Studiengang eine denkbar schlechte Grundlage dar. Wer Leistungskurse z.B. in Chemie und Biologie absolviert hat, bringt kaum physikalische Grundkenntnisse mit. Diese Mängel müssen durch individuellen Einsatz in studentischen Arbeitsgruppen kompensiert werden. Die Fachschaft der Technischen Fakultät führt ein Orientierungsprogramm für Studienanfänger durch, das auch auf diese Schwierigkeiten eingeht.

Ein solcher Studiengang wird am grauen Tisch geboren; Wirklichkeit wird er erst durch die Studenten und Studentinnen, die ihn durchleben und ausgestalten. Ihr Erfolg ist auch der seine.

Simulation von Modellen und Prozessen

I. Bausch-Gall
Bausch-Gall GmbH,
München

J. Mahrun
J. W. Goethe Universität,
Frankfurt

Die Simulation komplexer Systeme ist ein Fachgebiet, in dem die Zusammenarbeit von Informatik und Physik eine lange und fruchtbare Tradition hat. In vielen Forschungsgebieten der Physik stößt man an die Grenzen des Experimentierens oder die Experimente sind sehr aufwendig und damit teuer. Daher müssen die Systeme so weit wie möglich durch Simulationen untersucht werden. In der Physik werden, wie in sonst keinem anderen Gebiet, besonders komplexe Systeme simuliert. Mit Simulationsrechnungen werden außerdem sowohl Versuche geplant, also auch begleitend und unterstützt. Jedoch auch die Simulation dieser Systeme stößt oft an die Grenze des heute rechentechnisch Möglichen. Daher spielen vor allem neue Rechnerarchitekturen und effiziente Algorithmen zur Ausnutzung dieser Rechner eine besondere Rolle. Die Simulationstechnik als wissenschaftliche Methode zur Entwicklung und Überprüfung von Modellen, zur Organisation der Datenhaltung und der Modell- und Ergebnisverwaltung ist ebenfalls von gemeinsamen Interesse. Diese Aspekte wurden in den Vorträgen dieser Sitzung repräsentiert.

J. Honerkamp gibt einen Überblick über den methodischen Stellenwert der Simulation bei der Zeitreihenanalyse, der Lösung inverser Probleme und der Simulation stochastischer Modelle und deren Steuerung. Die Simulation dient hier vorwiegend zur Überprüfung von Modellen in anders nicht zugänglichen Systemen. Ein wichtiges Problem der Modellbildung ist dabei die Identifikation von Modellen, deren dynamisches Verhalten nur durch Messdaten, nicht jedoch durch Gleichungen beschrieben werden kann. Hier erweist sich die Erfassung der Daten und Erstellung der Modelle als gemeinsames Interessengebiet von Physikern, Informatikern und Ingenieuren.

A. Hübler schildert den Prozeß der Modellentwicklung am Beispiel der Dendritenbildung in Kupfergelen, die deterministisches Chaos aufzuweisen scheint. Die Differentialgleichungen des Modells enthalten jeweils eine große Zahl verborgener Variablen und müssen durch Optimierung, Fehlerminimierung oder "maximum-likelihood"-Analyse am Experiment überprüft werden.

W. von der Linden schildert Simulationen von Hochtemperatur-Supraleitern als klassisches Beispiel der Anwendung von Supercomputern, wobei in diesem Fall ein hochparalleles System verwendet wurde.

Als letzter Vortragender erläutert H. Fuss eine Anwendung der Theorie der Petri-Netze auf die Überprüfung von Simulationsergebnissen. Die in der Simulation allein nachweisbare Relation der Ununterscheidbarkeit darf nicht mit Gleichheit verwechselt werden, da ihr insbesondere die Eigenschaft der Transitivität fehlt. Auch deswegen sollte man bei Schlußfolgerungen aus Simulationen besonders vorsichtig vorgehen.

Die Sitzungsleitung übernahmen Dr. Ingrid Bausch-Gall und Professor Dr. Joachim Maruhn. Die Vorträge wurden aus 12 Vortragsanmeldungen ausgewählt.

SIMULATION VON ZELLULARAUTOMATEN

D.Stauffer
Institut für Theoretische Physik
Universität
D-W-5000 Köln 41

1. Definition

Gegeben sei ein sehr großes Gitter, also zum Beispiel ein Schachbrettmuster (Quadratgitter). Auf jedem Gitterplatz sitzt eine Variable, die von Physikern als Spin oder Besetzungszustand, von Mathematikern als Boolesche Variable, von Technikern als Bit bezeichnet wird und nur zwei Werte annehmen kann: Spin nach oben oder nach unten gerichtet, Platz besetzt oder unbesetzt, Boolesche Variable wahr oder falsch, Bit 1 oder 0. Die Orientierung eines Spins zum Zeitpunkt t ist eindeutig bestimmt durch die Orientierung seiner nächsten Gitternachbarn zum vorherigen Zeitpunkt t–1. Somit sind also Ort und Zeit von vornherein diskret und erleichtern die Simulation. Solche Systeme nennen wir nach John von Neumann *Zellular-Automaten*[1]. Natürlich kann man sie auch komplizierter definieren, mit Einfluß von mehr als nur den nächsten Nachbarn, mit mehr als einem Bit pro Gitterplatz, oder probabilistisch mit einer durch die Konfiguration der Nachbarn gegebenen Wahrscheinlichkeit. Ich empfinde eine Fahrkarten-Maschine, die nur mit einer Wahrscheinlichkeit von zehn Prozent mir nach Geldeinwurf einen Fahrschein herausrückt, nicht als funktionierenden Automaten, und beschränke mich daher hier auf völlig deterministische Automaten wie oben definiert.
Diese Definition entspricht einem rein parallelen und nicht sequentiellen Vorgehen bei der Behandlung der Gitterplätze: Kein Spin hängt zum Zeitpunkt t ab von anderen gerade vorher berechneten Spins des Zeitpunkts t. Manchmal hat ein Übergang zu sequentiellem Abarbeiten keinen Einfluß, manchmal schon. Die Parallelmethode eignet sich besonders für Vektor- und Parallelrechner.
Ein Beispiel ist der Automat des logischen ODER, wobei ein Platz besetzt wird, wenn mindestens einer seiner Gitternachbarn zum vorherigen Zeitpunkt besetzt war. Dann breitet sich die Menge der besetzten Plätze wie eine Infektion aus, und ein einzelner Prozessor eines NEC-SX3 Supercomputers bearbeitet in einer Dimension mehr als 10^{10} Plätze pro Sekunde[2]. Fraktale Bildchen ergeben sich mit der ENTWEDER-ODER Regel, wo ein Spin genau dann nach oben steht, wenn vorher eine ungerade Zahl seiner Nachbarn nach oben gerichtet war. Besonders populär war Conways Spiel des Lebens, wo nächste und übernächste Nachbarn im Quadratgitter berücksichtigt wurden. Zahlreiche verschiedene Aspekte der Zellularautomaten werden in einem allerdings rasch veraltenden Buch[1] behandelt, einfache Programmieraspekte in Ref.3 erklärt, und umfangreiche wissenschaftliche Rechnungen (multi-spin coding, da Silva-Herrmann Algorithmus) in Ref.2 zusammengefaßt. Zwei Konferenzen von 1989 sind in Ref.4 veröffentlicht.

2. Klassifikation, Chaos und Phasenübergänge

2.1 Wolfram-Klassen

Eine *Wissenschaft* wurden die Zellularautomaten nach meinem Gefühl erst, als Wolfram[1] eine systematische Klassifikation *aller* 256 eindimensionalen Zellularautomaten versuchte. Auf dem Quadratgitter hat jeder Platz vier Nachbarn, die $2^4 = 16$ Nachbarkonfigurationen bilden können. Zu jeder dieser Konfiguration gibt es eine Regel, die den Spin nach oben setzt, und eine andere Regel, die ihn nach unten setzt. Also kommen wir zu insgesamt $2^{16} = 65536$ verschiedenen Regeln. Ein Supercomputer schafft es, diese in einer Minute alle auf einem 64×64 Gitter zu simulieren, denn wegen Rotations-, Spiegelungs- und Rauf-Runter-Symmetrie braucht man nur 4856 von ihnen

zu betrachten. (Wenn jeder Spin nicht nur von seinen Nachbarn, sondern wie bei Wolfram auch von sich selbst abhängt, dann gäbe es im Quadratgitter sogar 2^{32} statt 2^{16} Regeln, und der Computer müßte sich auf eine sinnvolle Auswahl beschränken.) Zunächst initialisieren wir die Gitter rein zufällig, d.h. die Hälfte der Spins steht anfangs nach oben und die andere nach unten.
Beobachtet man nun das Verhalten des Systems nach vielen Iterationen (eine Iteration entspricht einer einmaligen Behandlung jedes Gitterplatzes, also dem Schritt von t−1 zu t), so ergeben sich asymptotisch verschiedene Möglichkeiten:

- Klasse 0: Alle Spins zeigen dauernd nach unten
- Klasse 1: Alle Spins zeigen dauernd nach oben
- Klasse 2: Alle Spins zeigen dauernd in eine Richtung: manche nach oben, der Rest nach unten
- Klasse 3: Das ganze Gitter oszilliert zwischen: alle Spins nach oben und alle Spins nach unten
- Klasse 4: Das ganze Gitter oszilliert mit Periode zwei, aber nicht alle Spins sind parallel
- Klasse 5: Es gilt keiner der obigen Fälle.

Leider ist auf dem Quadratgitter die Klasse 5 ("sonstige") die größte, während zu den Klassen 0 und 1 je 2,8 Prozent, zu Klasse 2 und 3 ein bzw. sieben Promille, und zu Klasse 4 etwa 2,1 Prozent der 65536 Regeln gehören. (Ein Fünftel der Mitglieder von Klasse 5 führen zu gleichmäßigen Verschiebungen der gesamten Spinkonfiguration.)
Gerling[5] hat systematisch die Zahl der Nachbarn von zwei bis sechs variiert, indem er auch die eindimensionale Kette, das Bienenwabengitter mit drei Nachbarn, sowie das Dreieckgitter und das kubische Gitter mit sechs Nachbarn untersuchte. Je größer die Zahl der Nachbarn is, um so kleiner war der Prozentsatz der Regeln in den Klassen 0 bis 4. Ähnlich wie bei manchen anderen Methoden der Physik ist auch diese Wolfram-Klassifikation in einer Dimension viel zweckmäßiger als in höheren Dimensionen. Eine andere Klassifikation[12] betrachtet nur die Regeln selbst, nicht die zufällige Anfangskonfiguration, und sucht nach Fixpunkten, die stabil gegen das Flippen eines einzigen Input-Spins sind; so wurden 4 Milliarden Regeln erfolgreich klassifiziert.

2.2 Stabilität oder Chaos ?

Eine alternative Klassifikation fragt, ob die Regeln zu *Chaos* führen. Genauer, wir stellen fest, ob eine anfangs lokalisierte Störung sich über das Gitter ausbreitet, räumlich beschränkt bleibt, oder nach einiger Zeit wieder verschwindet. Um dieses aus der Biologie stammende "damage spreading" zu untersuchen, simulieren wir gleichzeitig zwei Systeme mit den gleichen Regeln und der gleichen Anfangskonfiguration. Nur in der Mitte des Quadratgitters wird anfangs in einem der beiden Systeme ein einzelner Spin, oder eine ganze Gitterreihe von Spins, gegenüber dem anderen System umgedreht. Während der Simulation markieren wir als Schaden ("Hamming-Abstand") die Menge der Gitterplätze, auf denen sich bei einem Platz-für-Platz Vergleich der beiden Systeme die entsprechenden Spins unterscheiden. Wenn dieser Schaden im Lauf der Zeit wieder verschwindet oder wenigstens räumlich beschränkt bleibt, so sprechen wir von stabilem bzw. marginalen Verhalten. Erreicht der Schaden dagegen den Rand, so nennen wir das System instabil oder chaotisch.
Ähnlich wie der Flügelschlag eines Schmetterlings im brasilianischen Urwald später das chaotische Wetter in Europa beeinflussen kann, so kann bei chaotischen Zellularautomaten das Flippen eines einzigen Spins später die ganze Konfiguration beeinflussen. Falls die Spins mit Prozessoren eines massiv- parallelen Computers identifiziert werden, so wird also ein einziger Rechenfehler in einem chaotischen Netzwerk die gesamten Endergebnisse beeinflussen, während er in einem stabilen Netzwerk wieder verschwindet.
Die Simulationen[2,5] in der eindimensionalen Kette, dem Bienenwabengitter und dem Quadratgitter zeigen, daß etwa sechzig Prozent der Regeln zu chaotischem Verhalten führen; im Dreiecks- oder kubischen Gitter sind es mehr als 98 Prozent. Je mehr Nachbarn man hat, um so chaotischer ist das Leben.

2.3 Phasenübergänge

Bisher hatten wir anfangs stets die Hälfte der Spins nach oben und die andere nach unten orientiert; jetzt variieren wir diesen Prozentsatz p der anfänglich nach oben stehenden Spins. Von den zahlreichen dabei gefundenen Phasenübergängen zwischen verschiedenen Wolfram-Klassen beschränke ich mich hier auf die bei "Bootstrap-Perkolation"[6]. Hier bleibt ein Platz frei, wenn zum vorherigen Zeitpunkt mindestens m seiner z nächsten Nachbarn frei waren; andernfalls wird er dauernd besetzt. Im Quadratgitter mit z=4 ist die Lösung trivial bei m=1 (isolierte Löcher werden gefüllt, sonst passiert nichts) und m=4 (ganzes Gitter wird schließlich gefüllt, außer bei p=0), m=2 ist einem normalen Perkolationsproblem[7] ähnlich, und m=3 ist besonders wichtig: Für sehr kleine p bleibt ein stabiles Gemisch besetzter Rechtecke (getrennt durch unbesetzte Flächen) übrig, während für mittlere und große p sich schließlich das ganze Gitter füllt. Dieser Phasenübergang zwischen Unordnung und vollständig gefülltem Gitter bei m=3, z=4 passiert bei einem Schwellwert für die Anfangskonzentration p. Mathematische Theoreme zeigen nun, daß in einem Quadratgitter der Kantenlänge L der Schwellwert wie 1/log(L) verschwindet. Bei anderen Gittertypen treten auch andere Potenzen von log(L), oder gar eine Proportionalität zu 1/log(log L) auf; wenn bei m=3, z=4 besetzte Plätze auch wieder frei werden, falls gerade 3 oder 4 Nachbarn frei sind, dann soll der Schwellwert wie die Quadratwurzel von 1/log(L) verschwinden. Simulationen mit hunderten von Millionen Gitterplätzen bestätigen in einigen Fällen diese Theorien gut, in anderen Beispielen aber gelten numerisch eher andere Gesetze, oder der Vergleich mit der Theorie ist unklar.
Interessant ist dies nun deshalb, weil Physiker gerne behaupten, am Verhalten unendlich großer Systeme interessiert zu sein. In Wirklichkeit aber betrachten sie 10^{24} Teilchen, also $L = 10^8$ in drei Dimensionen. Wenn die Effekte endlicher Systemgröse mit einer vernünftigen Potenz von L variieren, ist $L = 10^8$ auch hinreichend nahe an Unendlich. Wenn aber ein 1/ln(L)–Gesetz gilt, wie im obigen Beispiel, dann ist ein Schwellwert 1/ln(L) nicht nahe an Null, sondern immer noch bei 5,4 Prozent. Für die Praxis könnte dann eine Computer-Simulation mit $L = 1000$ näher an der physikalischen Anwendung liegen als das mathematische Theorem für unendlich große Gitter, ganz im Gegensatz zur üblichen Physik-Propaganda. Biologisch relevant ist eventuell $L = 100$, wenn die Gitterplätze mit Rezeptor-Molekülen auf der Oberfläche der weißen Blutkörperchen unseres Immunsystems identifiziert werden (M.Kaufman, priv.Mitt.) Wichtig ist dann nicht der Grenzwert, sondern die Art, wie er erreicht wird.

3. Anwendungen: Fluß durch poröse Medien

Von den zahlreichen Anwendungen der Zellularautomaten ist die zweidimensionale Hydrodynamik[8] wohl die wichtigste[4]. Auf jedem Gitterplatz sitzen jetzt sechs Spins, und die Struktur ist ein Dreiecksgitter mit je sechs Nachbarn pro Gitterplatz. In grober Vereinfachung einer echten Molekulardynamik (die mit Kraft = Masse mal Beschleunigung arbeiten würde) fliegen die Moleküle, alle mit der gleichen Geschwindigkeit, von einem Gitterplatz zu einem seiner Nachbarn. In jede der sechs Richtungen, die von einem Gitterplatz ausgehen, kann entweder genau ein Teilchen fliegen, oder gar keins; daher genügen sechs Bits, um den Zustand eines Gitterplatzes zu charakterisieren. Wenn wir von Zeit t–1 zu Zeit t übergehen, so bewegen sich alle Teilchen von einem Platz zu dem entsprechenden Nachbarplatz. Dort stoßen sie dann eventuell auf andere Teilchen und gehorchen den üblichen Stoßgesetzen mit Erhaltung von Energie und Impuls.
Da nur mit Bits gearbeitet wird, sind Simulationen sehr schnell: Kohring[9] braucht nur 2 Nanosekunden pro Platz und Zeitschritt auf einem NEC-SX3 Prozessor. Es treten auch keine numerischen Instabilitäten durch Rundungsfehler auf, und komplizierte Geometrien lassen sich bequem einbauen. Allerdings sind die "finite-size" Effekte besonders groß und daher störend: Bei Kanälen mit Radius R sind die Korrekturen von der Form

$$1 + 7\lambda/R + ...$$

wobei λ die mittlere freie Weglänge ist[9] und zum Beispiel sieben Gitterkonstanten betragen kann. Alle numerischen Simulationen sollten also die systematischen Fehler durch endliches R berücksichtigen und sorgfältig auf den experimentellen Wert $\lambda/R \to 0$ extrapolieren. In der Literatur geschieht das bisher fast nie, natürlich auch, weil wohl erst seit einem Jahr 10^8 Gitterplätze bearbeitet wurden.

Von praktischer Bedeutung ist der Fluß von Öl und Wasser durch Sand oder Sandstein. Das poröse Medium kann in zwei Dimensionen durch zufällig plazierte Kreise (oder Rauten) simuliert werden, die das feste Material symbolisieren und zwischen denen die Flüssigkeit strömt. Wenn die Porosität p der Volumanteil der Flüssigkeit ist, so ist für nicht zu kleine p in den Simulationen[9] die mittlere Fließgeschwindigkeit zu

$$exp(6.5p)/(1-p)$$

proportional, eine handfeste Vorhersage der Zellularautomaten.

4. Computerprobleme

Die Rechengeschwindigkeit bei der Simulation solcher Zellularautomaten ist 0,0 Megaflops, denn Gleitkomma-Operationen kommen überhaupt nicht vor. Sinnvoller sind Mups (Mega Updates per Second), die Zahl der pro Mikrosekunde abgearbeiteten Gitterplätze. 1000 Mups sind durchaus normal für einen einzelnen Prozessor einer Cray oder NEC-SX3. Seit Jahren faßt man 32 oder 64 nicht benachbarte Spins in einem Computer-Wort zusammen und benutzt logische bit-per-bit Befehle, die zum Beispiel 64 ODER-Operationen auf einmal bearbeiten[2]. Moderne Computer und ihre Compiler sollten dies können und auch schnell die Zahl der in einem Wort gesetzten Bits bestimmen. Das parallele Abarbeiten der Spins gestattet leichte Vektorisierung, so daß ein Transputer etwa zwei Zehnerpotenzen[10] langsamer ist als ein Cray-Prozessor. Die Connection - Machine mit ihren 65536 Einzelbit-Prozessoren ist leider in der Praxis nicht so viel schneller als eine NEC-SX3.

Spezialrechner für Zellularautomaten wurden (und werden) in Paris und in Cambridge (MA) gebaut, sind aber bei Fertigstellung nicht unbedingt schneller als für Gleitkomma-Operationen gedachte Superrechner. Vielleicht hätte ein auf ganze Zahlen und Bits spezialisierter Rechner[11], der ansonsten allgemein und bequem programmierbar ist, eine bessere Chance.

Danksagung: Viele der vorgestellten Resultate stammen von R.W.Gerling und G.A.Kohring.

1. S.Wolfram, Rev.Mod.Phys. 55, 601 (1983) und: *Theory and Applications of Cellular Automata*, World Scientific, Singapore 1986
2. D.Stauffer J.Phys.A 24, 909 (1991); R.W.Gerling and D.Stauffer, Int.J.Mod.Phys. C 2, 799 (1991); F.Bagnoli, Int.J.Mod.Phys. C (1992); see also P.M.C. de Oliveira, *Computing Boolean Statistical Models*, World Scientific, Singapore 1991
3. R.W.Gerling, Mathem.Naturwiss.Unterr. 43, 451 (1990), D.Stauffer, Computers in Physics 5, 62 (Jan/Feb. 1991)
4. Physica D 45, 1-483 (1990) und 47, 1-339 (1991)
5. R.W.Gerling, Physica A 162, 186 und 196 (1990), 167, 611 (1990)
6. J.Adler, Physica A 171, 453 (1991)
7. D.Stauffer und A.Aharony, *Introduction to Percolation Theory*[2], Taylor and Francis, London 1992
8. U.Frisch, B.Hasslacher, und Y.Pomeau, Phys.Rev.Lett. 56, 1505 (1986)
9. G.A.Kohring, Int.J.Mod.Phys.C 2, 755 (1991), J.Stat.Phys. 66 (1992) und preprint
10. S.S.Manna, D.Stauffer und D.W.Heermann, Physica A 162, 20 (1989)
11. M.Neschen, Proc.of HICSS-25, ACM, Jan.1992, Hawaii
12. G.A.Kohring, Physica A (1992)

Kann die Analyse komplexer Systeme auch als eine Aufgabe der Informatik angesehen werden?

Josef Honerkamp
Fakultät für Physik
und
Freiburger Materialforschungszentrum (F·M·F)
Albert–Ludwigs–Universität Freiburg

Inhaltsverzeichnis

1 Informatik und die Analyse komplexer Syteme

In der Physik untersuchte man bisher vorwiegend Systeme, die als isoliert betrachtet werden können oder für die wenigstens der Einfluß von außen wohl definiert ist. Es ist dann häufig möglich, experimentell zugängliche Größen aus einer Theorie heraus zu berechnen; der Kontakt mit dem experimentellen Resultat besteht im wesentlichen in einem Vergleich oder in einer Anpassung von wenigen Parametern.
Komplexe Systeme andererseits sind gekennzeichnet durch viele Freiheitsgrade und dadurch, daß der Einfluß von außen nur pauschal als ein (statistisches) "Rauschen" formulierbar ist. Hier ist es oft nur noch möglich, aufgrund von Daten die Systeme zu charakterisieren und allgemeinere Modellansätze zu vervollständigen. Ziel ist neben der Charakterisierung oft auch eine Klassifizierung sowie eine Optimierung und Steuerung bei unvollständiger Kenntnis der Systeme oder Prozesse.
Die Schwierigkeiten der mathematischen Modellierung auf der Basis fundamentaler Gesetzmäßigkeiten führt bei komplexen Systemen zu einer größeren Bedeutung des inversen Problems, nämlich ausgehend von den Meßdaten Information über das System zu gewinnen. Dadurch erhält die mathematische Statistik und Stochastik eine Schlüsselrolle, denn

- die Daten sind stets durch Meßfehler verrauscht und man muß somit bei den Schlußfolgerungen aus den Daten auch die Fortpflanzung der Fehler (Vertrauensintervalle) studieren, um den Grad der Verläßlichkeit der Folgerungen zu kennen. Dabei wird der fundamentale Begriff der Zufallsvariablen benötigt.

- Daten, die zeitabhängigen Prozessen entsprechen, sind oft sehr irregulär und wenig reproduzierbar. Sie sollten dann eher als Realisierung eines stochastischen Prozesses interpretiert als durch ein deterministisches Modell beschrieben werden.

Die mathematische Statistik und Stochastik stellt die angemessenen Begriffe und Methoden bereit; in ihrer Sprache kann die Information, die man durch Messungen an den Systemen erhält, erst quantitativ formuliert werden.
Häufig auftretende Probleme, die nur in diesem Sinne befriedigend gelöst werden können, sollen kurz in den folgenden drei Abschnitten vorgestellt werden.

2 Schlecht gestellte inverse Probleme

Häufig interessiert man sich für Verteilungen einer physikalischen Größe. In einer Suspension z.B. sind viele Teilchen mit unterschiedlicher Größe vorhanden und man möchte für jede gegebene Grösse die Anzahl der Teilchen wissen. Die Verteilungsfunktion ist hier also eine Häufigkeit als Funktion der Größe. Fast immer sind aber die Verteilungen nicht direkt meßbar. Man erhält über diese Information, indem man die Materialien mit Lichtstrahlen, Röntgenstrahlen oder Neutronen beschießt und die gestreute Intensität mißt. Aus diesen Streudaten ist dann auf die gesuchte Verteilung zurückzuschließen.
Durch theoretische Überlegungen weiß man häufig, auf welche Streudaten eine gegebene Verteilung führt. Nennen wir diese z.B. $h(\tau)$, wobei τ für die Teilchengrösse steht, und bezeichnen wir die experimentell meßbare Größe mit $G(\omega)$, wobei ω den Streuwinkel charakterisiert, so gilt oft eine lineare Beziehung der Art

$$G(\omega) = \int d\tau K(\omega, \tau) h(\tau)$$

Die Kernfunktion $K(\omega, \tau)$ ist dabei durch die Theorie und durch die Meßanordnung bestimmt.
Es ist einfach, bei gegebener Verteilungsfunktion $h(\tau)$ die Streudaten $G(\omega)$ vorherzusagen, um sie mit den experimentellen Daten zu vergleichen. Das inverse Problem, aus Meßdaten für $G(\omega)$ die Verteilungsfunktion $h(\tau)$ verläßlich zu schätzen, ist allerdings viel schwieriger. Die Meßdaten werden notwendigerweise bei endlich vielen diskreten Werten $\{\omega_i, i = 1, \ldots, N\}$ gewonnen und sind immer mit einem experimentellen Fehler behaftet. Die Information aus dem Experiment besteht also aus den Daten $\{g_i^\sigma, i = 1, \ldots, N\}$, wobei sich g_i^σ darstellt als

$$g_i^\sigma = G(\omega_i) + \sigma_i \eta_i.$$

Dabei ist σ_i ein Maß für den experimentellen Fehler und η_i eine Realisierung einer normalverteilten Zufallsvariablen mit Mittelwert 0 und Varianz 1. Meßdaten sind also Realisierungen von Zufallsvariablen, jede Messung ist mit einem jeweils anderen Fehler behaftet. Für die Verteilung dieser Fehler nehmen wir an, daß sie einer Gaußverteilung entspricht.
Aus diesen endlich vielen, durch Fehler verrauschten Daten hat man also nun die Funktion $h(\tau)$ zu schätzen. Die Vermutung ist naheliegend, daß man ohne weitere Information keine eindeutige Antwort erhalten kann. Die Information ist zu gering, um eine kontinuierliche Funktion in wohl definierter Approximation zu bestimmen.
Solche Probleme treten aber sehr häufig auf, eigentlich bei jeder Bestimmung einer Verteilungsfunktion, denn nie kann man diese direkt messen. Wir kennen das aus der Rheologie, wo man z.B. polymere Schmelzen einer oszillatorischen Scherung unterwirft und die sich daraus ergebenden Spannungen mißt. Aus diesen Meßdaten möchte man auf die Verteilung der Relaxationszeiten schließen, die besonders gut die Eigenschaften des Materials charakterisiert. In Lichtstreuexperimenten möchte man aus den Streudaten u.a. auf die Molekularmassenverteilung schließen. In der zweidimensionalen NMR-Spektroskopie sind die Meßdaten Signale von abklingenden magnetischen Feldern und aus diesen Signalen möchte man auf die Verteilung von Sprungwinkeln schließen, so daß man die möglichen

Umordnungen der Molekülstruktur kennenlernt. In der Tomographie geht es um die indirekte Messung von Gewebedichten. Chromatographie und Bildrekonstruktion sind weitere Gebiete, in denen solche inversen Pobleme identifiziert sind, und es gibt sicher viele weitere angewandte Gebiete, in denen Probleme existieren, die man auf diese Formulierung zurückführen kann.
Verläßliche Methoden für die Behandlung solcher Probleme sind also äußerst wertvoll und vielseitig anwendbar. Es gibt solche Methoden auch schon relativ lange. Die Maximum-Entropie-Methode ist schon 1984 von Gull und Skilling[1] formuliert worden und geht zurück auf das Maximum-Entropie-Prinzip von Jaynes (1957)[2]. Die Regularisierungsmethode ist von dem russischen Mathematiker Thikonov 1963[3, 4] entwickelt worden.
Beide Methoden führen noch weitere Information über die gesuchte Funktion $h(\tau)$ ein, indem sie zusätzliche Bedingungen an diese stellen. In der Regularisierungsmethode werden bestimmte Glattheitseigenschaften verlangt, in der Maximum-Entropie-Methode wird, wie der Name andeutet, die Funktion $h(\tau)$ gesucht, die maximale Entropie besitzt und noch mit den Daten verträglich ist.
Wenn auch die Idee dieser Methoden schon relativ alt ist, so sieht man sich bei der Realisierung der Idee in den verschiedenen Gebieten immer wieder vor neue Probleme gestellt. Die Kernfunktion kann z.B. noch unbekannte Parameter enthalten, die auch durch die Meßdaten geschätzt werden müssen. Das ist in der zweidimensionalen NMR-Spektroskopie der Fall. Die Relation zwischen Verteilungsfunktion und Meßgröße kann nichtlinear sein, wie in der Lichtstreuung. Oder: die Relation kann zwar noch linear sein, aber die Kernfunktion unbekannt. Das ist in der Bildverarbeitung oft der Fall. Dann muß man mit bekannten Testverteilungen und mit Hilfe der gleichen Methoden erst einmal die Kernfunktion schätzen. Schließlich kann es durch die Menge der Daten, die ja auch zwei- und dreidimensionale Datenfelder sein können, und durch die Menge der Stützpunkte für die Verteilungsfunktion zu numerischen Problemen kommen und besonders schnelle Algorithmen können gefragt sein.
Die Behandlung aller dieser Probleme hat sich als äußerst fruchtbar und anregend erwiesen. Bei der Einführung in die Rheologie polymerer Schmelzen sind besonders robuste Versionen der Regularisierungs– und Maximum–Entropie– Verfahren entwickelt worden[5, 6, 7, 8, 9, 10]. Die Behandlung inverser Probleme in der zweidimensionalen NMR–Spektroskopie und in der Lichtstreuung mit solchen Methoden ist gerade formuliert worden[11, 12]. Weitere Gebiete, in die diese Methoden eingeführt werden sollen, sind die Chromatographie und die Pharmakokinetik.
Überall ist nicht nur eine Schätzung der entsprechenden Verteilungsfunktion zu bestimmen, unerläßlich ist dabei auch eine Quantifizierung der Ungenauigkeit in der Schätzung aufgrund des experimentellen Fehlers. Jede Methode, die die Fortpflanzung dieser Fehler nicht kontrolliert, ist suspekt.

3 Simulation stochastischer Prozesse

Im vorigen Abschnitt hatten wir zeitunabhängige Meßdaten betrachtet. Wir haben diese jeweils als eine Realisierung einer Zufallsvariablen auffassen müssen, um der Tatsache gerecht zu werden, daß jede Messung durch experimentelle Fehler verfälscht wird. Zufallsvariablen können aber auch zeitabhängig sein, und der Zufallscharakter muß nicht immer nur Folge einer experimentellen Unsicherheit sein. Oft auch fehlt die vollständige Information über die zeitliche Entwicklung eines Systems. Der Zustand, den ein System in einem nächsten Zeitschritt annimmt, ist oft nicht eindeutig bestimmt: es gibt mehrere Möglichkeiten und man kann die Dynamik nur dadurch charakterisieren, daß man die Wahrscheinlichkeiten angibt, mit denen die einzelnen möglichen Zustände angenommen werden.
Wir wollen das am Beispiel einer chemischen Reaktion verdeutlichen. Nennen wir die beteiligten Moleküle $X_j, j = 1, \ldots, J$, so ist der chemisch relevante Zustand des Systems von Molekülen durch

die Menge $(N_1, \ldots, N_J)$ der Teilchenzahlen N_j beschrieben. (Nicht immer ist die Spezifikation eines Zustands so einfach wie hier bei der chemischen Reaktion.) Durch eine chemische Reaktion der Form

$$\sum_j s_j X_j \xrightarrow{k} \sum_j r_j X_j \quad ,$$

kann sich der Zustand ändern, die Anzahl N_j der Moleküle vom Typ j geht über in $N'_j = N_j + r_j - s_j$. Aus der Theorie der chemischen Reaktionen kennt man auch die Übergangswahrscheinlichkeit pro Zeiteinheit, diese ist

$$W_{N'N} = \frac{\Omega k}{\prod_j s_j!} \prod_j \frac{N_j(N_j-1)\ldots(N_j-s_j+1)}{\Omega^{s_j}}.$$

Selbst wenn wir nur diese eine Reaktion berücksichtigen, gibt es also zwei mögliche Zustände, in denen das System nach dem Zeitschritt dt vorliegen kann: Der neue Zustand $(N'_1, \ldots, N'_J)$ mit der Wahrscheinlichkeit $W_{N'N}dt$ sowie der alte Zustand $(N_1, \ldots, N_J)$ mit der Wahrscheinlichkeit $1 - W_{N'N}dt$.

In der Regel liegen natürlich mehrere chemische Reaktionen vor; man hat dann die stöchiometrischen Koeffizienten $(s_j^\alpha, r_j^\alpha), j = 1, \ldots, J$, sowie die Übergangswahrscheinlichkeiten $W^\alpha_{N'N}$ für $\alpha = 1, \ldots, \alpha_R$ zu unterscheiden. Entsprechend größer ist die Anzahl der Zustände, in denen das System zur Zeit $t + dt$ vorliegen kann, wenn es zur Zeit t in dem Zustand $(N_1, \ldots, N_J)$ war, und für jeden Übergang läßt sich die Wahrscheinlichkeit ausrechnen.

Solch ein stochastischer Prozeß für einen Zustandsvektor $(N_1, \ldots, N_J)$ wird im allgemeinen durch eine Mastergleichung beschrieben. Diese ist eine Differentialgleichung für die Größe $p(N_1, \ldots, N_J; t)$, die die Wahrscheinlichkeit dafür angibt, daß zum Zeitpunkt t der Zustand $(N_1, \ldots, N_J)$ vorliegt. Wir haben aber den Prozeß schon so definiert, daß es näher liegt, nicht die zeitliche Entwicklung dieser Wahrscheinlichkeit zu betrachten, sondern direkt die von $(N_1, \ldots, N_J)$. Das bedeutet, bei einer numerischen Behandlung dieses Prozesses wollen wir nicht die Differentialgleichung für die Wahrscheinlichkeiten lösen, sondern Realisierungen des stochastischen Prozesses $(N_1(t), \ldots, N_J(t))$ erzeugen. Das ist direkt, numerisch sehr einfach und führt auch viel schneller auf die Größen, die eigentlich interessieren.

Die wichtigste Größe, die auch häufig nur betrachtet wird, ist offensichtlich die Konzentration c der einzelnen Molekülsorten. Diese ist üblicherweise als die Anzahl der Teilchen pro Volumen definiert, genauer ist sie durch

$$c_j(t) = \frac{< N_j(t) >}{\Omega}$$

bestimmt, wobei Ω das Reaktionsvolumen sei. Den Erwartungswert der Teilchenanzahl $N_j(t)$ erhält man einfach, indem man über mehrere Realisierungen mittelt.

Hier nun ist eine wichtige Erklärung am Platze. Die übliche Weise, chemische Reaktionen quantitativ auch in ihrem zeitlichen Verlauf zu beschreiben, ist die der Reaktionskinetik: man stellt nach gewissen Regeln Gleichungen für die Konzentrationen $c_j(t)$ auf. Man erhält so ein System von J Differentialgleichungen für $c_j(t), j = 1, \ldots, J$. Im stochastischen Ansatz andererseits simuliert man den Prozeß $(N_1(t), \ldots, N_J(t))$, d.h. man erzeugt mehrere Realisierungen des stochastischen Prozesses und erhält daraus durch Mittelung $< N_j(t) >$. Es zeigt sich, daß die Simulation des stochastischen Prozesses nicht nur elementarer ist, sondern auch allgemeiner, flexibler und leichter durchführbar, denn:

- Die Gleichungen der Reaktionskinetik für die Konzentrationen $c_j(t)$ sind wiederum ableitbar aus der Mastergleichung für die Wahrscheinlichkeit $p(N_1, \ldots, N_J; t)$. Man hat dabei allerdings einige Näherungen zu machen, die bei großen Teilchenzahlen unproblematisch sind. Man erkennt dabei aber: die Mastergleichung ist die Grundgleichung für die chemische Reaktion, die Gleichungen für $c_j(t)$ sind abgeleitete Gleichungen.

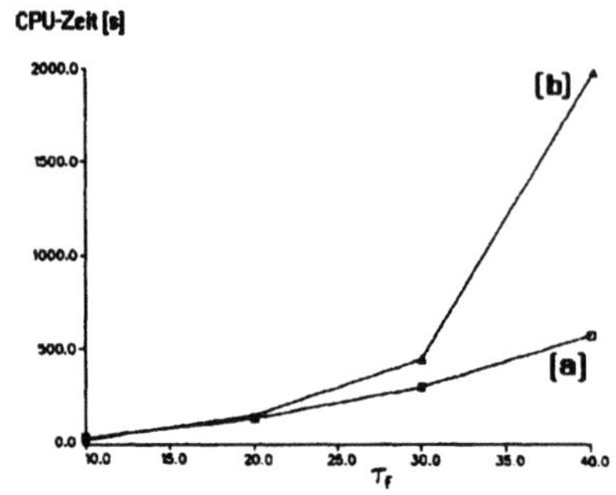

Abbildung 1: Vergleich der CPU–Zeiten, dargestellt in Abhängigkeit von der Zeit τ_f, für die man die Reaktion simuliert: (a) für die Erzeugung von 10 Realisierungen des stochastischen Prozesses; (b) für die Berechnung mittels der Differentialgleichung

- Bei großen Systemen verlangt die Lösung der entsprechenden Differentialgleichungen für $c_j(t)$ große mathematische Vorarbeit wie die Berechnung eines speziellen Systems von Basisfunktionen. Die Programme sind daher sehr aufwendig und manche Modifikationen in den Reaktionsraten sind nur mit großer Mühe oder gar nicht zu berücksichtigen. Die Simulation des stochastischen Prozesses ist immer einfach und es macht keinen Unterschied, in welcher Form die Reaktionsraten auftreten.

- Dadurch, daß man die chemischen Reaktionen auf einer elementareren Ebene betrachtet, erhält man auch mehr Information. Man kann nicht nur die Konzentrationen $c_j(t)$ betrachten, sondern auch noch die Fluktuationen $\mathrm{Var}(N_j(t))$. Die Konzentration entspricht ja nur der mittleren Teilchenanzahl, die wahre Teilchenanzahl kann um diese mittlere fluktuieren und die Stärke dieser Fluktuationen ist durch die Varianz bestimmt. Das ist interessant, wenn die Teilchenzahlen klein sind.

- Die Anzahl der Realisierungen bei der Simulation bestimmt den Schätzfehler für die Konzentrationen $c_j(t)$. Aus dem Experiment kennt man diese auch nur mit einer bestimmten Ungenauigkeit. Man kann die Anzahl der Realisierung gerade nur so groß wählen, daß die Schätzfehler von der Größenordnung der experimentellen Fehler sind. Dadurch verkürzt sich oft die Rechenzeit für die Simulation erheblich.

All dieses ist lange bekannt und auch schon in Lehrbüchern zu finden[13, 14]. Praktiziert wird es aber nur sehr selten. Neu ist allerdings, wie leistungsfähig diese stochastische Simulation auch für sehr große Systeme von chemischen Reaktionen ist, z.B. für Polymerisationsreaktionen . Das ist u.a. für die Koagulations-Reaktion

$$X_s + X_r \xrightarrow{k_{r,s}} X_{s+r}\ , \quad r, s = 1, 2, 3, \ldots$$

gezeigt worden[15, 16]. Dabei ist X_s ein ein Molekül der Länge s und $k_{s,r}$ die Reaktionsrate, die im allgemeinen von r und s abhängt. Die entsprechende Differentialgleichung für die Konzentrationen $c_j(t)$ ist analytisch lösbar nur in Ausnahmefällen, z.B. im Fall $k_{r,s} \equiv k = const.$ Der allgemeine Fall ist hier numerisch sehr aufwendig. Mit der Simulationsmethode sind beide Fälle nicht nur gleich einfach, der Vergleich der Rechenzeiten fällt mit zunehmender Dauer, für die man die Reaktion betrachtet, immer günstiger für die Simulationsmethode aus (siehe Abb.1).

4 Klassifikation stochastischer Prozesse

Die Simulation eines stochastischen Prozesses ist möglich, wenn dieser mathematisch wohl definiert ist, und zwar mit allen seinen möglichen Zuständen und mit allen Übergangswahrscheinlichkeiten. Die Realisierungen des Prozesses sind dann Trajektorien im Raum der Zustände. Der Beobachter oder Experimentator begegnet einem stochastischen Prozeß aber in einer ganz anderer Weise. Er notiert z.B. in festen Zeitabständen eine Meßgröße und erhält auf diese Weise eine Zeitreihe. Solche Zeitreihen sind fast nie vollständig reproduzierbar, sie sind oft irregulär und man muß meistens davon ausgehen, daß man mit der Zeitreihe eine Realisierung eines stochastischen Prozesses beobachtet.
Fast in jedem Bereich der Naturwissenschaft werden Zeitreihen aufgenommen. Am bekanntesten sind sie in der Meteorologie und Klimatologie, als Monatsmittel der Temperatur oder als Luftdruckdifferenzen zwischen prominenten Meßstationen. Heutzutage kommen die stündlichen Messungen der Schadstoffkonzentrationen und des Ozons hinzu. In der Medizin werden Zeitreihen in fast jedem Teilgebiet in großen Mengen produziert, sei es in der Neurophysiologie, in der Schlafforschung oder in der Chronobiologie. Was kann man aus diesen Zeitreihen herauslesen, welche Information ist in ihnen enthalten und mit welchen Methoden ist diese Information zu entdecken?
Zeitreihenanalyse, die über eine einfache Beschreibung der Zeitreihe durch Schätzung der Varianz und Bestimmung der Fourier-Transformierten hinausgehen will, muß natürlich einen Rahmen für mathematische Modelle benutzen. Eine Zeitreihe kann stationär oder instationär sein, stochastisch oder doch deterministisch (vielleicht dann sogar chaotisch), linear oder nichtlinear sein. Das sind natürlich alles mathematische Begriffe, und wenn man sagt, eine Zeitreihe sei stationär, stochastisch und linear, so meint man, daß sie als Realisierung eines solchen Prozesses angesehen werden darf. Um solches zu entscheiden, und zwar bei gegebener Wahrscheinlichkeit für einen Irrtum, sind die Methoden der Mathematischen Statistik, insbesondere der Testtheorie heranzuziehen. Diese Charakterisierung ist außerordentlich wichtig und ist Vorbedingung für eine mathematische Modellierung, denn die Struktur des mathematischen Modells, das der Zeitreihe angepaßt werden soll, sieht in jedem Fall anders aus.
Eine solche Charakterisierng geht weit über das hinaus, was heutzutage mit realistischen Zeitreihen in der Praxis angestellt wird. Sie ist aber auch dann oft schon das Ende der Möglichkeiten, wenn man ausschließen muß, daß die Zeitreihe durch ein *lineares* Modell beschrieben werden kann. Die Identifikation *linearer* Modelle aus entsprechenden Zeitreihen ist allerdings ein mathematisch gut untersuchtes Gebiet, aber leider erweisen sich die interessanten realistischen Zeitreihen gerade als nichtlinear.
Wir wollen eine solche Charakterisierung an zwei Beispielen vorführen:

- Tremor-Zeitreihen bestehen aus Messungen der Beschleunigung einer ausgestreckten Hand. Bei Normalpersonen findet man nur ein leichtes Zittern, bei Patienten mit der Parkinson-Krankheit ein starkes Schütteln (Abb.2). Es gibt aber viele Zwischenformen, die mittlere Amplitude allein ist kein verläßliches Maß für die Einteilung in Klassen "gesund", "Parkinson", usw. Im klinischen Alltag werden einfache Analyseergebnisse aus solchen Zeitreihen und die klinische Erfahrung dazu benutzt, eine bestimmte Diagnose bei den untersuchten Patienten zu stellen. Man kann so von einer klinischen Klassifikation sprechen, von einer bestimmten Art, Information über den Patienten zu gewinnen und diese auszuwerten. Die oben beschriebene Charakterisierung durch statistische Tests auf Stationarität, Stochastizität und Linearität stellt eine andere Klassifikation dar, die sich weniger auf die Intuition des Klinikers stützt, aber objektiv und auch automatisierbar ist. Sie steht natürlich aber in Gefahr, daß wesentliche Gesichtspunkte, die der Kliniker mit seiner Erfahrung intuitiv erfaßt, nicht berücksichtigt werden. Es wird aber mehr Information aus den gegebenen Messungen herausgeholt, insofern werden die Meßdaten

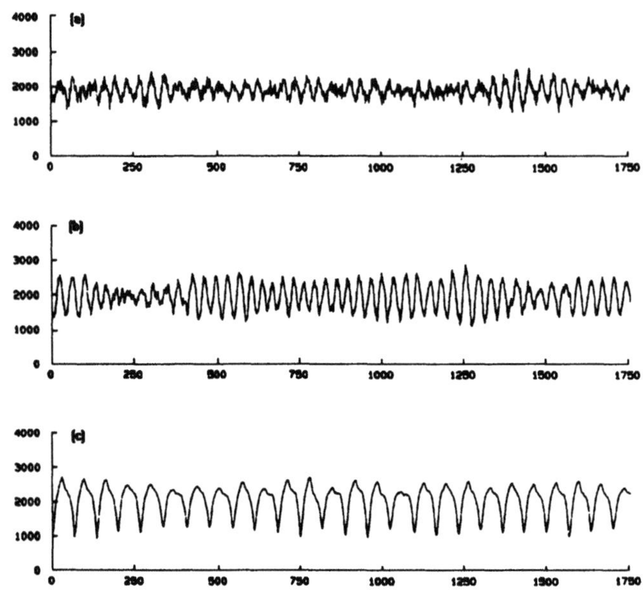

Abbildung 2: Zeitreihen der Beschleunigung der ausgestreckten Hand bei (a) einer Normalperson, (b) einem Patienten mit "essentiellem Tremor", (c) einem Patienten mit Parkinson-Tremor. Die Zeitreihen sind mit einem Skalenfaktor versehen, so daß das Augenmerk mehr auf die unterschiedliche Dynamik hingewiesen wird.

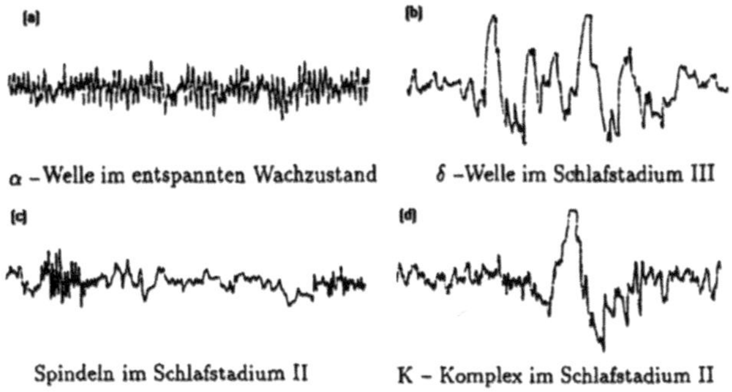

Abbildung 3: Häufig vorkommende Abschnitte in EEG–Zeitreihen, die als spezielle Segmente identifizierbar sind: (a) α–Wellen, (b) δ–Wellen, (c) Spindeln, (d) K–Komplexe

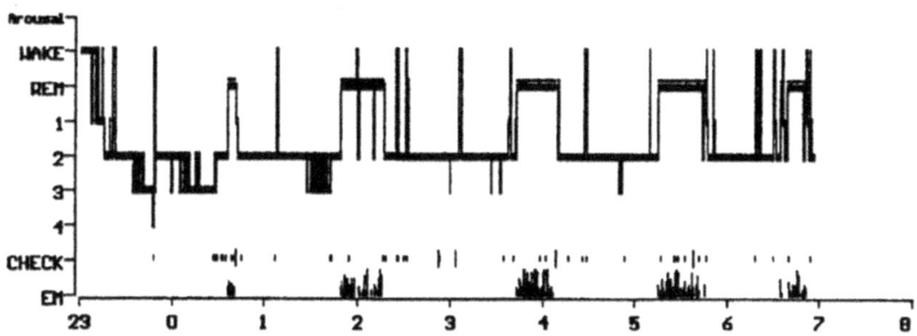

Abbildung 4: Schlafstadien in Abhängigkeit von der Zeit, Beispiel bei einem Probanden

besser verwertet. Zu prüfen ist, ob in den Meßdaten überhaupt genügend Information für eine angemessene Diagnose bzw. Klassifikation enthalten ist.

- In der Schlafforschung versucht man, verschiedene Schlafstadien zu identifizieren. In den entsprechenden Labors werden bei Probanden die Encephalogramm (EEG)–Signale, das Elektrokardiogramm (EKG), das Elektromyogramm (EMG), das Elektrookulogramm (EOG, eine Maß für die Stärke der Augenbewegungen) und die Atmung während einer Nacht gemessen. Das liefert Daten im Umfang von 108 Megabyte, deren graphische Darstellung zu einem Buch von etwa 8 cm Dicke führt. Eine einzelne EEG–Spur besteht aus etwa 3 Millionen Datenpunkten.

  In EEG–Zeitreihen erkennt man in allgemeinen verschiedene typische Segmente (Abb.3a–d). Es gibt Abschnitte, in denen ein α-Rhythmus vorherrscht (Abb.3a), oder ein δ–Rhythmus (Abb.3b), oder auch Abschnitte, in denen verstärkt Spindeln (Abb.3c) oder K–Komplexe (Abb.3d) auftreten. Die Häufigkeit, mit der diese EEG–Segmente erscheinen, sind mit gewissen Schlafstadien verbunden. Die Klassifikation von Rechtschaffen und Kales[18] sieht sechs Schlafstadien vor: Wach, REM, I, II, III, IV. Für jede Klasse werden charakteristische Eigenschaften des EEG, EOG, EMG und der Atmung angegeben.

  Während des Schlafes durchwandert man so verschieden Stadien (Abb.4), und die Form, in der das geschieht, sagt etwas aus über die "Güte" des Schlafes. Die Bestimmung der Schlafstadien aufgrund der aufgezeichneten Daten geschieht "per Hand", Seite für Seite des Datenbuches wird inspiziert und ein erfahrener Auswerter kann die entsprechende Klassifikation relativ sicher durchführen.

Die bisherigen Versuche, die Klassifikation der Daten zu automatisieren, benutzen das Kurzzeitspektrogramm einer einzelnen EEG–Zeitreihe. Daraus erhält man einen ersten Eindruck von der Abfolge der Schlafstadien. Weiterhin bestimmt man die gleitende Varianz, die ein Erkennen der δ–Wellen, besonders im Schlafstadium IV, erlaubt.

Wünschenswert wäre eine Charakterisierung der Daten eines bestimmten zeitlichen Abschnittes durch einen Merkmalsvektor und eine darauf folgende automatische Klassifizierung mit den Methoden der mathematischen Statistik. Für die Charakterisierung bietet sich die Wavelet–Transformation an, die einer Fourier–Transformation auf verschiedenen Zeitskalen entspricht, jedoch schneller als das Kurzzeitspektrogramm zu erhalten ist. Die Klassifikation soll mit einem Hidden–Markov–Modell versucht werden. Dabei werden die beobachteten Zeitreihen je nach Stadium bzw. Klasse einem Modell für einen stochastischen Prozeß angepaßt, die im Hintergrund ablaufende Folge von Schlafstadien ist dabei aber aus den Daten gleichzeitig zu schätzen.

5 Zusammenfassung

In den vorhergehenden Abschnitten haben wir drei große Gebiete betrachtet, in denen stochastische Begriffe und Methoden eine grundlegende Rolle spielen. In allen drei Gebieten ist die Relevanz der Probleme und der Problemlösungen für die Anwendung unmittelbar evident. Verläßliche Schätzungen von Verteilungen aus deren indirekter Messung sind notwendig, um überhaupt solche Messungen nutzbar zu machen für die Charakterisierung von Materialien. Simulation von chemischen Reaktionen oder anderer Prozesse ist eine Vorbedingung für deren Steuerung und gezielte Beeinflussung. Schließlich müssen effiziente Methoden zur Analyse und Charakterisierung von Zeitreihen gemeinsam mit den mathematischen Stochastikern angewandt und weiterentwickelt werden, damit uns die Datenfriedhöfe nicht über den Kopf wachsen.

An der Freiburger Universität gibt es seit vier Jahren ein interdisziplinäres Seminar, an dem sich Mathematiker, Theoretische Physiker, Medizinische Biometriker, aber auch Forstwissenschaftler und Ökonomen beteiligen. Inzwischen wird geplant, ein Zentrum für Datenanalyse und Modellbildung zu gründen. Das Konzept für dieses Zentrums sieht folgendes vor:

- Die Mitglieder des Zentrums sollen in exemplarischer Weise die Nützlichkeit der Methoden der Stochastik und Statistik in Anwendungsgebieten (Neurophysiologie, Umweltforschung, Sensortechnik, usw.) demonstrieren. Besonderer Wert soll dabei auf solche Fragestellungen gelegt werden, die Mathematiker zur Weiterentwicklung der Methoden anregen.

- Mitglieder des Zentrums können in der Regel Theoretiker (Mathematiker, Statistiker, theoretische Physiker) werden, die sich mit Anwendern zu einer Projektgruppe verabreden, um Daten aus der Anwendung gemeinsam zu analysieren und zu interpretieren. Somit sind die Projektgruppen zwangsläufig interdisziplinär besetzt. Eine Ausnahme von dieser Regel ist nur in der Abteilung für mathematische Grundlagen möglich. Mathematische Probleme, die bei den Projektgruppen entstehen, können in dieser Abteilung eingehender untersucht werden.

- Das Zentrum soll ein Kompetenzzentrum für die wissenschaftliche Datenanalyse und Modellbildung werden, das von Forschungseinrichtungen und von der Industrie und Wirtschaft angesprochen werden und damit kooperieren kann.

- Das Zentrum soll im universitären Rahmen Aus- und Weiterbildung betreiben wie auch nach Möglichkeit Schulungen für einen breiteren Kreis durchführen, um das Wissen über Grundkonzepte und Basismethoden bei den Anwendern zu vermehren.

Die Probleme, die in diesem Zentrum analysiert werden sollen, sind nur interdisziplinar zu lösen. Es sind keine Probleme der traditionellen Physik, auch keine Probleme, die man sofort der Informatik zuordnen würde. Man untersucht komplexe Systeme, und zwar in der Denkweise eines Physikers, mit Methoden der Stochastik und mit dem Ziel, möglichst viel der Information, die in den Meßdaten vorhanden ist, zu entdecken. Bei der Einordnung dieses Forschungszweiges stehen wir vor einem Klassifikationsproblem, dessen Lösung uns allerdings nicht besonders drängend erscheint.

Literatur

[1] Gull, S.F., Skilling, J., *Maximum entropy method in image processing*, (1984) IEEE Proc. **131**, 646-659

[2] Jaynes, E. T., *Information Theory and Statistical Mechanics*, (1957) Phys. Rev. **106**, 620-630

[3] A.N. Tikhonov, *On the solution of ill-posed problems and the regularization method*, Dokl. Akad. Nauk SSSR, **151**, (1963) 501–504

[4] A.N. Tikhonov, V.Y. Arsenin, *Solutions of ill-posed problems*, Wiley, New York, 1977

[5] J. Honerkamp and J. Weese, *Determination of the relaxation spectrum by a regularization method*, Macromol. **22** (1989) 4372-4377.

[6] J. Honerkamp and J. Weese, *Tikhonovs regularization method for ill-posed problems*, Cont.Mech.and Therm. **2** (1990) 17-30.

[7] J. Honerkamp, *Ill-posed problems in Rheology*, Rheol. Acta **28** (1989) 363-371.

[8] C. Elster, J. Honerkamp, *A modified maximum entropy method and its application to creep data*, Macromolecules **24** (1991) 310-314.

[9] C. Elster, J. Honerkamp, J. Weese, *Using Regularization methods for the determination of relaxation and retardation spectra*, 1991, wird veröffentlicht in Rheol.Acta

[10] J. Weese, *A reliable and fast method for the solution of Fredholm integral equations of the first kind based on the Tikhonov regularization*, 1991, wird veröffentlicht in "Comp. Phys. Commun."

[11] D. Grabowski, J. Honerkamp, *The determination of the reorientational angle distribution in two-dimensional NMR–spectroscpy on powder samples*, 1991, wird veröffentlicht in J.Chem.Phys.

[12] J. Honerkamp, D. Maier, J. Weese, *Nonlinear Regularization Method for Light Scattering*, 1991, preprint Freiburg

[13] N.G. van Kampen, *Stochastic Processes in Physics and Chemistry*, 1981, North–Holland

[14] J. Honerkamp, *Stochastische Dynamische Systeme*, 1990, VCH–Verlag

[15] H.P. Breuer, J. Honerkamp, F. Petruccione, *Stochastic Simulation of Polymerization Reactions*, 1991, wird veröffentlich in "Computational Polymer Science"

[16] H.P. Breuer, J. Honerkamp, F. Petruccione, *A Stochastic Approach to Complex Chemical Reactions*, 1991, preprint Freiburg

[17] C. Gantert, J. Honerkamp, J. Timmer, *Analyzing the Dynamics of the Hand Tremor Time Series*, 1991, preprint Freiburg

[18] A. Rechtschaffen, A. Kales, *A manual of standardized terminology, techniques, and scoring system for sleep stages of human subjects*, 1968, Department of Health, Education and Welfare, Washington DC

Complex Systems Research: Interfacing Supercomputers with Experiments

A. Hübler
Center for Complex Systems Research,
Department of Physics, Beckman Institute,
University of Illinois, Urbana, IL 61801, USA

Some new paradigms in complex systems research indicate a basic need for a direct interface between supercomputers and real experiments. We present some of those those paradigms, in particular the paradigms on the dynamical key and on dynamical reconstruction and discuss their impact on future experiments.

1. Introduction

Complex systems research has made a tremendous progress during the last few years. Concepts like synergetics and slaving[1], neural nets[2], fuzzy logics[3], genetic algorithms[4], artificial life[5], deterministic chaos[6], fractal geometries[7], solitons[8], catastrophes[9], and self organized criticality[10] have been used to model typical phenomena of complex systems, i.e. high dimensional evolving systems. The progress was triggered by advances in computer technology, since digital computers are good chaos generators. This progress is expected to have a strong impact on industrial technology in the near future since the chaotic signals produced by computers are ideal tools to interact with real complex systems[11, 12] and to control them[13, 14, 15, 16]. That results from the fact that nonlinear systems react most sensitively if the period of the perturbation fits to their actual internal rhythm[17, 18, 19]. This is the principle of the dynamical key. We present applications of this principle which show that dynamical properties of some complicated experimental systems can be determined orders of magnitude more accurately if sine generators or sources for monochromatic electromagnetic waves are replaced by chaos generators. In addition, we show that it is possible to get sharp diffraction peaks from disordered materials if the incoming wave is chaotic. This makes it possible to determine the structure of disordered materials accurately. A better understanding of disordered materials could make it easier to operate industrial processes in strong nonequilibrium, since physical systems tend to be disordered in strong nonequilibrium. A large flow of heat, a large current or a large flow of any other substance penetrating a physical system usually causes disorder, for example amorphous or fractal structures or turbulent dynamics. Some of those physical systems in strong nonequilibrium have the capability to repair minor damages themselves. They are regenerative. The investigation of regenerative materials may trigger the production of self repairing electronic equipment. The principle of minimum resistance and the principle of optimal interaction explain properties of regenerative materials.

In order to calculate an optimal driving force for a certain complex systems a model of the natural dynamics of the system is needed. However in many cases only a certain fraction of the variables of an complicated experimental system can be measured. The paradigm on dynamical reconstruction states in this context: for an accurate modeling, the number of unobserved variables should not exceed the number of measured variables and the dynamics should by of the experimental systems should cover a certain region of the state space homogeneously. This means that the modeling unit has to measure many variables simulateously and control at the same time many actuators. In addition the dynamics of the experimental system should cover a region of the state space homogeneously. This means in most cases, it has to be chaotic or even random. Only super computer can handle those requirements.

Those principles and, in particular, the various control techniques for chaotic systems may be applicable in social and economic systems too. Computers are speeding up all business interactions, whereas management techniques are still based on personal experience and judgement. In order to match the speed determined by computer based business, management has to be computer assisted, too. Since economic systems are complex nonequilibrium systems[20] and time scale arguments are most important in physical systems as well as in economic systems, the principle of the dynamical key in combination with the existing knowledge in complex systems research should provide the right tool for an efficient control of turbulence as well as computer assisted management.

A dramatic change in experimental equipment and materials paired with a broad applicability of the new concepts could produce a substantial technological breakthrough. Many systems in everyday life and in industry exhibit chaos, turbulence, and catastrophes. Chaos, turbulence and catastrophes are frightening people because they are generally assumed to be uncontrollable and unpredictable. People try to avoid them or to limit the "negative" influences of chaos, turbulence and catastrophes to a minimum. The method of dynamical control enables us now to handle some of the most sophisticated, unpredictable systems, including some systems exhibiting chaos, turbulence and catastrophes and to use them for scientific and industrial purposes. Therefore it is necessary to replace the labels "negative" and "unacceptable" for features, like chaos, turbulence, catastrophes and slaving, by a careful scientific classification. The goal of this article is to present some new theorems, paradigms and conjectures in the framework of existing methods. In the context of some physical examples we present a mathematical background for these paradigms and colloquial interpretations. We hope that this facilitates the transfer to other disciplines. In addition, we emphasize possible industrial applications. The next few sections provide an overview of paradigms, theorems and conjectures on modeling and control of complex systems. The last section gives a summary and a perspective on future developments.

2. Simple Models for Complex Systems

2.1 Slaving, Aims, Order Parameters, Evolving Systems

Synergetics and the theory on dynamical systems are concerned with the long-term behavior of the solutions of time dependent systems; in particular, the aim is to determine the level of self organization or chaos that the system can generate. Since the basic methodologies of synergetics and dynamical systems theory are well known and have been applied to different problems [21], here only the main results are recalled, in order to illustrate the differences between the approach in the present paper and the usual method.

A dynamical system in continuous time is a differential system,

$$d\tilde{x}(t)/dt = f(\tilde{x}(t), \tilde{p}(t) + \tilde{F}(t)), \tag{1}$$

where t is the time variable and $\tilde{x}(0) = \tilde{x}_0$. $\tilde{p}$ is a set of control parameters. Usually, some of those control parameters are experimentally accessible, i.e. can be changed in the course of an experiment. This is indicated by $\tilde{F}(t)$. $\tilde{p}$ and $\tilde{F}$ have the same dimension, but in practice most of the components of $\tilde{F}$ may be zero. $\tilde{x}(t)$ represents the actual state of the system. The corresponding state space can be finite or infinite dimensional. In a chemical system the components of $\tilde{x}$ might represent the concentrations of several species as well as temperature or the flow of heat at a certain location in the container. In a physical system, for example, a simple pendulum made from a metal bar which is fixed on one end and oscillates under the influence of a gravitational field, the components of $\tilde{x}$ might represent the motion of the center of mass of the bar and all vibrational states of the bar. For some dissipative systems it has been illustrated that all solutions of Eq.1 converge to a compact invariant set, the global attractor. For the pendulum we mentioned before, as well as for most other macroscopic mechanical setups, this means that all vibrational modes are slaved, i.e. decay rapidly to a limiting value which is uniquely determined by the motion of the order parameters, in this case the motion of the center of mass. For example, if we start out from a 2-dimensional system with second order nonlinearities but no driving forces $\tilde{F}(t)$:

$$\begin{aligned}\dot{\tilde{x}}_1(t) &= \tilde{p}_1\tilde{x}_1(t) - \tilde{x}_1(t)\tilde{x}_2(t) && (2)\\ \dot{\tilde{x}}_2(t) &= -\tilde{p}_2\tilde{x}_2(t) + \tilde{x}_1^2(t) && (3)\end{aligned}$$

where $\tilde{p}_1$ and $\tilde{p}_2$ are two control parameters, and assume that $\tilde{x}_1(t)$ is slow compared to $\tilde{x}_2(t)$, i.e. $\tilde{p}_1 \ll \tilde{p}_2 \approx 1$ then one can solve Eq.3 by assuming that $\tilde{x}_1 \approx const.$:

$$\tilde{x}_2(t) = \frac{\tilde{x}_1^2(t)}{\tilde{p}_2} + \frac{1}{\tilde{p}_2}(\tilde{x}_2(0) - \tilde{x}_1^2(t))e^{-\tilde{p}_2 t}. \tag{4}$$

In Eq.4, the first term describes a slow motion given by $\tilde{x}_1(t)$ and the second term a transient which rapidly decays. After this rapid decay of the transient, $\tilde{x}_2$ is entirely determined by $\tilde{x}_1$:

$$\tilde{x}_2 = \frac{1}{\tilde{p}_2}\tilde{x}_1^2, \tag{5}$$

$\tilde{x}_2$ is "slaved" by $\tilde{x}_1$[1]. Since Eq.5 is invertible it is not obvious who slaves whom. However a certain asymmetry is given by the fact that a small sudden change of $\tilde{x}_1$ decays slowly or persists, whereas any change of $\tilde{x}_2$ decays fast, by definition. In a geometrical representation, the trajectory converges rapidly to a one dimensional manifold given by Eq.5, and then performs a slow motion given by Eq.5 and Eq. 2:

$$\dot{x}_1 = p_1 x_1 - \frac{1}{p_2} x_1^3. \tag{6}$$

where we neglected the ~ in order to label x_1 as an order parameter and p_1 and p_2 as significant control parameters. This procedure is called adiabatic elimination of fast variables. Synergetics provides a systematic approach to estimate the location of those attractive, low dimensional limiting sets(Eq.5), the approximate inertial manifolds (AIM) and to determine the motion of the state vector on the AIM(Eq.6). In the above example we did not include any driving forces. In section 4 we study the same system including driving forces.

In most physical experiments the significant control parameters are not exactly constant but change at a slow rate compared to the order parameters. An example is the temperature or humidity in a laboratory which might change slightly during the experiment. Usually an experimentalist tries to fix the significant control parameters, for example through a temperature control or by putting the experiment in a special container. But one can never fix all of those control parameters. If one of those control parameters is changing at a noticeable rate, we call the system evolving. In an evolving system, the location of the AIM as well as the flow on the AIM may change as a function of time. The distinction between slaved variables, order parameters, and time dependent control parameters in an evolving system is based on time scale and stability arguments: slaved variables approach rapidly a limiting value which is determined by the order parameters. The dynamics of the order parameters emerges from the structure of the experimental system and may be complicated, but has usually a much longer time scale than the slaved variables. The significant control parameters change, if at all, on an even longer time scale which is typically longer than the period of the experiment and are usually set by the experimentalist. There may be experimental systems where this classification does not fit.

Given that the dynamics of a dissipative system always collapses rapidly onto one of those inertial manifolds, several questions arise: (i) Is the AIM smooth? Eq.6 contains a third order nonlinearity whereas the original equation (Eq.2) was second order. A successive elimination of more and more slaved variables in a high dimensional systems might produce higher and higher nonlinearities and converge to a non-smooth flow vector field. (ii) How attractive is the AIM? (iii) Does the systems stay on the AIM when an external force is applied in order to control the system? (iv) What are the properties of the equation of motion of the state vector on the AIM. In particular is the motion deterministic?

These questions are very important from a practical point of view. If the AIM is not smooth at a certain location, and the state vector reaches this location, it could mean rapid change of the concentration of chemical species and, for example, an explosion. (iii) addresses the question of whether we can use simplified chemical rate equations or the motion of the center of mass of a macroscopic mechanical system for a control. And (iv) means for mechanical systems: Can we expect that the dynamics of macroscopic quantities such as, for example, the center of mass of a pendulum to be deterministic and smooth? We discuss this problem of determinism in the context of classical mechanics in the next section. In section 4 we discuss the issue of controlling complex systems based on equations on the AIM.

2.2 Deterministic Chaos and Fractal Strange Attractors

For many complex systems it seems to be true that the high dimensional system is attracted to a low dimensional AIM and visits only a small, bounded region of this manifold: it approaches regular or strange attractors. Usually the adiabatic elimination procedure gives highly nonlinear flow vector fields, but if the experimental trajectory visits only a very small bounded region of the AIM without singularities, then it is often possible to get a good approximation by an appropriate expansion with cut off, for example a Taylor series expansion where all third order and higher order terms are neglected around the center of mass of the trajectory or a singular point. In this case the model of the complex system is low dimensional and has only nonlinearities of low order(LL-system). Nonlinear Dynamics has shown that LL systems can exhibit irregular motion which is sensitive to initial conditions: deterministic chaos. Therefore, deterministic chaos is important in complex systems research. Since the flow vector field of LL systems is smooth, their trajectories have often a simple geometry, for example they are fractal. Though this fact is generally only conjectured, FEIGENBAUM[22] proved it for the logistic map dynamics. Therefore, fractal strange attractors and deterministic chaos can be found in complex systems, but may not be generic for real systems, as we illustrated in the last section. But why are the Feigenbaum scenario and fractal strange attractors found on so many physical systems? The reasons are historical. The Feigenbaum scenario is typical for a sinusoidal driving force resonantly perturbing a damped, linear oscillator with a third order nonlinearity and very small higher order nonlinearities. Spectroscopy with small sinusoidal driving forces is the main topic in many physics labs, and most investigated are systems weakly damped and weakly nonlinear because then the response to sinusoidal perturbations, i.e. the signal to noise ratio, is as large as possible. For this equipment and those systems the Feigenbaum scenario is a generic route to chaos when the amplitude of the sinusoidal perturbation is increased.

2.3 Dynamical Reconstruction

The paradigm of dynamical reconstruction is the inverse of the paradigm of deterministic chaos. In this case a complicated time series is given and possible sources for such a time series are sought. In addition, we have to deal with the problem that in most experimental situations not all order parameters are observable. The paradigm on dynamical reconstruction states that *the most accurate model for prediction and control is obtained if the dynamics of the investigated system is highly chaotic and all preknowledge about the system is used, where it is unimportant whether the chaotic dynamics is imposed via external perturbations or is natural.*

From the point of view of an experimentalist the desired features of reconstructed models are:

- Simplicity: The model should encode an observed time series or spatial structure out of N data, with only a few parameters K, i.e. $K \ll N$, where the encoding itself should not be too complicated.
- Reproducibility: Various runs of the experiment must reproduce the parameters of the model.
- Physical Meaning of the Parameters: There should be a smooth relation, between the significant control parameters of the experiment and the parameters of the model in order to associate the parameters of the model with a physical meaning. In addition, a smooth relation between model parameters and experimentally significant control parameters is useful for predicting evolving systems.
- Prediction: The model should predict the future [23, 24] up to a certain accuracy and predict the dynamics for neighboring settings of the significant control parameters [25] and initial conditions.
- Accuracy and significance of the parameters: The neglect of a parameter of small magnitude or the neglect of a parameter which is smaller than its error bar should not decrease the quality of the model [26].
- Estimation of hidden order parameters: In many experimental systems not all order parameters are experimentally observable. The model should provide an estimate for those hidden order parameters.

Since noise affects the experimental system, it is never possible to reproduce the parameters of the equation of motion exactly. A measurement is considered to be reproducible if statistical tests indicate that the deviations of the parameters between individual measurements are due to the influence of noise. We will discuss such a test.

The modeling may be considered as a special type of filter for the data. If the dynamics of an experimental system is periodic, it turns out that a fast Fourier transformation very frequently satisfies all those conditions with just a few nonzero Fourier components. These Fourier amplitudes together with the corresponding Fourier frequencies are the parameters of a model. But as soon as transient dynamics, resulting from an initial condition which does not belong to the limit cycle, is included in the time series, the dynamics is no longer periodic, i.e. the Fourier spectrum gets complicated and the number of nonzero parameters may equal the number of points in the time series [28]. In particular, if the limiting dynamics is chaotic, nearly all those parameters get large. Since in this case the condition $K \ll N$ is violated, a Fourier transformation is not an appropriate modeling process for the chaotic time series.

Assume the observed time series originates from a system whose dynamics can be described by an N_d-dimensional system of differential equations of first order $\dot{\vec{y}} = \vec{f}(\vec{y})$ ($\vec{y} \in S = N_d$-dimensional state space [29, 23]) or from a map $\vec{y}_{n+1} = \vec{f}(\vec{y}_n)$ ($n = 1, 2, \ldots$) with a flow vector field absent of singularities. Then we can model the dynamics by an analytical estimation of its flow vector field.

In certain cases this analytical estimate of the flow vector field may have just a few fit parameters. In this case the analytical estimate of the flow vector field is a simple model of the aperiodic time series. In the following we will focus on this situation.

The problem of dealing with hidden variables in dynamical systems was first addressed by PACKARD et al.[29] and RUELLE in 1980. They noted that a state space representation of the dynamics could be reconstructed from a single time series through the use of delay coordinates. This delay-coordinate reconstruction would then be topologically equivalent to the dynamics of the true system. WHITNEY had shown much earlier that any compact manifold with dimension m can be embedded in R^{2m+1}[30]. TAKENS extended this in proving that an embedding can be obtained from any system from only a time series by using $2m+1$ delay coordinates[31]. While this combination of ideas thus far has been extremely useful in studying nonlinear systems, several difficulties arise in their application.

The most obvious difficulty in using delay coordinates is the issue of interpreting the results physically. If one is concerned only with forecasting, the method used for the modeling is irrelevant, as long as it works. Successful modeling techniques based upon delay coordinates and/or partitioning the state space to generate local fits have been developed[23, 32]. However, relating these models to physical principles or existing theories is often difficult if not impossible. For example if the two variables of a three dimensional ODE are hidden, and the time series of the third variable is represented in state space using delay coordinates, the simplest model for the flow is typically an integro-differential equation, i.e. substantially more complicated than the ODE. Another important consideration is that Takens's theorem does not apply to systems with noise, i.e. experimental data. And Whitney's theorem as well as Takens' modification apply only to smooth and stationary manifolds, whereas smooth and stationary AIMs may not be generic in experimental situations.

The dynamical reconstruction[33] which we present in the next few paragraphs addresses the problems associated with an optimal state space reconstruction as well as desirable features for the modeling process. To illustrate the basic idea we use a simple example. Given is a system of two coupled maps:

$$x_1(t+1) = f(x_1(t), x_2(t), p_1 + F^1_{noise}(t) + F_1(t)) \tag{7}$$

$$x_2(t+1) = f(x_1(t), x_2(t), p_2 + F^2_{noise}(t) + F_2(t)) \tag{8}$$

where $x_1(t), t = 1, 2, \ldots$ is assumed to be measurable and given, whereas $x_2(t)$ is unobservable, i.e. a hidden variable. p_1 and p_2 are two sets of parameters, for example, the coefficients of a Fourier expansion of f. $F^1_{noise}(t)$ and $F^2_{noise}(t)$ represent small uncorrelated noise and F_1 and F_2 are control forces. F_1 and F_2 are assumed to be known and may be zero. If p_1 and p_2 are known, we use a maximum likelihood method to estimate the value of the hidden variable[34], i.e. $x_2^{guess}(t)$ is estimated by the x_2 value of the most probable trajectory of length i_{max}:

$$\begin{aligned} x_2^{guess}(t) &\in \{\hat{x}_2^o(t) | \prod_{i=0}^{i_{max}} P(t, t_i, t_{i+1}, \hat{x}_2^o(t+t_i), p_1, p_2) \\ &\geq \prod_{i=0}^{i_{max}} P(t, t_i, t_{i+1}, \hat{x}_2(t+t_i), p_1, p_2)\} \end{aligned} \tag{9}$$

where $t_0 = 0$, $t_{i+1} > t_i$, $\hat{x}_2(t+1) = f(x_1(t), \hat{x}_2(t), p_2 + F_2(t))$, and $\hat{x}_2^o(t+1) = f(x_1(t), \hat{x}_2^o(t), p_2 + F_2(t))$. $P(t, t_i, t_{i+1}, x_2, p_1, p_2)$ is the probability of the transition from the state $(x_1(t+t_i), x_2)$ to a state with a given $x_1(t+t_{i+1})$ within $\Delta t = t_{i+1} - t_i$ steps in time. In principle one may choose $t_{i+1} = t_i + 1$ with a large i_{max}, but if x_2 is estimated numerically this would need a lot of computer power. Systematic numerical studies indicate that the estimate for x_2 does not substantially degrade if $i_{max} > 1$ is small, for example equal to two, and if Δt equals a quarter of the typical time scale of the system. If x_2^{guess} cannot be determined uniquely it is sometimes useful to increase i_{max}. In particular, if the number of hidden variables is larger than the number of measured variables i_max has to be two or even larger. Therefore it is advantageous to measure the dynamics of at least half of all order parameters of the experimental system.

If P is Gaussian, the optimum value is given by the minimum value of Q, the mean square difference between the predicted and the observed time series $Q(t, \hat{x}_2(t), p_1, p_2) = \sum_{i=1}^{i_{max}}(x_1(t+t_i) - f_1(x_1(t+t_i-1), \hat{x}_2(t+t_i-1), p_1 + F_1(t+t_i-1)))^2$:

$$x_2^{guess}(t) \in \{\hat{x}_2^o(t) | Q(t, \hat{x}_2^o(t), p_1, p_2) \leq Q_j(t, x_2(t), p_1, p_2)\} \tag{10}$$

If only rough estimates of p_1 and p_2 are known, they can be improved by a maximum likelihood estimation, too:

$$\begin{aligned}&(p_1, p_2)^{opt} \in \\ &\{(p_1^o, p_2^o) | \prod_{i=0,j=1}^{i_{max}, j_{max}} P(t_j, t_i, t_{i+1}, x_2^{guess}(t_j), p_1^o, p_2^o) \\ &\geq \prod_{i=0,j=1}^{i_{max}, j_{max}} P(t_j, t_i, t_{i+1}, x_2^{guess}(t_j), p_1, p_2)\}\end{aligned} \tag{11}$$

and if P is Gaussian through a minimization of Q:

$$\begin{aligned}(p_1, p_2)^{opt} &\in \{(p_1^o, p_2^o) | \sum_{j=1}^{j_{max}} Q(t_j, x_2^{guess}(t_j), p_1^o, p_2^o) \\ &= \min \sum_{j=1}^{j_{max}} Q(t_j, x_2^{guess}(t_j), p_1^o, p_2^o)\}\end{aligned} \tag{12}$$

where t_j is a set of times, for example $t_j = 2j$. In section 4.2 we illustrate how to estimate Q algebraically for the case where all order parameters are observable. It seems that this approach can be generalized to the modeling process with hidden variables.

There are many methods to search for the extrema of the likelihood (Eq.11) or Q (Eq.11) numerically[33]. A simple approach is to search the minimum with a gradient search and an appropriate choice of initial conditions. As an initial condition one picks the rough estimate and a certain number of neighboring positions in order to check for local minima. In particular if the system is evolving, i.e. $p_1 = p_1(t)$ and $p_2 = p_2(t)$, then the estimates from the previous modeling process are well suited as initial conditions. For evolving systems only data from the most recent history should be used for the modeling. A detailed discussion of this issue was given by OHLE et al.[26]. If no rough estimates for p_1 and p_2 are available, one has to define a region of interest and start the search from a large number of initial conditions within this region.

There is never a unique model for the whole state space, since it is impossible to collect data from all regions of the state space. However, it may be possible to cover a certain region of the state space with data. If the measured variables and the hidden variables cover this region of interest then the model parameters as well as their accuracy can be determined uniquely[26]. The parameters of the model are most accurate if the dynamics covers this region homogeneously[27]. In order to obtain such a homogeneous coverage, intrinsic noise in the system as well as external driving forces, which force the system to cover the region of interest homogeneously may be useful[33] . Therefore it is advantageous for the modeling process to impose a random motion or a highly chaotic motion onto the experimental system. However, there is a big problem associated with such a control: slaved variables may be stimulated and make the reconstruction of a simple model impossible. To do a control without stimulating slaved variables is the main issue of dynamical control and the principle of the dynamical key, which we will discuss later.

It turns out that the proposed modeling process optimizes most of the requirements which we mentioned above. It is not confined to smooth dynamics on the AIMs or smooth AIMs since f in Eq.7 may be singular or even multivalued. The obtained models tend to be simple, since foreknowledge about the system can be used. Since the model parameters can be determined uniquely, including their error bars, there are usually no problems with reproducibility or interpolations between neighboring parameter settings. In the next few paragraphs we discuss how to use such models for a control.

3. Controlling Complex Systems

3.1 Nonlinear Resonances and the Principle of the Dynamical Key

The principle of the dynamical key states: *Nonlinear systems react most sensitively to a special class of aperiodic driving forces. Typically, the characteristic time scales of the optimal driving force match at all times the characteristic time scales of the system. In some cases the optimal driving force as well as the resulting dynamics are similar to the transients of the unperturbed system.*

The following theorem illustrates this paradigm in a physical context[35]. Given is the velocity of the oscillation $\dot{x}(t)$ and the driving force $F(t)$ in terms of amplitude $A(t) \geq 0$, $B(t) \geq 0$, time scale $T_x(t) = 2\pi/|\omega_x(t)|$, $T_F(t) = 2\pi/|\omega_F(t)|$ and phases ϕ_x, ϕ_F:

$$\dot{x}(t) = A(t) \sin(\omega_x(t) * t + \phi_x) \tag{13}$$

$$F(t) = B(t) \sin(\omega_F(t) * t + \phi_F) \tag{14}$$

where $x(t)$ is a state vector and $\omega_x(t)$ and $\omega_F(t)$ are real or imaginary scalars. There is a a 100% energy transfer $P(t)$, i.e. 0%of the transferred energy are reflected and

$$P(t) = \dot{x}(t)F(t) \geq 0 \tag{15}$$

for all times t, if the two time scales and phases are equal. In this theorem we used the efficiency of the energy transfer as a measure of the size of the response. This definition is useful for physical and chemical systems and might help to understand phenomena such as catalysis[40]. But the paradigm applies for other measures, too. We will show some examples later. There is a certain arbitrariness in the definition of ω_x and ω_F. Since the theorem applies for any definition of ω_x and ω_F, it means that all time scales have to match. However, it provides useful physical interpretations only if the definitions are appropriately chosen for a given dynamical system. Another critical question concerns the type of similarity between the driving force and the transient dynamics needed to get an optimal response. No general answer is known yet, but we will give some examples. In the next few paragraphs we compare the impact of sinusoidal driving forces and optimal driving forces on low dimensional nonlinear oscillators and complex systems and study some examples.

Sinusoidally driven nonlinear oscillators have complex, in many cases chaotic, dynamics. The response to externally applied forces is relatively small and does not match any well defined resonance condition. E.g., when a nonlinear mechanical pendulum is perturbed by a sinusoidal force the response is comparatively small in amplitude[35], is in many cases chaotic[36] and does not satisfy any well defined resonance condition, even when the frequency of the driving force coincides with a peak (resonance) in the power spectrum of the dynamics of the unperturbed system[37]. In order to obtain a large response, the frequency of the driving force must be shifted in such a way that it coincides at all amplitudes with the characteristic frequency of the oscillator[35]. But even when the frequency of the driving force coincides at all amplitudes with the basic resonance of the unperturbed system, the resonance condition is not exactly satisfied, since all the other peaks[37] of the power spectrum of the unperturbed system have to be taken into account as well. Recently a method has been presented to calculate driving forces which satisfy the resonance condition exactly[35]. We will show that they are in agreement with the principle of the dynamical key. In general, resonant driving forces are those which reach a certain target, break a special bonding of a molecule, or satisfy a given constraint with the smallest driving force possible.

Perturbations of small amplitude are most important for the stimulation of complex systems. Its has been shown[1] that due to the slaving principle a large variety of complex systems can be modeled by low dimensional systems of differential equations. If one uses this system of differential equations to calculate driving forces which force the experimental system to perform a special dynamics (goal dynamics) and applies these forces to the real system, then generally some slaved variables which are not included in the model will be excited, too. If we take for example the system given by Eq.2 and investigate the influence of an additive driving force we would expect that the response is given by:

$$\dot{x}_1 = p_1 x_1 - p_2 x_1^3 + F(t). \tag{16}$$

In order to check this expectation we go back to the full set of equations and assume, for the sake of simplicity, that the driving force $F(t)$ acts on both variables in the same way:

$$\dot{\tilde{x}}_1(t) = \tilde{p}_1 \tilde{x}_1(t) - \tilde{x}_1(t)\tilde{x}_2(t) + \tilde{F}(t) \tag{17}$$

$$\dot{\tilde{x}}_2(t) = -\tilde{p}_2 \tilde{x}_2(t) + \tilde{x}_1^2(t) + \tilde{F}(t) \tag{18}$$

where $\tilde{p}_1$ and $\tilde{p}_2$ are two control parameters. If we assume that $\tilde{x}_1(t)$ is slow compared to $\tilde{x}_2(t)$, i.e. $\tilde{p}_1 \ll \tilde{p}_2 \approx 1$, then one can solve Eq.3, by assuming that $\tilde{x}_1 \approx const.$ and $F(t) \approx const$ in Eq.18:

$$\tilde{x}_2(t) = \frac{1}{\tilde{p}_2}\tilde{x}_1^2(t) + \tilde{F}(t) + (\tilde{x}_2(0) - \frac{1}{\tilde{p}_2}(\tilde{x}_1^2(t) + \tilde{F}(t)))e^{-\tilde{p}_2 t} \tag{19}$$

In Eq.19 the first two terms describe a slow motion given by $x_1(t)$ under the influence of a driving force and the third term a transient which rapidly decays. After this rapid decay of the transient, $\tilde{x}_2$ is entirely determined by $\tilde{x}_1$ and $\tilde{F}$:

$$\tilde{x}_2 = \frac{1}{\tilde{p}_2}\tilde{x}_1^2 + \tilde{F}, \tag{20}$$

$\tilde{x}_2$ is "slaved" by $\tilde{x}_1$ and $\tilde{F}$. In a geometrical representation, the trajectory converges rapidly to a two dimensional manifold given by Eq.20 and then performs a slow motion given by Eq.20 and Eq. 17:

$$\dot{x}_1 = (p_1 + F(t))x_1 - \frac{1}{p_2}x_1^3 + F(t). \tag{21}$$

where we neglected the $\tilde{}$ in order to label x_1 as a order parameter. It turns out that Eq.21 is in agreement with Eq.16 if we assume that $F(t) \ll p_1 \ll p_2$. The result of this little illustration is that F has to be very small and slow, whereas there is no constraint on the size of the response. Very similar results can be obtained for a macroscopic physical pendulum. If an external force is applied to the pendulum, the response of the pendulum is given by the equation of motion of the center of mass which contains the driving force as an additive term. However, if the driving force is large and has a short time scale, as for example a kick by a hammer, this equation does not give a correct description of the response, since all kinds of vibrational modes will be stimulated. Only if the typical time scale of the driving force is much larger than the typical time scale of all slaved variables can the response of the pendulum be predicted with the low dimensional model.

In order to calculate resonant driving forces $F(t)$ for nonlinear systems we start out with a damped motion in a nonlinear potential:

$$m\ddot{x} + \eta_1\dot{x} + \frac{\partial V(x)}{\partial x} = F(t) \tag{22}$$

where m is the mass, η_1 a friction coefficient and where the first term represents inertial force, the second term a velocity dependent friction force, the third term a nonlinear potential force and $F(t)$ a driving force.

A strong resonance is defined through the following set of conditions: $x(t=0) = x_0$, $x(t=T) = x_T$, T, $H(t=0) = H_0$, $H(t=T) = H_T$ where $H = \dot{x}^2/2 + V(x)$ are given, i.e.

$$\Delta H = \int_0^T F\dot{x} - \eta_1\dot{x}^2 dt = fixed \tag{23}$$

and where the mean square average of the driving force

$$\bar{F}^2 = \int_0^T F(t)^2 dt \tag{24}$$

is as small as possible. The corresponding Lagrange equation reads:

$$L = F(t)^2 - \lambda_1(t)(m\ddot{x} + \eta_1\dot{x} + \frac{\partial V(x)}{\partial x} - F(t)) - \lambda_2(F\dot{x} - \eta_1\dot{x}^2) \tag{25}$$

where $\lambda_1(t)$ and λ_2 are the Lagrange parameters for the constraints given by Eq.22 and Eq.23. The equations of motion $\partial L/\partial x - d/dt\partial L/\partial\dot{x} + d^2/dt^2\partial L/\partial\ddot{x} = 0$ and $\partial L/\partial F = 0$ yield the equations

$$-\lambda_1\frac{\partial^2 V(x)}{\partial x^2} + \eta_1\dot{\lambda}_1 + \lambda_2\dot{F} - 2\lambda_2\eta_1\ddot{x} - \ddot{\lambda}_1 = 0 \tag{26}$$

$$2F + \lambda_1 - m\lambda_2\dot{x} = 0 \tag{27}$$

in addition to the constraints given by Eq.22 and Eq.23. A solution of these four equations is given by

$$F(t) = 2\eta_1 \dot{y} \tag{28}$$

$$\lambda_1(t) = -4\eta_1 \dot{y} \tag{29}$$

$$\lambda_2 = 0 \tag{30}$$

where the dynamics of the auxiliary variable $y(t)$ if given by

$$m\ddot{y} - \eta_1 \dot{y} + \frac{\partial V(y)}{\partial y} = 0 \tag{31}$$

Generally $y(t)$ is not periodic if $\frac{\partial V(y)}{\partial y}$ is nonlinear. Therefore the optimal driving force $F(t)$ is aperiodic. In addition, the dynamics of $x(t)$ is given by

$$x(t) = y(t) \tag{32}$$

For many systems the solution given in Eq.32 is stable and has a certain basin of attraction. In this case Eq.32 provides an estimate of the response $x(t)$ even if $x(t=0) \neq y(t=0)$, but is in the basin of attraction. Since Eq.31 is autonomous, i.e. has no explicit time dependence, in contrast to Eq.22, it is sometimes possible to solve Eq.31 in a closed form. In this case the response due to the aperiodic, optimal driving force is given in a closed form[17], i.e. is simple in terms of standard algebraic concepts.

Eq. 31 results from unperturbed experimental dynamics (Eq. 22) through a reflection of time. This fact provides a physical interpretation: The nonlinear oscillator reacts most sensitive if perturbed by a driving force which mimics the time reflected transient dynamics of the unperturbed system. In this case the driving force matches at all times the actual rhythm of the nonlinear oscillation and the absorption of energy is 100%, i.e. the energy transfer is always positive(principle of the dynamical key):

$$P(t) = F(t)\dot{x}(t) = 2\eta_1 \dot{x}^2(t) \geq 0 \tag{33}$$

This is an example of the similarity relation mentioned in the principle of the dynamical key: the response of the experimental systems equals the time reflected transient dynamics of the unperturbed system (Eq. 32)and the optimal driving force is proportional to the time derivative of the time reflected transient dynamics of the unperturbed system (Eq. 28). And, of course, the typical time scales of the driving force F and the velocity $\dot{x}$ match, since, for the solution given by Eq.32, they are directly proportional to each other: $F(t) = 2\eta_1 \dot{y} = 2\eta_1 \dot{x}$

3.2 Dynamical Control

The problem of finding nonlinear resonances is closely related to the problem of optimal control of nonlinear systems. Whereas extensive literature exists on the subject of linear controls with feedback and without feedback [43], little is known about their nonlinear counterparts [35, 15, 41]. Recently, progress in chaos theory [35, 23] has led us to a new approach to the control process [11, 17, 42]. We assume that the dynamics of the order parameters of a complex system can be modeled by a differential equation of the type

$$\begin{aligned} &\dot{x}\left(t, r_1, r_2, \ldots\right) = \\ &f\left(x_1, x_2, \ldots, t, r_1, r_2, \ldots, \frac{\partial}{\partial r_1}, \frac{\partial}{\partial r_2}, \ldots, F\left(t, r_1, r_2, \ldots\right) + p_m\left(t, r_1, r_2, \ldots\right)\right) \end{aligned} \tag{34}$$

where y is a n-dimensional vector of order parameters and f is a n-dimensional flow vector field which depends on the order parameters $x_1, x_2, \ldots$, spatial coordinates $r_1, r_2, \ldots$, time t, and spatial derivatives. F is a driving force which acts on a set of significant control parameters p which is assumed to be close to the true parameter setting $p_m \approx p$. The goal of the control process is to make the limiting behavior of $x(t- > \infty)$ and a given goal dynamics $y(t)$ equal, by using an appropriate F.

Now we have to distinguish between two cases: (i) dynamical control, where no information about the actual state of the experimental system is required[11, 44], and (ii) feedback based controls [15]. For dynamical control complete entrainment occurs if the driving force makes the flow vectors of the experimental dynamics and the goal dynamics equal at the location of the goal dynamics, *i.e.*

$$\begin{aligned} &\dot{y}\left(t, r_1, r_2, \ldots\right) = \\ &f\left(y_1, y_2, \ldots, t, r_1, r_2, \ldots, \frac{\partial}{\partial r_1}, \frac{\partial}{\partial r_2}, \ldots, F\left(t, r_1, r_2, \ldots\right) + p\left(t, r_1, r_2, \ldots\right)\right) \end{aligned} \tag{35}$$

and if the special solution $\mathbf{x}(t) = y(t)$ is stable [13]. If $y(t)$ is not differentiable, Eq.35 must be analyzed in terms of distributions. Now we present an example for the control of a chaotic motion with dynamical control. In this case the experimental dynamics is a chaotic logistic map dynamics:

$$x(t+1) = 4x(t)(1 - x(t)) + F(t) \tag{36}$$

and the goal dynamics is given by another logistic map dynamics

$$y(t+1) = 3/8 + z(t)/4 \tag{37}$$

$$z(t+1) = bz(t)(1 - z(t)) \tag{38}$$

where $0 < x(t) < 1$, $0 < z(t) < 1$, $0 < b < 4$, and $t = 1, 2, \ldots$. In order to apply Eq.34 and Eq.35 to that problem we have to replace $\dot{y}(t)$ by $y(t+1)$ to do a Taylor expansion of Eq.35. A comparison of coefficients in the expansion yields:

$$F(t) = y(t+1) - 4y(t)(1 - y(t)) \tag{39}$$

This driving force is independent of the actual states of the experimental system and can be calculated before the experiment is done. The stability of the control can be calculated as follows

$$\epsilon(t+1) = x(t+1) - y(t+1) \approx \frac{\partial x(t+1)}{\partial x(t)}\Big|_{x(t)=y(t)} \epsilon(t) \tag{40}$$

where $\epsilon(t) = x(t) - y(t)$ is the distance between goal and real dynamics. $\epsilon(t)$ tends to zero if

$$\lambda_c = lim_{T\to\infty}\frac{1}{T}\sum_{t=1}^{T}\log|\frac{\partial x(t+1)}{\partial x(t)}|_{x(t)=y(t)}| \leq 0 \tag{41}$$

where λ_c is called control coefficient. For the logistic map in Eq.36 the control coefficient is smaller than zero for any goal dynamics $y(t)$ where $3/8 < y(t) < 5/8$. A generalization of this stability analysis for higher dimension systems and continuous systems is straightforward. The definition of the control coefficient is similar but not equal to the definition of Liapounov exponents

$$\lambda_x = lim_{T\to\infty}\frac{1}{T}\sum_{t=1}^{T}\log|\frac{\partial x(t+1)}{\partial x(t)}|_{x(t)}| \tag{42}$$

$$\lambda_y = lim_{T\to\infty}\frac{1}{T}\sum_{t=1}^{T}\log|\frac{\partial y(t+1)}{\partial y(t)}|_{y(t)}| \tag{43}$$

where λ_x and λ_y are the Liapounov exponents for the unperturbed experimental system and the goal dynamics respectively. This means that the goal dynamics and the experimental dynamics can both be chaotic, i.e. $\lambda_x > 0$ and $\lambda_y > 0$, and nevertheless the control is stable since $\lambda_c < 0$. This happens, for example, for $b = 4$ in the example given above.

For feedback controls[15] the flow vectors of the experimental dynamics and the goal dynamics have to be made equal, too, and therefore they are very similar to dynamical control. However, since information about the actual state of the system can be used one has

$$\begin{aligned}&\dot{y}\,(t, r_1, r_2, \ldots) = \\ &f\left(x_1, x_2, \ldots, t, r_1, r_2, \ldots, \tfrac{\partial}{\partial r_1}, \tfrac{\partial}{\partial r_2}, \ldots, F\,(t, r_1, r_2, \ldots) + p\,(t, r_1, r_2, \ldots)\right)\end{aligned} \tag{44}$$

the special solution $x(t) = y(t)$ is usually stable. This is a distinct advantage of feedback controls over dynamical controls. Some targets are not reachable with dynamical controls since they are unstable. It is not possible to get a stable dynamical control if the goal dynamics has a stable limit cycle at a location where the experimental system has an unstable limit cycle, whereas feedback controls can deal with this situation[15]. However, feedback controls need the information about the actual state of the system, which in many physical systems, as for example quantum systems, is not accessible. Any delay between the measurement and the resulting adjustment of the control force causes the instabilities. All feedback controls need continuous access to sensors and a control unit which has a typical time scale much faster than the typical time scale of the experiment, whereas dynamical control needs no feedback after an initial modeling period where an estimate of p is determined, and no feedback at all, if a model of the experimental dynamics is known. In addition, the time scale of the control unit does not need to be faster than the typical time scale of the experiment and is in most applications much slower. For example, in order to control the motion of a hot nucleus, a spatial aperiodic sequence of magnetic and electric fields which the nucleus is traveling through, could be preassigned. In this case the time constant of the control unit is zero. Since the control force is calculated for both types of control in the same way, the size of the control force is the same.

Therefore, in the following we focus on dynamical control. As soon as the goal dynamics can be integrated in a closed form, one also has a closed form for the response of the driven experimental system, even if the driving force is aperiodic.

A driving force is called a resonant perturbation if it forces the experimental system to satisfy a certain set of constraints like, for example, to transfer a certain energy or to reach a certain target and be extraordinary small. In this case, the variational principle establishes a link between the goal dynamics and the experimental dynamics, which is the similarity mentioned in the principle of the dynamical key. The type of similarity which is established between the two dynamics depends on the constraints. In general, the goal equation is equivalent to the equation of motion resulting from the variation. In addition, in some cases solvability conditions of the variational calculus determine reachable targets.

3.3 General Resonance Spectroscopy

In section 3 we presented a method to extract a model from a given time series: dynamical reconstruction. Dynamical reconstruction provides accurate models for the natural dynamics of the system. However, these models are only useful for a control if there is an overlap between the natural dynamics of the system and the goal dynamics. In particular, if the experimental dynamics is simple, for example a fixed point dynamics, this is not the case. In this case it is necessary to explore the state space through appropriate perturbation. These perturbations have to be small in order to avoid the stimulation of slaved variables, as we argued earlier. A systematic procedure for such an active modeling is general resonance spectroscopy.

In linear resonance spectroscopy, one applies a sinusoidal driving force to a harmonic oscillator. Such an oscillator has only one parameter, the characteristic frequency. The amplitude of the resulting stimulated oscillator is a function of the driving frequency and a maximum when the driving frequency is equal to the oscillator's characteristic frequency. One determines the characteristic frequency of the oscillator, the model parameter, by systematically varying the driving force until this resonance, i.e. large response is observed.

In other words, one is seeking to match two quantities - the driving frequency and the characteristic frequency - using the amplitude of the oscillation as a measure of how good the match is.

This procedure can be generalized. The central concept is simple: If our model of some experimental dynamics is a good one, then we should be able to predict the response of the driven experimental system. Using the deviation $Q = \overline{|z(t) - x(t)|}$ between the prediction $z(t)$ and the real trajectory of the driven system $x(t)$, we find how good the model is. By varying the parameters in the model to minimize this deviation, we find the "general resonance" and thus fine-tune the model. $1/Q$ becomes infinitely large at this resonance and indeed CHANG et al.[39] have shown that for harmonic oscillators $1/Q$ as a function of the characteristic frequency is closely related to standard resonance curves.

The goal dynamics must be chosen with some care in order to make the modeling process successful. The control must be stable for a perfect model, so that small perturbations in the initial conditions remain small. Also, for modeling purposes, one wishes the goal dynamics to cover a large area of the state space, as the model derived by this procedure is valid only over the range of the goal dynamics. For the same reason, one would prefer a chaotic goal dynamics to a periodic one.

Finally, for experimental considerations, one would like (i) to keep the driving force as small as possible to avoid exciting modes that one is not interested in and (ii) a goal dynamics which is significantly different from the natural dynamics in terms of the experimental noise level, since in most experimental systems Q can be accurately measured rather than $1/Q$ we investigate the resonance curve determined by Q. In this case a rough estimate of the signal to noise ratio of the resonance curve is given by $Q = |\overline{|z(t)|} - \overline{|x(t)|}|/\sigma_D$ where σ_D is the standard deviation of the noise level.

4. A Virtual Environment for Matter: Science Fiction?

The principle of the dynamical key states that only very special external driving forces provoke a large response in a nonlinear oscillator, which seems to make it very unlikely that a given natural system could stimulate a large response in another natural system or get any strong interaction with the other natural system. However, the principle of optimal interaction asserts that just those states with a complicated dynamics but an optimal interaction are stationary or even attractive[38]. Therefore, we conclude that interacting physical systems are very likely to be observed or adapt to complicated states, where their interaction, for example, the size of the energy exchange, is as large as possible or as small as possible (principle of matching). The term "principle of matching" originates from Biology, where is has a very similar meaning[45].

According to the definition given in section 2.1, control parameters are those variables which have a time scale much longer than the typical time scale of the order parameters. The energy content of the two oscillators is changing on a slow rate compared to the period of oscillation of each oscillator[38]. Therefore, the energy content of each oscillator may be considered a slowly evolving control parameter. In this terminology the principle of matching reads: *Interacting physical systems adjust those control parameters which are not fixed by external constraints, in such a way that their interaction is extremal.*

The principle of matching would suggest that modules emerge if more than two physical systems interact, where a module is a set of strongly interacting systems and where the boundary of a module is marked by weak interaction. Little is known about those systems.

In the context of the principle of matching we discussed adaptation of two physical systems. Now we want to pose the question about what happens if one of these physical systems is replaced by a computer. Digital computers are ideal chaos generators, since they are deterministic and react sensitively to any change in the computer code. As we have shown earlier, those chaotic signals can be used to provoke strong responses from physical systems. Sensors make it possible for computers to interact directly with matter and other physical systems. With these sensors the computer could establish an adapting virtual environment for physical systems. The number of sensors ans actuators should equal at least half of the number of order parameters in the experimental system, according to the paradigm on dynamical reconstruction (sect. 2.2). Therefore the number of sensors and actuators may be large, for a complicated experimental system. We can only guess in which way such an adapting virtual environment surrounding an adapting physical system would evolve. Maybe it would make it possible to produce completely relaxed, stress free, super clean materials or with a special type of disorder. Three things would be new: (i) the material is exposed to very different types of external nonequilibria as for example, acoustic and optic perturbations as well as a flow of heat and an electric current simultaneously, (ii) the external nonequilibria are not stationary but adapting, according to the response of the material and (iii) the external nonequilibria may have a complicated space and time dependence. Massive parallel supercomputers might provide the necessary compute power.

Even if such a virtual environment for matter is not visible in the near future, we expect that the paradigms which we presented in this paper might trigger a major change of scientific instrumentation and materials in the near future. Due to the principle of the dynamical key, function generators for aperiodic driving forces seem to be most useful. Since most function generators are microprocessor based, only minor changes of the hardware are needed to upgrade a sine generator to a chaos generator. However, the replacement of this equipment in all scientific and industrial labs, for all frequency ranges, could be a major business.

Most nonlinear systems have many significant parameters. In order to determine those parameters through general resonance spectroscopy, it is necessary to scan through a high dimensional parameter space in search for the resonance, i.e. the measurement instrument needs to behave as a generic physical system, namely, adapt to the state of strongest interaction (principle of matching). It should be no technological problem to add an analog input to a standard function generator in order to use the feedback from the experimental system for a fast readjustment of the parameter setting of the function generator. This would upgrade a function generator to an adaptive system. Those fast adaptive function generators would make it possible to do an automatic search in high dimensional parameter spaces for a modeling of nonlinear systems.

References

1. H. Haken, *Synergetics, an introduction* (Springer, Berlin, 1983), chapt. 7.

2. R.P. Lippmann, IEEE ASSP Mag., p. 4, April (1987).

3. L.A. Zadeh, *Fuzzy sets and applications* (Wiley, New York, 1987).

4. J.H. Holland, *Adaptation in natural and neural systems* (Univ. of Michigan Press, Ann Arbor, 1975).

5. C.G. Langton in *Artificial life*, edited by C.G. Langton. (Addison-Wesley, Redwood City, CA 1989).

6. H.G. Schuster, *Deterministic Chaos* (Physik-Verlag, Weinheim, 1984).

7. B.B. Mandelbrot, *The fractal geometry of nature* (Freeman, San Francisco, 1983).

8. D.K. Campbell, A.C. Newell, R.J. Schrieffer, H. Segur, Physica **18D**, 1(1986); R.K. Bullough and J.P. Caudrey eds., *Solitons* (Springer, New York, 1980).

9. R. Thom, *Structural stability and morphogenesis* (W.A. Benjamin, Reading, Mass. 1975).

10. P.C. Bak, C. Tang, and K. Wiesenfeld, Phys. Rev. A **38**, 364 (1988).

11. A. Hübler, E. Lüscher, Naturwissenschaften **76**, 67 (1989);A. Hübler, Helv.Phys.Acta **62**, 543 (1989).

12. K. Chang, A. Hübler, N. Packard in *Quantitative measures of complex dynamical systems*, edited by N.B. Abraham. (Plenum Press, New York, 1989), p. 225; K. Chang, S. Kodogeoriou, A. Hübler, E.A. Jackson, Physica D **44**, 407 (1990).

13. E.A. Jackson, A. Hübler, Physica D **44**, 407 (1990).

14. R. Shermer, A. Hübler, N. Packard, Phys. Rev. A **43**, 5642 (1991).

15. E. Ott, C. Grebogi, J.A. Yorke, Phys. Rev. Lett. **64**, 1196 (1990).

16. W.L. Ditto, S.N. Rauseo, M.L. Spano, Phys. Rev. Lett. **65**, 3211 (1990).

17. B. Plapp, A. Hübler, Phys. Rev. Lett. **65**, 2302 (1990).

18. T. Meyer, A. Hübler, N. Packard, *Quantitative measures of complex dynamical systems*, edited by N.B. Abraham. (Plenum Press, New York, 1989), p. 253.

19. G. Mayer-Kress, A. Hübler, in *Quantitative measures of complex dynamical systems*, edited by N.B. Abraham. (Plenum Press, New York, 1989), p. 155.

20. W.A. Brock in *Beyond belief: randomness, prediction, and explanation in science*, edited by J. Casti and A. Karlqvist. (CRC Press, Boca Raton, FL, 1991), p. 230.

21. H. Haken, *Advanced Synergetics* (Springer, New York, 1983).

22. M.J. Feigenbaum, J. Stat. Phys. **19**, 25 (1978); J. Stat. Phys. **21**, 669 (1979).

23. J.D. Farmer, S. Sidorowich, Phys. Rev. Lett. **59**, 845 (1987).

24. J.P. Crutchfield, B.S. McNamara, Complex Syst. **1**, 417 (1987).

25. J. Cremers, A. Hübler, Z. Naturforsch. **42a**, 797 (1987).

26. A. Hübler, Helv. Phys. Acta **62**, 343 (1989); F. Ohle, F. Dinkelacker, A. Hübler, Technical Report CCSR-90-13, Center for Complex Systems Research, Univ. of Illinois at Urbana-Champaign.

27. A. Hübler, J. Miller, D. Pines, N. Weber, preprint 1991.

28. E. Roesch, H. Eckelmann, A. Hübler, Construction of Differential Equations from Experimental Data of Kármán Vortex Streets, report #11/1988, Max-Planck-Institut für Strömungsforschung, Göttingen, ISSN 0436-1199; F. Ohle, P. Lehmann, H. Eckelmann, A. Hübler, Phys. Fluids **A** **2**, 479 (1990).

29. N.H. Packard, J.P. Crutchfield, J.D. Farmer, and R.S. Shaw, Phys. Rev. Lett. **45**, 712 (1980).

30. H. Whitney, Ann.Math. **37**, 645 (1936); **45**, 220 (1936), **45** 247 (1936).

31. F. Takens, in *Dynamical systems and turbulence*, edited by D.A. Rand and S. Young. (Springer, New York, 1981), p. 366.

32. T.P. Meyer, F.C. Richards, N. Packard, Phys. Rev. Lett. **63**, 1753 (1989).

33. J. Breeden, A. Hübler, Phys. Rev. A **42**, 5817 (1990); J. Breeden, F. Dinkelacker, A. Hübler, Phys. Rev. A **42**, 5827 (1990).

34. W. Eberl, T. Aschenbrenner, M.Athelogou, A. Hübler, preprint 1991.

35. G. Reiser, A. Hübler, E. Lüscher, Z. Naturforsch. **42a**, 803 (1987).

36. B.A.Huberman and J.P.Crutchfield, Phys. Rev. Lett. **43**, 1743 (1979); D.D. Humieres, M.R. Beasley, B.A. Huberman, and A. Libchaber, Phys. Rev. A **26**, 2483 (1982).

37. D.Ruelle, Phys. Rev. Lett. **56**, 405 (1986); U.Parlitz and W. Lauterborn, Phys. Lett. **107A**, 351 (1986).

38. T. Eisenhammer, A. Hübler, T. Geisel, E. Lüscher, Phys. Rev. A **41**, 3332 (1990).

39. K. Chang, A. Hübler, preprint 1991.

40. S. A. Kauffman, private communication.

41. S. Shi, H. Rabitz, J. Chem. Phys **92**, 2972 (1990); S. Shi, H. Rabitz, J. Chem. Phys **92**, 364 (1990).

42. A. Hübler, R. Georgii, M. Kuchler, W. Stelzl, E. Lüscher, Helv. Phys. Acta **61**, 897 (1988); R. Georgii, W. Eberl, E. Lüscher, A. Hübler, Helv. Phys. Acta **62**, 290 (1989).

43. see for example B.D.O. Anderson, *Stability Analysis of Adaptive Systems: Passivity and Averaging Analysis*, Cambridge, MA, MIT Press 1986.

44. A.Hübler, Ph.D. thesis at Technische Universität München (1987).

45. J. E. Mittenthal, A. B. Baskin, in *Principles of organization in organisms* edited by J.E. Mittenthal and A. B. Baskin (Addison-Wesley, Redwood City, CA, 1992).

QUANTEN-MONTE-CARLO SIMULATIONEN

Wolfgang von der Linden
Max-Planck-Institut für Plasmaphysik
D-8046 Garching bei München

1. Einleitung

Quanten-Monte-Carlo (QMC) Verfahren sind in der theoretischen Festkörperphysik zu einem nützlichen und wichtigen Hilfsmittel geworden. Insbesondere die Entdeckung der Hochtemperatur-Supraleitung durch K.A. Müller und J.G.Bednorz [1] – und das damit verbundene Interesse an Modellen für stark wechselwirkende Elektronen – hat zu einem vermehrten Einsatz von Quanten-Monte-Carlo-Verfahren geführt. Ihnen kam nicht zuletzt die enorme Leistungsfähigkeit der heutigen Großrechner zugute, und sie werden sicherlich mit der neuen Generation, den Teraflop-Maschinen, weiter an Bedeutung gewinnen.
Viele interessante physikalische Eigenschaften eines Festkörpers werden in entscheidendem Maße durch die gegenseitige Wechselwirkung der Elektronen bestimmt. Man spricht in diesem Fall von starker Elektronen-Korrelation. Typische Phänomene Antiferromagnetismus, Metall-Isolator-Übergang und – so wird seit einigen Jahren spekuliert – die Hochtemperatur Supraleitung [2]. Hierbei soll die eigentlich abstoßende Coulomb-Wechselwirkung, aufgrund von Quanteneffekten, zu einer effektiven attraktiven Wechselwirkung führen.
Um solche Phänomene zu untersuchen, separiert man i.d.R. die interessierenden Effekte und studiert sie anhand von vereinfachten Modellen. Ein typisches Modell, das im Rahmen der Hochtemperatur-Supraleitung intensivst untersucht wird, ist das Hubbard Modell.

2. Das Hubbard Modell

Das Hubbard-Modell [3] ist das einfachste Modell, das bereits wesentliche Eigenschaften und Schwierigkeiten eines typischen Vielteilchenproblems enthält. John Hubbard schlug dieses Modell vor ungefähr 30 Jahren vor, um den Metall-Isolator-Übergang und den Magnetismus frei beweglicher Elektronen zu studieren.
Es wird in diesem Modell vereinfachend angenommen, daß den Elektronen pro Gitterplatz nur ein Orbital zur Verfügung steht. Für die Hochtemperatur-Supraleitung ist man insbesondere an zwei-dimensionalen Quadratgittern interessiert. Die Bewegung der Elektronen findet in Sprüngen zwischen benachbarten Gitterplätzen statt. Pro Gitterplatz kann sich maximal ein Elektron einer Spinrichtung aufhalten. Die abstoßende Coulomb-Wechselwirkung ist dahingehend vereinfacht, daß Elektronen nur miteinander wechselwirken, wenn sie sich am selben Platz befinden. Eine solche "Doppelbesetzung" ist nur für Elektronen mit entgegengesetztem Spin möglich und führt zu einer Wechselwirkungsenergie ($E_{WW} = U \times$ Zahl der Doppelbesetzungen).
Trotz der scheinbaren Einfachheit des Hubbard-Modells sind bislang nur wenige Eigenschaften analytisch exakt bekannt, und man ist auf numerische Methoden angewiesen, um rigorose Ergebnisse zu erhalten.

3. Quanten-Monte-Carlo Verfahren

Wo liegt eigentlich das Problem einer Vielteilchenrechnung, und was ist der Nutzen der Monte-Carlo-Verfahren? Die Antwort auf beide Fragen hängt mit der extrem großen Zahl möglicher Konfigurationen zusammen.
Ein Zahlenbeispiel soll dies veranschaulichen. Wir betrachten das Hubbard Modell auf einem

Quadratgitter mit 10x10 Gitterplätzen und je 50 Elektronen pro Spinrichtung. Somit gibt es $L = \binom{100}{50}^2 \approx 10^{58}$ Konfigurationen. In einer quantemechanischen Rechnung müßten nun LxL-Matrizen diagonailsiert werden. An eine exakte Diagonalisierung von Matrizen dieser Größe ist natürlich nicht zu denken. Dennoch ist es mit Quanten-Monte-Carlo-Verfahren möglich, akurate Ergebnisse für solche und gar noch größere Systeme zu erlangen. Dies ist möglich, da nur ein geringer Anteil der Zustände für ein zuverlässiges Ergebnis notwendig ist. Analytisch sind die relevanten Zustände allerdings schwer zu ermitteln, und man ist auf stochastische Methoden angewiesen.
In der Regel ist man am thermodynamischen Erwartungswert einer Grösse O interessiert. Es läßt sich zeigen, daß das quantenmechanische Problem immer auf die Berechnung eines klassischen thermodynamischen Erwartungswertes für gewisse Hilfsfelder abgebildet werden kann

$$\langle O \rangle = \sum_{S_i(\tau_j)=\pm 1} O(S) \cdot \rho(S) \quad . \tag{1}$$

Die Hilfsfelder $S_i(\tau_j)$ sind im Falle des Hubbard Modells besonders einfach. Sie können nur die Werte ± 1 annehmen und stellen somit Ising-artige Spins dar. Die quantenmechanischen Vertauschungsrelationen führen allerdings dazu, daß die Hilfsfelder nicht nur vom Gitterplatzindex i, sondern auch von einem Zeitenindex τ_j abhängen. Für eine allgemeinverständliche Einleitung in die QMC-Verfahren verweise ich den Leser auf [4].
Der Summand in (1) besteht aus zwei Anteilen, von denen $\rho(S)$ das Gewicht des Zustandes angibt und die Eigenschaften einer Wahrscheinlichkeitsdichte besitzt ($\rho(S) \geq 0, \sum_S \rho(S) = 1$). Die Größe $O(S)$ ist der Erwartungswert der Observablen O im Zustand S.
Es liegt in der Natur des wechselwirkenden Vielteilchenproblems, daß die meisten Zustände, aufgrund der Funktion $\rho(S)$ ein verschwindend kleines Gewicht $\rho(S)$ besitzen und in der Summe vernachlässigt werden können. Um in (1) nur relevante Konfigurationen zu berücksichtigen, verwendet man das sogenannte "Importance Sampling", das auf Metropolis, Rosenbluth und Teller zurückgeht [5,6]. Hierzu wird eine Sequenz von Zuständen "$S_1, S_2, .., S_L$" (Markov-Kette) erzeugt, die der Verteilung $\rho(S)$ genügen, und mit denen man einen Schätzwert $\langle \tilde{O} \rangle$ für den Erwartungswert (1) angeben kann:

$$\langle \tilde{O} \rangle = \frac{1}{L} \sum_{l=1}^{L} O(S_l) \quad . \tag{2}$$

Darüber hinaus liefert der "Zentrale Grenzwertsatz" den statistischen Fehler des Schätzwertes $\langle O \rangle = \langle \tilde{O} \rangle \pm \sqrt{\sigma^2/L}$, wobei σ^2 die Varianz der Funktion $O(S)$ ist. Das bedeutet, daß der stochastische Fehler mit $1/\sqrt{L}$ gegen Null strebt und nicht explizit von der Dimension des Problems abhängt.
Der Hauptanteil der Rechenzeit steckt in der Berechnung der Grössen $\rho(S)$ und $O(S)$. Ohne näher auf die Details eingehen zu können, soll hier zumindest skizziert werden, welche numerischen Operationen notwendig sind.
Zunächst wird jedem Elektron ein fester Vektor $(c_1, c_2, \ldots, c_N)$ der Länge N (Zahl der Gitterplätze) zugeordnet. c_i^2 gibt hierbei die Wahrscheinlichkeit an, dieses Elektron am Gitterplatz i anzutreffen. Die quantenmechanische Bewegung der Elektronen in dem durch die Ising-Spins vorgegebenen Potential wird mittels Matrizen $M_\alpha(S)$ beschrieben

$$\vec{c}_l\,' = M_m \cdots M_2 \cdot M_1 \cdot \vec{c}_l \quad . \tag{3}$$

Die Elektronen (Vektoren $\vec{c}_l$ mit $l = 1, 2, \ldots, N_e$) propagieren unabhängig von einander. Es ist deshalb möglich, den Rechenzeit-intensivsten Anteil des Verfahrens massiv zu parallelisieren, indem die Propagation (3) auf verschiedenen Prozessoren durchgeführt wird.

4. Anwendung auf ein Elektron-Phonon Modell

Ich möchte nun noch auf QMC-Simulationen zu sprechen kommen, die im Rahmen des Großprojektes für Hochtemperatur-Supraleitung am HLRZ in Jülich durchgeführt worden sind. Hierbei wurde ein Elektron-Phonon Modell untersucht, bei dem zusätzlich zum Hubbard-Modell noch Gitterschwingungen berücksichtigt worden sind.
Zur Demonstration nehmen wir ein besonders einfaches Elektron-Phonon Modell, bei dem sich oberhalb jeden Gitterplatzes der "Kupfer-Oxyd" Ebene noch ein weiteres Atom befindet.
Diese sogenannten Apex-Sauerstoff-Atome schwingen senkrecht zur Ebene in einem Doppelmulden-Potential, wobei sie im wesentlichen nur zwei Positionen, die der Potentialminima, annehmen können. Die Position der Apex-Atome beeinflußt das Potential der Elektronen in der Kupfer-Oxyd-Ebene. Dieses Modell, wurde von K.A. Müller zur Untersuchung der anharmonischen Schwingungen des Apex-Sauerstoffs in den Hochtemperatur-Supraleitern vorgeschlagen [7]. Eine detaillierte Diskussion befindet sich in [8].

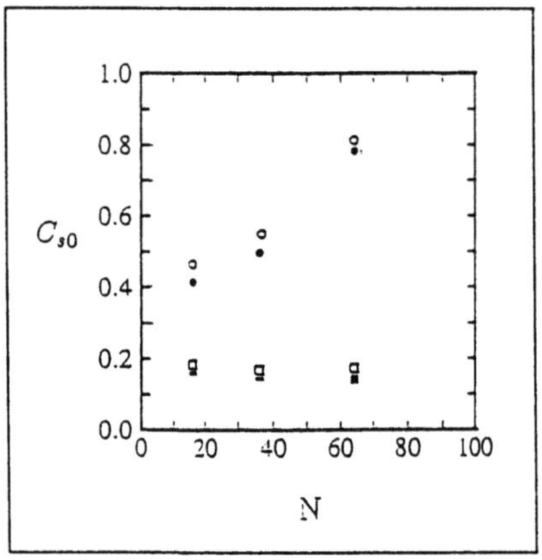

Abb.1 Besetzungszahl der lokalen Cooper-Paare im k=0 Zustand für verschieden Systemsgrössen N und einer Elektronenkonzentration von ≈ 31%. Kreise repräsentieren die Resultate für $U = 0$ ($U_{eff} = -1.6|t|$) und Quadrate die für $U = 4|t|$ ($U_{eff} = 2.4|t|$). Ausgemalte Symbole stellen die Ergebnisse des vollen Problems dar und offene die des effektiven Hubbard-Modells mit den angegeben U_{eff}.

Im Rahmen einer Störungstheorie ($\Omega >> g, t$) kann das Elektron-Phonon-Modell in ein effektives Hubbard-Modell, mit $U_{eff} = U - \frac{g^2}{\Omega}$, transformiert werden [8], wobei g die Elektron-Phonon-Kopplung und Ω die Phonon-Frequenz angibt. Für $U_{eff} < 0$ ist die Wechselwirkung attraktiv und kann zu Supraleitung führen.
Supraleitung, als makroskopisches Quantenphänomen, tritt nur dann auf, wenn ein Quantenzustand makroskopisch besetzt ist. Aufgrund der lokalen Wechselwirkung bilden sich Cooper-Paare mit vorwiegend lokalem Charakter aus, die mit dem Operator $\Delta^+ = \frac{1}{\sqrt{N}} \sum_i c_{i\uparrow}^+ c_{i\downarrow}^+$ erzeugt werden. Der Cooper-Paar-Zustand kann, wie bei der Bose-Kondensation der durch $b_{k=0}^+$ erzeugte Zustand,

makroskopisch besetzt werden. In diesem Fall wächst die Besetzungszahl $C_{s0} = \langle\Delta^+\Delta\rangle$ linear mit der Systemsgröße an. In Abbildung 1 ist C_{s0} als Funktion der Systemsgröße aufgetragen. Es wurden zwei Sätze von Parametern verwendet, die einen Vergleich mit störungstheoretischen Ergebnissen erlauben und die auf ein effektives Hubbard-Modell mit $U_{eff} < 0$ und $U_{eff} > 0$ führen. Man erkennt deutlich, daß die Besetzungszahl im ersten Fall mit der Systemsgröße anwächst. Das Elektron-Phonon Modell erfüllt für diese Parameter eine notwendige und hinreichende Voraussetzung für Supraleitung. Im Fall des repulsiven Hubbard-Modells hingegen ($U_{eff} > 0$), bilden sich keine lokalen Cooper-Paare aus. Das ist in Anbetracht der lokalen abstoßenden Wechselwirkung nicht weiter verwunderlich. Man macht jedoch dieselbe Beobachtung auch für ausgedehnte Cooper-Paare mit s-Wellen Symmetrie [8]. Aufgrund dieser und ähnlicher QMC-Ergebnisse ist es sehr unwahrscheinlich, daß das repulsive Hubbard-Modell ein angemessenes Modell für die Hochtemperatur Supraleiter ist, wie es ursprünglich von P.W.Anderson vermutet wurde [2].
Die gute Übereinstimmung der störungstheoretischen Ergebnisse mit denen des Elektron-Phonon Modells, wie sie aus Abbildung 1 hervorgeht, macht deutlich, daß die Vermittlung der attraktive Wechselwirkung, durch Austausch von Phononen, in der Simulation korrekt stattfindet.
QMC-Simulationen für eine realistischeres Elektron-Phonon Modell haben gezeigt, daß viele der supraleitenden Eigenschaften der Hochtemperatur-Supraleiter darin erklärt werden können [8,9].

5. Schlußbemerkung

Quanten-Monte-Carlo-Verfahren, sowie das gesamte Gebiet der *computational physics*, sind in der Festkörperphysik zu einem hilfreichen Werkzeug geworden, und es konnten in den letzten Jahren deutliche Fortschritte erzielt werden.
Allerdings ist die Entwicklung der QMC-Verfahren noch lange nicht abgeschlossen und es gibt noch einige fundamentale Probleme, auf die hier nicht eingegangen wurde, die es zu bewältigen gilt.
Dennoch haben Quanten-Simulationen bereits entscheidend zum Verständnis stark korrelierter elektronischer Systeme beigetragen. Ihre Bedeutung wird, parallel mit der Verbesserung der Leistungsfähigkeit der Großrechner, weiter zunehmen.

Danksagung: Für die anregende und interessante Zusammenarbeit an dem hier vorgestellten QMC-Problem sei Herrn Prof.I.Morgenstern, Prof.H.deRaedt und Dr.M.Frick gedankt. Es sei insbesondere auch dankend erwähnt, daß die QMC-Rechnungen im Rahmen eines Großrojektes am HLRZ der KFA Jülich durchgeführt worden sind.

[1] J.G.Bednorz and K.A.Müller, Z.Phys.B **64**, 189 (1987).
[2] P.W. Anderson, Science **235**, 1196 (1987).
[3] J. Hubbard, Proc. Roy. Soc. A **276**, 238 (1963).
[4] W. von der Linden, Phys.Bl. **47**, 747 (1991).
[5] K. Binder , Phys. Bl. **40**, 53 (1984).
[6] K. Binder, in K.Binder (Hrsg.): Monte-Carlo-Methods in Statistical Physics, Topics in Current Physics 7, Springer Verlag, Berlin 1979.
[7] K.A. Müller, Z.Phys.B **80**, 193 (1990).
[8] M.Frick, I.Morgenstern, and W. von der Linden, Z.Phys.B, **81**, 327 (1990); ibid. **82**, 339 (1991).
[9] M.Frick, I.Morgenstern und W. von der Linden, Int.J.Mod.Phys. **xx**, xx (1992).

Systembeschreibung und –Simulation mit Petri-Netzen

H. Fuss
Institut für Methodische Grundlagen (F 1)
Gesellschaft für Mathematik und Datenverarbeitung (GMD)
Bonn – Birlinghoven

0. Einleitung

Petri-Netze sind inzwischen weltweit bekannt und werden häufig angewandt, doch wird in einem Großteil der Anwendungen lediglich die dazugehörige Graphik als eine Art Flußdiagramm-Sprache benutzt. Die zugrundeliegende Idee der Netze – und das ist das Phänomen der Nebenläufigkeit (engl: concurrency) – ist aber viel umfassender, und damit bedeutender.

Nebenläufigkeit tritt in fast allen realen Systemen auf. Sie ist aber meist erst dann von Bedeutung, wenn das System räumlich verteilt ist und die Übertragungsgeschwindigkeiten von Nachrichten und Kommandos im System von der gleichen Größenordnung sind wie seine (materielle) interne Arbeitsgeschwindigkeit, d.h. wie die Reaktionszeiten des Systems.

1. Systembeschreibung: Nebenläufigkeit als Systemeigenschaft

Reale Systeme sind verteilt im Raum und agieren in der Zeit; jede Aktion zwischen den System-Teilen, auch die Informationsübertragung innerhalb des Systems, benötigt dafür eine gewisse Zeitdauer; jedoch werden diese physikalischen Eigenschaften eines Systems oftmals bei seiner Beschreibung und Simulation, häufig gar bei seiner Konstruktion, nicht genügend beachtet.

Bei der Beschreibung eines realen Systems durch ein DGL-System z.B. werden die Zustandsübergänge, d.h. die neuen Werte für *alle* Variablen für *denselben Zeitpunkt* berechnet. Das unterstellt jedoch dem Original-System (s.u. bei 'Bi-Simulation'), daß die Systemzustände aller Teilsysteme jedem anderen Systemteil zu jeder Zeit (insbesondere zur Taktzeit) gleichzeitig bekannt sind, es unterstellt ihm also eine unendlich große Informations- bzw. Signal-Übertragungsgeschwindigkeit.

Schon Minkowski hat in seinem Raum-/Zeit-/Ereignis-Modell[1] durch die Idee der 'Vorkegel' bzw. 'Nachkegel' gezeigt, (daß nicht nur nicht alle Information immer überall sein kann, sondern darüberhinaus auch:) daß die kausale Ordnung der Ereignisse in Raum und Zeit keine Vollordnung (Kette) bildet, wie man sie für die reellen Zahlen kennt und wie man sich das auch für die Zeit gedacht hatte, sondern daß Ereignisse eine *Halbordnung* bilden.

Diese Erkenntnis führte schließlich auch zu einer Auffassung von einer *Halbordnung der Zeit* und ermöglichte es, von *lokaler Zeit* sprechen zu können. Die Halbordnung (*'partial order'*) impliziert, daß 'Gleichzeitigkeit' *nicht transitiv* ist; somit ist 'Zeit' ein ungeeigneter Sortierbegriff.

Halb-geordnet-sein, und damit teilweise Unvergleichbarkeit, gilt überall da, wo genau genug gemessen, verglichen und sortiert wird, also u.a. *in der gesamten experimentellen Physik*; denn es handelt sich hier um ein generelles Phänomen aus der Theorie des Messens. Es ist nämlich ein konzeptioneller Irrtum davon auszugehen, daß beim Messen (von Länge, Zeit oder anderen Größen) ein Vergleich zwischen Objekt und Meßlatte/Meßskala auf *Gleichheit* gemacht wird. In Wirklichkeit geht es beim Meßvorgang um das (Nicht-mehr-)Wahrnehmen von Differenzen, also um Unterscheidbarkeit, genauer: um (reale, experimentelle) *Un-Unterscheidbarkeit*, und nicht um (ideelle) Gleichheit.

Zur Veranschaulichung mögen die (paarweise) 'Un-Unterscheidbarkeits-Balken' dienen:

[1] Ein Ereignis ist 'früher', wenn es auf ein anderes eine kausale Wirkung haben kann

2. System-Simulation; Prozesse, Abläufe; Verhalten

Zunächst scheint eine Präzisierung der Ausdrucksweise angebracht: obwohl vielfach so (verkürzt) genannt, sind es nicht Systeme, die simuliert werden, sondern die in ihnen ablaufenden Vorgänge, ihre Prozesse, d.h. das *Verhalten* des Systems im Verlaufe der Zeit.

Eine Verhaltens-Simulation sollte *kongruent in beiden Richtungen,* Realität $\rightarrow$ Simulations-Modell ($\mathcal{R} \longrightarrow \mathcal{S}$) und Simulations-Modell $\rightarrow$ Realität ($\mathcal{S} \longrightarrow \mathcal{R}$) sein.

Gemeint ist folgendes: Wenn das reale System $\mathcal{R}$, in der Situation (im 'Zustand') r befindlich, durch eine Aktion ϱ in die Situation r' gerät, und dieses modelliert wird mit Φ in ein Simulations-Modell $\mathcal{S}$, in dem der Übergang von $\mathcal{S}$ aus der Situation ('Zustand') s durch die modellierte Aktion σ in die Situation s' abgebildet wird, dann soll auch selbstverständlich r' durch die Modellierungsabbildung Φ in s' übergehen, und die Abbildung Φ^{-1} führt s in r und s' in r' zurück, d.h. die Rollen von $\mathcal{R}$ und $\mathcal{S}$ könnten genausogut vertauscht sein: **Bi-Simulation**.

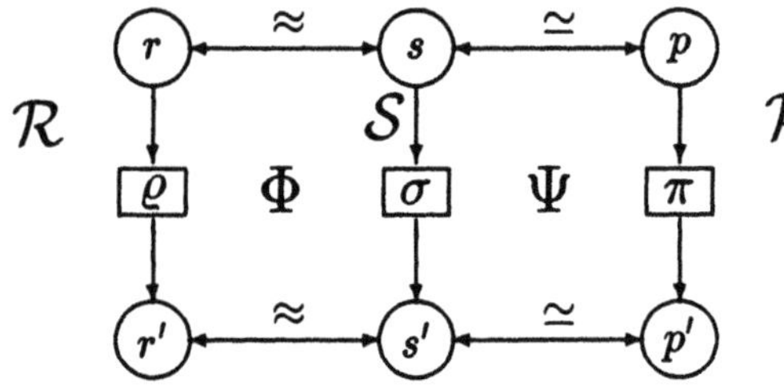

Desgleichen: Wenn das Simulations-Modell $\mathcal{S}$, in der Situation ('Zustand') s befindlich, ... und die dazugehörige numerische Berechnung durch die Abbildung (Programmierung) Ψ des Modells $\mathcal{S}$ in ein Programm $\mathcal{P}$ aus der Situation ('Zustand') p durch den Programmablauf π die Folgewerte p' errechnet, dann auch $\Psi(s') = p'$, und $\Psi^{-1}(p) = s$ und $\Psi^{-1}(p') = s'$.

Da meist noch mehr Abbildungsschritte bei der Simulation dazwischengeschaltet sind (z.B. Zielvorstellungen über die Simulationsergebnisse, graphische Darstellungen etc), muß man diese Forderung für alle Zwischenschritte erheben. (Tri-Simulation ...).

Durch ungeeignete Wahl der Simulationssprache wird bewirkt, daß diese Forderung – zumindest für die in der Einleitung angegebenen Fälle – garantiert *nicht* erfüllt werden kann.

3. Petri-Netze

In der *concurrency*-Theorie sind die Petri-Netze (PN) das mathematisch bzw. systemtheoretisch am weitesten bearbeitete Modell. Sie sind – zumindest ihre graphischen Symbole □ (Kästchen) und ○ (Kreise), die durch $\longrightarrow$ (Pfeile) verbunden werden – allgemein bekannt genug, so daß sich eine längere Beschreibung hier erübrigt.

$\mathrm{PN} = \{\mathrm{S}, \mathrm{T}, \mathrm{F}; \mathrm{M}_0; \mathcal{R}\}$ (Stellen, Transitionen, Flußrelation; Anfangsmarkierung; Schaltregel)

Dabei stehen die Kästchen für dynamische Objekte (Veränderungen, Ereignisse, System- bzw. Zustandsübergänge), die Kreise für statische Objekte (Dinge, Variablen, ihre Werte, Zustände...).
Die Pfeile geben Richtungen an: a) Flüsse von physischen Objekten (Materialien, Substanzen ...),

von ideellen Objekten (Geld, Zeit) - oder b) Einflüsse, Wirkungen, Informationsweitergabe (Unterschrift, Autorisierung, Kenntnisnahme, Geheimnummer/Code/Schlüssel etc.).
Die Marken auf einer Stelle bezeichnen das momentane Vorhandensein eines Objektes (auf diesem Platz), die Gültigkeit einer Bedingung, den Wert u.ä. (Ggf. ist bei der Darstellung von Nachrichten zu unterscheiden zwischen dem materiellen Datenträger und der immateriellen Information selbst.)
Die Schaltregel beschreibt die dem Netz innewohnende Dynamik, die Veränderungs-Automatik.

Zwar ist die formal korrekte Anwendung der PN-Graphik schon allein für sich eine nützliche Bereicherung der System-Beschreibungs- und -Darstellungsmöglichkeiten, erzwingt sie doch i.a. eine korrekte Darstellung von Flußstrukturen und kausalen Abhängigkeiten (der Relationen nacheinander/nebeneinander). Aber die gezielte inhaltliche Anwendung der *concurrency*-Theorie, darstellbar durch diese PN-Graphik, kann noch zu einigen Verbesserungen bei der Konstruktion von Systemen (realen und Simulations-Systemen) beitragen. Das wäre der Fall, wenn man die logisch bzw. räumlich verteilte Struktur des Original-Systems genauso verteilt durch Teil-Netze darstellt, und wenn man nebenläufige Abläufe im Original gezielt in ebensolche des Modell-Systems abbildet.

Unter den PN gibt es verschiedene Klassen und Spezialisationen, in denen unterschiedliche Eigenschaften gelten; je nach beabsichtigten Zweck werden geeignete Netz-Typen benutzt.

4. Konkurrente Systeme, Parallel-Verarbeitung, Sequentialisierung

Nebenläufigkeit prägt das Verhalten vieler Systeme in signifikanter Weise; ohne ihre Betrachtung bleiben bestimmte Verhaltensweisen mancher realer Systeme unerklärlich. Sie ist allerdings aus den eingangs erwähnten Gründen (zentrale/hierarchische Steuerung, Steuerungs-Geschwindigkeit ähnlich der Arbeitsgeschwindigkeit) lange unbemerkt geblieben. Das änderte sich schlagartig, als die Objekte, die in einem System verarbeitet wurden, selber Nachrichten und Kommandos waren, also mit der Erfindung der Computer, und wurde noch eimal erheblich wichtiger mit der Erfindung der Mehrprozessor-Maschinen. Vorher konnten meistens noch fehllaufende Verarbeitungsmechanismen durch immer schnellere Steuerungssignale eingeholt und korrigiert werden.
Im Computer jedoch wurde mit derselben (Höchst-)Geschwindigkeit gesteuert wie verarbeitet.

Konkurrente Systeme sind dadurch gekennzeichnet, daß sie keine zentrale Steuerung enthalten (das sind z.B. alle Systeme, in denen Menschen bestimmend mitarbeiten, ferner viele Mehrprozessorsysteme, Datennetze, mailing-Systeme). Die system-typische partielle Autonomie bedeutet lokale Entscheidungsbefugnis und Verfolgen eigener Interessen, die allerdings denen anderer Systemteile zuwiderlaufen können. Zum Funktionieren eines konkurrenten Systems ist erheblicher organisatorischer Aufwand neuerer Art nötig, damit nicht unerwünschte Zustände wie Systemstillstand, unendliche Schleifen, unbeabsichtigtes Auslöschen von später noch benötigter Information u.dergl. eintreten.

Besonderes Augenmerk ist der Sequentialisierung von nebeneinander ('parallel') ablaufenden Prozessen zu widmen; da ja das Ergebnis in beeindruckender Weise von der Reihenfolge (Beispiel: update/lesen) abhängen kann. Unglücklicherweise werden im Umgang mit Simulations-Programmen viele dieser Konflikt-Entscheidungen dem jeweiligen Programm bzw. Betriebssystem überlassen, sodaß der Kunde nicht einmal von der Existenz, geschweige denn von der Lösung solcher Konflikte erfährt. Damit hat der Benutzer die Kontrolle über die Simulationsergebnisse verloren.

5. Zusammenfassung

Halb-Ordnungen treten in realen Systemen viel häufiger auf, als man sich weithin bewußt ist, z.B. in der ganzen experimentellen Physik. Wohlgemerkt: hier wird nicht behauptet, daß die zu messenden *Objekte* bzgl. der zu messenden Eigenschaft nicht vollständig geordnet *sind* bzw. sein könnten, sondern, daß man eine Voll-Ordnung *nicht feststellen kann*, d.h. daß die *Meßwerte* halb-geordnet sind. Diese Halb-Ordnungen werden i.a. auf eine Voll-Ordnung (z.B. auf die rationalen Zahlen) abgebildet. Inwieweit dadurch Simulationsergebnisse verfälscht werden, oder gar bei der Systemkonstruktion etwas Unmögliches versucht wird, muß von Fall zu Fall geprüft werden.

Hier scheint es noch einen erheblichen Forschungs- und Aufklärungs-Bedarf zu geben.

LITERATUR:

H.Fuss: *Reversal Simulation with Place-Transactor-Nets.*
in: H.Wedde (Ed.): *Adequate Modeling of Systems.* pp.222-232, Springer (1983)

H.Fuss: *Zur Simulation von Zufall und Verläßlichkeit.*
in: D.P.F.Möller (Hrsg.): *Simulationstechnik.* pp.141-146; Inf.FB 109, Springer

H.Fuss: *Simulating 'Fair' Random Numbers.*
in: Proc. 2nd Europ. Simulat. Congress ESC '86 (Antwerp), pp.250-257. SCS (1986)

H.Fuss: (Hrsg.): *Fachgespräch Simulation mit Petri-Netzen und verwandten Methoden.*
in: R.Valk (Hrsg.): Proc. 18. GI-Jahrestg., pp. 533-606; IFB 187, Springer (1988)

H.Fuss: *Is Parallel Processing a Temporal Issue Only?*
in: D.Murray-Smith/D.Stephenson/R.Zobel (Eds.) Proc. 3rd Europ. Simulat. Congress ESC '89 (Edinburgh), pp.35-40. SCS Simulation Councils Inc., (1989)

H.Fuss: *Wie lang ist ein Meter? – Reflexionen über Daten in Simulationsmodellen*
in: Breitenecker/Troch/Kopacek (Hrsg.): *Simulationstechnik.* Proc. 6.ASIM-GI-Sympos. (Wien), pp.102-106, Fortschritte in der Simulationstechnik, Bd.1, Vieweg (1990)

B.Baumgarten: *Petri-Netze* Grundlagen und Anwendungen. 369 S., B.I.Verlag 1990

C.A.Petri: *Kommunikation mit Automaten.* IIM, Uni Bonn; Schrift Nr. 2, 1962

C.A.Petri: *Fundamentals of a Theory of Asynchronous Flow.*
in: Proc. IFIP Congr. 62, München, pp.386-390. N.Holl. 1963

C.A.Petri: *Concepts of Net Theory.* Proc. H.Tatr. pp.137-146. 1973

C.A.Petri: *Interpretations of Net Theory.* GMD-ISF Bericht 75-07

C.A.Petri: *Nichtsequentielle Prozesse.* IMMD Erlangen Ber. Vol.8, Nr.8 pp.57-82 (1976)
GMD-ISF 76-06.3 (Revd. Aufl. 1977) Engl. transl.: GMD-ISF 77-05

C.A.Petri: *General Net Theory.*
in: B.Shaw (ed.) Proc. IBM & Newcastle Seminar Sept.76, pp.131-169

C.A. Petri: *Concurrency* in: W.Brauer (ed.): *Net Theory and Applications,* Proc. Advanced Course, Hamburg LNCS 84, pp.251-260. Springer (1980)

C.A.Petri: *State-Transition Structures in Physics and in Computation.*
in: Int.J.Th. Physics, Vol. 21, No. 12, pp.979–992, (1982)

C.A.Petri: *Concurrency Theory* in: W.Brauer/W.Reisig/G.Rozenberg (eds.): *Advances in Petri Nets 1986,* Proc. Adv.Course, Bad Honnef LNCS 254, pp.4-24. Springer (1987)

C.A.Petri, E.Smith: *Concurrency and Continuity.*
in: LNCS Vol. 266, pp.273-292. Springer (1987)

W.Reisig: *Petri-Netze,* Eine Einführung. 158 S. Springer 1982
W.Reisig: *Petri Nets;* An Introduction. EATCS Vol.4, 161 p., Springer 1985

W.Reisig: *Systementwurf mit Netzen.* 125 S. Springer 1985

E.Smith: *Zur Bedeutung der Concurrency-Theorie für den Aufbau hochverteilter Systeme*
(Dissertation) GMD-Bericht Nr. 180. (164 S.) R.Oldenbourg (1989)

K.Voss/H.J.Genrich/G.Rozenberg (eds.) *Concurrency and Nets.* 622 p., Springer 1987

H.Plünnecke, W.Reisig: *Bibliography of Petri Nets 1990*
(4099 Einträge) in: *Advances in PN 1991,* S.317 ff, LNCS 524, Springer 1991

Petri Net Newsletter der GI-FG 1.1.2 'Petri Nets and Related System Models', 3 Hefte/Jahr

Diverse Modell-Generatoren und Simulatoren von Petrinetzen, z.T. als Tools mit ganzen Entwicklungs-Umgebungen und großer Software-Ausstattung, sind vielfältig erhältlich. Übersicht in:

F.Feldbrugge: *Petri Net Tools Overview.* in: LNCS 424, pp.151-178; Springer 1990

Echtzeit-Modellierung instationärer binomialer bzw. Poissonscher Ereignisfolgen

Günter Dörfel und Verena Schöps
Institut für Festkörper- und Werkstofforschung i. G.
8027 Dresden,Helmholtzstraße

1 Motive und Anwendungen

Jedes informationstragende physikalische Signal weist stochastische Merkmale auf. Dieser seit SHANNON auch quantitativ faßbare Sachverhalt ist nicht nur bei der Messung, Übertragung, Aufbereitung und Deutung physikalischer Signale zu berücksichtigen, sondern auch bei der Vorbereitung, Testung, Kalibrierung und Wartung der Meß- und Auswerteeinrichtungen und -prozeduren. Letzteres insofern, als die zum Test erforderlichen Signale den natürlichen Meßsignalen immer ähnlicher sein müssen, je stärker das reale Meßsystem von idealisierenden Modellen abweicht, und je komplexer sich die Meß- und Auswerteprozeduren gestalten. Dieser Gesichtspunkt rückt besonders dann ins Blickfeld, wenn die Messung mit einer signifikanten Veränderung bzw. einer Zerstörung der natürlichen stochastischen Signalquelle einhergeht, und damit die Wiederholbarkeit des Meßvorganges zur Ergebnissicherung nicht möglich ist. Anderseits sollte der Test – bei aller wünschenswerten Ähnlichkeit zwischen natürlichem stochastischen Meßsignal und synthetischem Testsignal – auch im Detail wiederholbar sein. Die Überbrückung dieses Widerspruchs führt auf die Anwendung *pseudo*-stoachastischer Testsignale – d.h. vollständig determinierter Signale mit Häufigkeitseigenschaften, die den jeweils maßgeblichen Wahrscheinlichkeitseigenschaften der nachzubildenden stochastischen Prozesse entsprechen oder zumindest nahekommen.

Die Fragen, die diese Vorgehensweise aufwirft, sind evident: Sie betreffen, was die Qualität der Modellierung eines komplexen Eigenschaftsbildes betrifft, den strukturellen und apparativen Bereich, und sie erheischen weiter Antwort, ob und wie das physikalische Meßkonzept, welches das Modellsignal in die Auswertung einbezieht, den prinzipiellen Widerspruch zwischen der Streuungsfreiheit determinierter Abläufe und der zum Eigenschaftsbild zwingend gehörenden Streuung der nachzubildenden natürlichen Prozesse aufzulösen vermag. Der letztgenannte Gesichtspunkt erlangt, wie noch gezeigt werden soll, bei der Synthese instationärer Prozesse besondere Bedeutung. Die im folgenden thesenhaft und ohne Beweisführung diskutierten Überlegungen und Ergebnisse wurden stimuliert und erlangt bei methodischen, programmtechnischen und apparativen Entwicklungen zur Modellierung belastungsinduzierter Geräuschemissionen in Werkstoffproben [1],[2] und anderer mit dem Ziel einer Schadensanalyse auszuwertender Signale [3]. Sie wurden weiter angeregt durch die Frage, ob man die einen mit kurzer Halbwertzeit ablaufenden radioaktiven Zerfallsprozeß kennzeichnenden Signale mit deterministischen Generaturstrukturen so gut nachbilden könne, daß es einem nachgeschalteten technischen Auswertesystem mit Gedächtnis (Vielkanalanalysator) nicht mehr möglich sei, zwischen natürlichem und pseudostochastisch erzeugtem Signal zu unterscheiden. Eine bejahende Antwort sollte dann die Möglichkeit eröffnen, dem natürlichen Meßsignal ein synthetisches Referenz-Signal vergleichbarer Zeitcharakteristik so zu überlagern, daß vom beobachtbaren Schicksal des Referenzsignals auf die Verfälschung, die das natürliche Meßsignal im Auswertesystem erfährt, zu schließen ist. Zählverluste, bedingt durch Aufstock- und Totzeiteffekte im Meß- und Auswertesystem, wären auf diese Weise weitgehend zu eliminieren [4]. Allgemeiner dargestellt, realisiert ein so beschaffenes physikalisches Meßsystem den relativ selten möglichen Fall, daß die (für jedwedes Meßsystem in irgend einer Form unumgängliche!) Kalibrierung zeitgleich mit dem Meßvorgang und absolut realitätsbezogen abläuft, mit allen positiven Konsequenzen für die quantitative Ausdeutung der Meßergebnisse.

2 Stationäre pseudo–stochastische Ereignisfolgen

2.1 Realisierungsprinzipien

Alle im o.g. Zusammenhang realisierten Generatoren und Prinzipien bauen auf den Eigenschaften linearer Maximalfolgen auf. Diese besagen bekanntlich, daß linear rückgekoppelte diskrete Schiebestrukturen unter bestimmten Voraussetzungen Symbolfolgen mit maximaler Länge und "günstigen" Korrelationseigenschaften erzeugen. "Linear" heißt, daß das Rückkopplungssignal durch Wichtung und *additive* Verknüpfung bestimmter Elemente des Systemzustandes erzeugt wird. In den Voraussetzungen ist weiter eingeschlossen, daß Gewichte und Zustandsgrößen Elemente eines endlichen q-wertigen (q-Primzahlpotenz, $q=p^m$) Zahlenkörpers ("Galois-Feld GF(q)") sein müssen, und daß das die Rückkopplungsstruktur des Schieberegisters beschreibende charakteristische Polynom der Systemmatrix bestimmten Teilbarkeitsbedingungen genügt, d.h. "primitiv" ist. Wichtung (=Multiplikation) und Addition sind nach den für diesen Zahlenkörper geltenden Rechenregeln zu realisieren [5],[6]. Die Simulation POISSONscher Punktprozesse beruht dann auf der Auskopplung bestimmter Systemzustände und deren Vergleich mit einem vorgegebenen stabilen oder variablen Muster [7]. Wir erläutern die Vorgehensweise anhand eines Schemas (Abb.1) und beschreiben dabei auch, wie im Fall 2^m-wertiger Signale und Systeme die Strukturierung der Generatoren auch ohne Kenntnis der Rechenregeln und primitiven Polynome im $GF(2^m)$, d.h. ausschließlich auf den vollständig tabellierten Polynomen für das GF(2) aufbauend und ausschließlich die Operationen KONJUNKTION und ANTIVALENZ nutzend, gelingt [8],[9]. Das ist insofern von entscheidender Bedeutung, als für extrem hohe Wertigkeiten, wie sie die eingangs genannten Anwendungen erfordern, weder die Rechenregeln noch die darauf aufbauenden geeigneten primitiven Generator–Polynome bekannt sind. (Und es darf bezweifelt werden, ob sie – falls bekannt – im Echtzeitbetrieb gut handhabbar wären!)

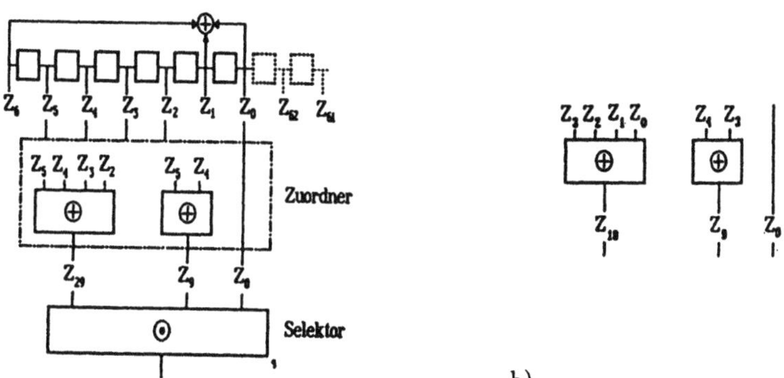

a) b)

Abb. 1a) Vereinfachter Modellgenerator zur Erzeugung binomialer Ereignisfolgen b) Zuordner, zu dem in Abb.a) angegebenen gleichwertig, aber mit im Raster Q=9 äquidistanter Auskopplung

2.2 Entwurfsregeln und Eigenschaften

Ein $(r = m \cdot n)$ bit breites, nach

$$z_{r-1}(i+1) = z_r(i) = z_0(i) \oplus \sum_{\alpha=1}^{r-1} c_\alpha z(i) \qquad c_\alpha \in GF(2) \tag{1}$$

rückgekoppeltes Schieberegister durchläuft dann lückenlos und wiederholungsfrei alle

$$L = 2^r - 1 \tag{2}$$

möglichen Zustände, wenn das der Rückopplungsschaltung entsprechende charakteristische Polynom

$$F(D) = D^r \oplus \sum_{\alpha=1}^{r-1} c_\alpha D^\alpha \oplus D^0 = 0 \tag{3}$$

irreduzibel ist und die Periodizitätsgleichung $D^{L'} \oplus D^0 = 0$ für kein $L' < L = 2^r - 1$ ohne Rest teilt. Solche Polynome heißen "primitiv" oder werden als"Maximalexponent-Polynome"bezeichnet. Von insgesamt L durch nach Takten gemessene Phasendifferenzen unterscheidbaren MF $z_{\varphi\nu}(i)$ werden m ausgewählt,

$$\mathbf{z}^T(i) = (z_{\varphi 0}, z_{\varphi 1}, \ldots, z_{\varphi m-1}) \quad , \tag{4}$$

durch Linearkombination der am Register abgreifbaren Folgen erzeugt und auf das Auftreten eines bestimmten Musters, z.B. des Alles-Eins- Musters, geprüft. Die so gewonnene Ereignisfolge

$$\mathbf{w}(i) = \{z_{\varphi 0}(i), z_{\varphi 1}(i), \ldots, z_{\varphi m-1}(i)\} \tag{5}$$

ist unter bestimmten Voraussetzungen als Muttertakt für die Erzeugung binomialer,im Grenzfall $m \gg 1$ Poissonscher Impulsfolgen optimal geeignet:

a) Die Phasendifferenzen zwischen den m ausgekoppelten Folgen $z_{\varphi_\nu} = z_{\varphi 0 + jQ}$ sind ganzzahlige Vielfache von Q,

$$Q = \frac{2^{m \cdot n} - 1}{2^m - 1} \quad , \quad j \in (0, 1, \ldots, 2^{m-1}) \quad . \tag{6}$$

b) Alle ausgekoppelten Folgen z_{φ_ν} sind linear unabhängig. I.a. bedarf das der Überprüfung, etwa anhand der automatentheoretischen Darstellung des Generierungsprozesses:

$$\mathbf{z}(i) = \mathbf{C}\mathbf{A}^i z(0) \quad ,$$

wobei nach Abb.1

$$\mathbf{A} = \begin{pmatrix} 0 & 0 & 0 & 0 & 1 & 1 \\ 1 & 0 & 0 & 0 & 0 & 0 \\ \vdots & \vdots & \ddots & & & \\ 0 & 0 & 0 & 0 & 1 & 0 \end{pmatrix} ; \mathbf{C}_{a)} = \begin{pmatrix} 1 & 1 & 1 & 1 & 0 & 0 \\ 1 & 1 & 0 & 0 & 0 & 0 \\ 0 & 0 & 0 & 0 & 0 & 1 \end{pmatrix} ; \mathbf{C}_{b)} = \begin{pmatrix} 0 & 0 & 1 & 1 & 1 & 1 \\ 0 & 1 & 1 & 0 & 0 & 0 \\ 0 & 0 & 0 & 0 & 0 & 1 \end{pmatrix}$$

mit Rang $\mathbf{C} = m = 3$ gilt.

Sind die ausgekoppelten Folgen z_{φ_ν} im Phasenraum jQ benachbart, so ist lineare Unabhängigkeit a priori gegeben! Die vorteilhaften Konsequenzen aus dieser Bemessung sind:

I: Alle (existierenden) Distanzen zwischen beliebigen (also nicht notwendig benachbarten !) Ereignissen einer Periode treten gleich häufig auf.

II: Die Muster, die $w(i)$ nach (5) realisieren, treten mit einer auf L bezogenen Häufigkeit

$$p_m = 2^{-m} \tag{7}$$

auf.

III. Auch alle höheren Häufigkeiten, d.h. die für das Auftreten und/oder Nichtauftreten von Ereigniskombinationen in $a \leq n$ nicht notwendig aufeinanderfolgenden Takten verantwortlichen , gehorchen dem binomialen Verteilungsgesetz. Das schließt insbesondere auch das Auftreten von Vielfachimpulsen (s-fach) mit der erwarteten Häufigkeit

$$p_s = 2^{-m \cdot s}, \qquad s \leq n \tag{8}$$

ein.

Abweichungen von diesem Verhalten treten nur insofern auf, als Impulsabstände

$$\Delta = j \cdot Q \quad , \quad j \text{ ganzzahlig,} \tag{9}$$

nicht existieren.
Folgerichtig werden Abweichungen vom idealen Zweiniviau-Verhalten der (normierten) Autokorrelationsfolge R(i) nur für Verschiebeparameter nach (9) beobachtet:

$$R_{ww}(i) = \begin{cases} 2^{m(n-1)} & \text{für } i = \pm l \cdot L, \quad l = 0,(1),\ldots \\ 0 & \text{für } i = \pm l \cdot L \pm jQ\ , \quad j \in [1,2,\ldots,2^{m-2}] \\ 2^{m(n-2)} & \text{sonst.} \end{cases} \tag{10}$$

2.3 Das Streuverhalten

Alle vorstehend genannten Eigenschaften gelten für die vollständige Generatorperiode und sind unter dieser Voraussetzung bei beliebigen Startbedindungen rechnerisch und experimentell streuungsfrei nachweisbar. Deutet man den instationären Prozeß als eine Aneinanderreihung (ggf. sehr kurzer) stationärer Prozeßabschnitte, so ist dieser aber nur dann durch geeignete Modulationsverfahren aus einem stationären Prozeß ableitbar, wenn sich alle erwarteten Eigenschaften auch in jedem beliebigen kurzen Prozeßabschnitt auffinden lassen, und zwar mit einer dem natürlichen Vorbildprozeß entsprechenden Streuung. Der Widerspruch zwischen determinierter Erzeugung und dem Begriff der Streuung wird dabei behoben, indem man alle $L = 2^r - 1$ denkbaren Anfangsbedingungen als gleichberechtigt und gleichwahrscheinlich ins Kalkül zieht. Die exakte Analyse [10] ergibt für N in ν Takten erzeugte Ereignisse

$$var\overline{N_\nu} = \overline{N_\nu^2} - \overline{N_\nu}^2 \quad ,$$

$$var\overline{N_\nu} = \frac{1}{2^r - 1}\left\{\nu 2^{r-m} + [\nu(\nu-1) - 2(\nu[|\frac{\nu}{Q}|] - Q\sum_{\varrho=1}^{[|\frac{\nu}{Q}|]}\varrho)]2^{r-2m}\right\} - \frac{\nu^2}{(2^r-1)^2} - 2^{2r-2m} \quad ,\tag{11}$$

$[|\frac{\nu}{Q}|]$ - größte ganze Zahl $\leq \frac{\nu}{Q}$,

und damit für $\nu < Q$ Takte

$$var\overline{N_\nu} \approx \nu 2^{-m} \quad . \tag{12}$$

Gl.(11) steht also für ein Streuverhalten, welches für kleine Prozeßabschnitte dem binomialen bzw. POISSONschen Verteilungsgesetz gehorcht und für lange Abschnitte schrittweise auf die für die vollständige Generatorperiode geltende (und gewollte!) Streuungsfreiheit hinzielt (Abb.2). Damit sind die Voraussetzungen für die Ableitung instationärer Ereignisfolgen aus stationären gegeben.

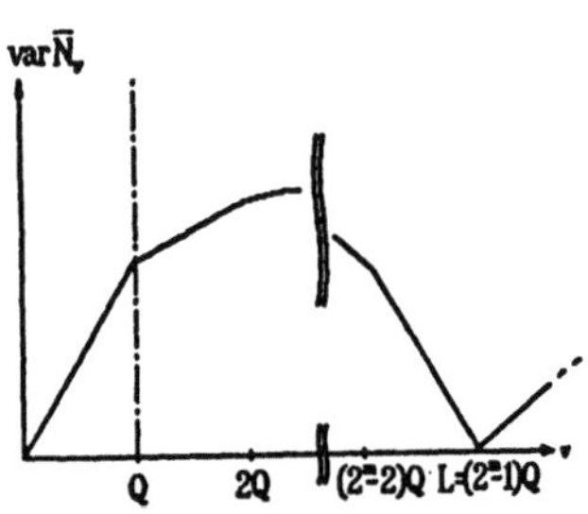

Abb.2: Varianz über der Taktzahl

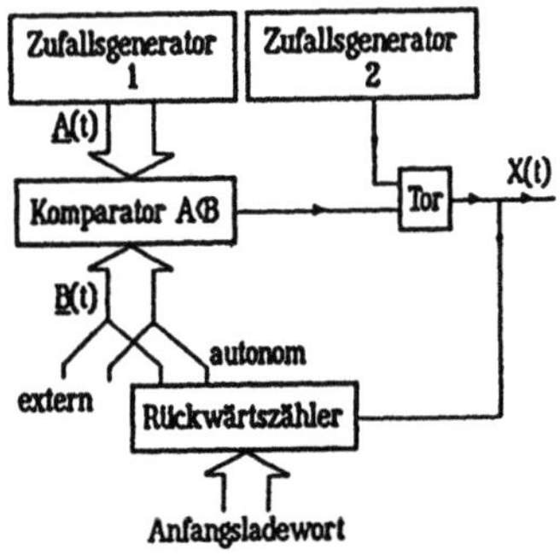

Abb.3: Instationärer Generator

3 Instationäre Ereignisfolgen

3.1 Gesteuerte und rückgekoppelte Prozesse

Ein Grundschema zur Erzeugung instationärer Folgen zeigt Abb.3. Der Zerfallsgenerator 2, er möge die im Abschnitt 2 entwickelten Eigenschaften besitzen, arbeitet auf ein Tor, über dessen Öffnung ein Komparator entscheidet. Letzterer wird von einem breitbandigen gleichverteilten Zufallsprozeß angesteuert. Die Ereignisdichte des ausgegebenen Pulsprozesses $X(t)$ folgt dem (inversen) Zeitgesetz, mit welchem die Komparatorschwelle geführt wird. Der instationäre Prozeß kann als durch "zufälliges Stehlen" entstanden interpretiert werden, wobei die Zufälligkeit des Stehlens eine ungewollte Strukturierung des Prozesses $X(t)$ ausschließen soll. Im rückgekoppelten Fall, wenn die Komparatorschwelle von der Zahl der akkumulierten Ausgangspulse abgeleitet wird, ergibt sich ein exponentielles Schwächungsgesetz für die Ereignisdichte:

$$\overline{x}(t) = \overline{x_0} e^{-\lambda t} \tag{13}$$

Dabei bestimmen die durch das "Anfangsladewort"festgelegte Zählkapazität N und die Anfangsimpulsdichte $\overline{x_0}$ den Schwächungskoeffizienten

$$\lambda = \frac{\overline{x_0}}{N} \tag{14}$$

Auch hier erweist sich für die strukturelle Ausgestaltung ein Konzept, welches auf dem "$m \cdot n$ – Feeback Shift Register Generator" [8] aufbaut, als vorteilhaft. Die diesbezüglichen Überlegungen werden in drei Schritten entwickelt und führen zu folgenden Eigenschaften:

IV: Werden im Selektor lediglich $m' < m$ von m durch den Zuordner ausgekoppelten Folgen bewertet, so erfährt das im Abschnitt 2 dargestellte Eigenschaftsbild keine grundsätzliche Veränderung. Die Ereignisdichte wird nunmehr von m' bestimmt; die unvermeidlichen Störungen liegen weiter im Abstand Q.

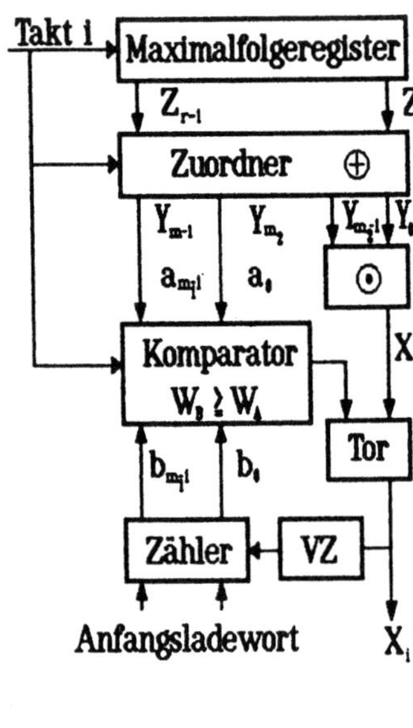

Abb4.: Autonomer instationärer Generator auf der Grundlage des $m \cdot n$-FSRG-Konzepts

V: Entgegen der Annahme, daß die Speisung der beiden Pseudo-Zufallsgeneratoren 1 und 2 (Abb.3) aus einem gemeinsamen Schieberegister mit m Auskopplungen — $m = m_1 + m_2$ — zu Verkopplungen und zusätzlichen Störungen führt, erweist sich dieser Weg, sofern die im Abschnitt 2 beschriebenen Grundregeln eingehalten werden, als der optimale (Abb.4). Eine Erklärung dafür liegt in dem Umstand, daß das Produkt zweier maximallanger Perioden selbst keine maximallange Periode sein kann:

$$(2^{m_1} - 1)(2^{m_2} - 1) < 2^{m_1+m_2} - 1 \tag{15}$$

Zwei unabhängige Generatoren 1 und 2 erzeugen also mehr Fehlstellen in der Abstandsverteilung als ein optimierter gemeinsamer.

VI: Man kann zeigen, daß die binär eingegebenen Komparatorschwellen, welche Abschwächungen entsprechend einer 2er-Potenz $2^{-\nu}$ erzeugen, exakt auf Ausgangsprozesse führen, die entstünden, wenn der Selektor weitere Abgriffe $> m_2$ in die Musterfindung einbezöge. Und zwar nicht nur bezüglich der integralen Eigenschaften, sondern auch bezüglich der schrittweisen Signalentwicklung. Damit scheint – ähnlich wie bei Eigenschaft IV –sicher, daß der Abschwächungsprozeß den Charakter der Ereignisfolge nicht ändert. Damit sind die wesentlichen strukturellen Merkmale und Signaleigenschaften für eine Klasse von Generatoren, die dem "$m \cdot n$ - FSRG" – Konzept folgt, genannt. Bezüglich einer technischen Ausgestaltungsmöglichkeit – orientiert an Forderungen aus der Kernstrahlungsmeßtechnik – sei auf [11] verwiesen.

3.2 Zerfallsketten

In der rückgekoppelten Struktur nach Abb.3,4 repräsentiert das "Anfangsladewort" des Rückwärtszählers z.B. die Zahl der zerfallsbereiten Atome einer Probe eines radioaktiven (kurzlebigen) Nuklids. Ist das Zerfallsprodukt seinerseits radioaktiv, so läßt sich dieser Vorgang modellieren, indem der Prozeß $X(t)$ dem Leerzählen des Zählers eines zweiten Generators, welcher den Tochterprozeß generiert, entgegenwirkt (Abb.5).

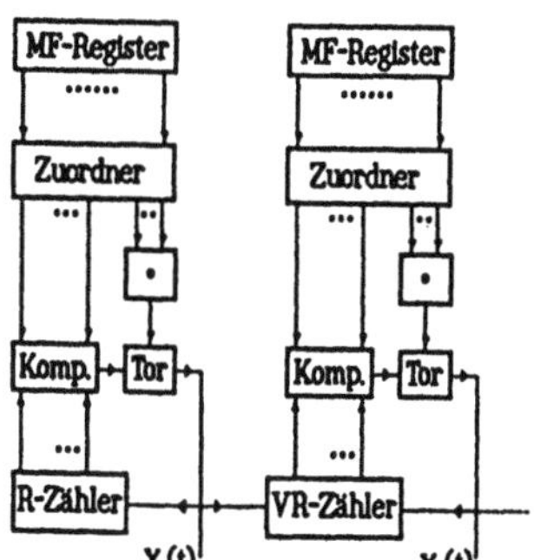

Abb.5: Modellierung von Zerfallsketten

4 Weiterungen

Die vorstehend beschriebenen Strukturen und ihre vorteilhaften Eigenschaften wecken weitere Wünsche. Sie betreffen informationstechnische und mathemathische Fragen zur Optimierung der Ausgestaltung der vorgestellten Strukturen, sie fragen nach der Nutzung anderer physikalischer Phänomene bei der Realisierung der Strukturen und bei der Erschließung neuer (höherer) Frequenzbereiche, und sie berühren übergreifende methodische Probleme:

- Hohe Ansprüche an die Qualität des Signalmodells und eine Annäherung an die Vielfalt der konkreten physikalischen Abläufe erfordern breite Register und damit auch eine große Zahl von logischen Verknüpfungen im Zuordner. Die Folgen dieses Aufwandes sind in erster Linie physikalisch-technischer Natur: sie wirken taktfrequenzbegrenzend. Nun lassen sich Wahrscheinlichkeitsaussagen machen, mit welcher Erfolgsaussicht man unter den fast $2^{m \cdot n}$ denkbaren gleichwertigen Realisierungen einer vorgegebenen Strukturklasse - sie ergeben sich aus der Verschiebung des lediglich durch Phasen*differenzen* festgelegten Systems von Auskopplungen über die gesamte Generatorperiode - aufwandsminimale Realisierungen, also solche mit einer vorgegebenen minimalen Anzahl logischer Verknüpfungen auffindet. Die Frage geht an den Mathematiker bzw. Informatiker, wie man auf effektive Weise solche Minimalstrukturen findet und welche Gesetzmäßigkeiten diesen Minimalstrukturen und ihrer Auffindung zugrunde liegen.

- Im Zusammenhang mit dem Streben nach höheren Grenzfrequenzen ist zu überlegen, ob und mit welchem Aufwand die klassische Schiebekette durch Laufzeitelemente ersetzt werden kann, die auf der Wellennatur akustischer oder elektrischer oder anderer Signale aufbauen. Die Probleme bestehen hier in der Realisierung der Steuerbarkeit der Laufzeit - die Taktsteuerung der Schieberegister ist da noch unübertroffen - und in der sauberen Realisierung der logischen Funktionen beim Verknüpfen von Wellenpaketen.

- Das hier vorgestellte Signalmodell und dessen Implementierung fußen auf algebraischen Methoden, die auch zur Beschreibung kollektiver physikalischer Phänomene selbst erfolgreich benutzt werden. Es wäre zu überlegen, welche (Echtzeit-)modelle physikalischer Prozesse möglich werden, wenn die dem hier beschriebenen Anwendungsfall zugrundeliegenden gravierenden Einschränkungen (Autonomie (Ausnahme: Zerfallskette), Primitivität des charakteristischen Polynoms, Linearität der Operationen) aufgegeben werden.

Die nachfolgend aufgelisteten Zitate markieren etwa den Teil des Fachgebietes, an dessen Entwicklung die Autoren Anteil hatten. Eine Würdigung weiterer insbesondere durch [5] und [6] angeregter Publikationen findet sich in [8] und [9].

Literatur

[1] Pompe,W.; Morgner,W: Sitz.-Ber. der AdW Nr.10 N, 1981

[2] Liewers,P.: Rauschdiagnostik, Berlin: Akademieverlag 1985

[3] Dörfel,G; Kirchhoff,G.; Schöps,V.:
Simulation spannungsinduzierter Schallemission durch ein Gerät mit pseudostochastischer Signalerzeugung.
Feingerätetechnik 32(1983) 10, 457-459

[4] Dörfel,G.; Kluge,W.; Kubsch,M.:
A pseudorandom pulser technique for the correction of dead-time and pile-up losses in γ-ray spectrometry.
Nucl.Instrum.Meth. 214 (1983), 435-440

[5] Zierler,N.: Linear recurring sequences.
J.Soc.Indust.Appl.Math. 7 (1959), 31-48

[6] Golomb,S,W.: Shift register sequences.
San Francisco: Holden Day 1967

[7] Pangratz,H.: Elektronischer Generator zur Nachbildung der Poisson-Statistik.
Frequenz 33(1979) 7/8, 204-207

[8] Dörfel,G.: The concept of the $m \cdot n$-Feedback Shift Register Generator with Optimum (Pseudo-)Randomness Properties.
A.E.Ü. 40 (1986)4, 213-218

[9] Dörfel,G.: Pseudostochastische Impulsgeneratoren - Strukturen und Anwendungen.
Exp.Techn.Phys. 35(1987)1, 1-12

[10] Dörfel,G.: Variance Properties of the $m \cdot n$-FSRG
(in Vorbereitung)

[11] Dörfel,G.; Schöps,V.; Zimmermann,J.:
Steuerbare und instationäre Impulsquellen zur Simulation Poissonscher Nutz-und Störsignale.
atp-msr 33/34(1991)8, 332-335

Einsatz der Informationstechnik im Air Space Management

H. Scheck
Dornier GmbH
Postfach 14 20
7990 Friedrichshafen

1. Einleitung

Eine strikte Trennung von Informatik, Natur- und Ingenieurwissenschaften hat es (solange wie sie als selbständige Disziplinen existieren) und konnte es nie geben. Ihre Verflechtungen, insbesondere wenn man anwendungsbezogen denkt, nehmen jedoch in immer stärkerem Maße zu, wobei insbesondere die Informationstechnik eine zunehmende Rolle eingenommen hat. Ein Beispiel hierfür findet man im Bereich Air Space Management (Luftraumüberwachung und Flugsicherung).

2. Historische Entwicklung

Bereits kurz nach dem Nachweis elektromagnetischer Wellen durch H.R. Hertz [1] im Jahre 1888 folgten die ersten Anwendungen dieser bedeutsamen Entdeckung. So gehen auch die Ursprünge der Funkmeßtechnik auf die Anfänge dieses Jahrhunderts [2] zurück, jedoch wurde ihre Brauchbarkeit lange verkannt. Eine der ersten bekannt gewordenen Anwendungen war die berühmte "Home Chain" der Engländer im zweiten Weltkrieg, einer Reihe von 18 Türmen bestückt mit Radarantennen zur Vorwarnung gegen deutsche Luftangriffe. Während in der Anfangszeit des Radars nur die Annäherung von Objekten erfaßt werden konnte, hat sich der Funkmeßtechnik heute ein wesentlich weiteres Feld eröffnet. Die Verarbeitung der hohen Informationsdichten, ihrer Verknüpfung mit zusätzlichen Daten sowie die Verfügbarmachung für den Nutzer wurde erst durch Einsatz informationstechnischer Mittel ermöglicht.

3. Informationselemente im Air Space Management

3.1 Das Primär-Radar

Die wichtigste Informationsquelle des Air Space Management besteht auch heute noch im ursprünglichen Radar, auch Primär-Radar genannt.

Moderne Radare liefern ca. alle 10 Sekunden Entfernung, Azimuth und Höhe eines Flugobjekts, wobei die Reichweite bis zu mehreren hundert Kilometern beträgt [3]. Die kritische Größe ist hierbei die Flughöhe, die meist nur bis auf einige hundert Meter genau bekannt ist. Da die Flugsicherung den Flugobjekten unterschiedliche Höhen (Flight Levels) zuweist um Kollisionen zu vermeiden, wird zusätzlich zum Primär-Radar ein Sekundär-Radar eingesetzt.

3.2 Das Sekundär-Radar

Das Sekundär-Radar ist in der Regel auf der Antenne des Primär-Radars montiert (Balken-Antenne). Das Sekundärradar kann ein im Flugzeug installiertes Antwortgerät, den sogenannten Transponder, direkt ansprechen. Das Flugzeug liefert als Information einen ihm zugewiesenen Identifizierungscode und seinen Flight Level. Der Flight Level ist eine relative Höhenangabe, die aus dem aktuellen Luftdruck ermittelt wird. Da alle Flugzeuge in einer Region denselben Luftdruck messen, ist der Flight Level eine brauchbare Größe, die Flugzeuge höhenmäßig zu separieren.

3.3 Sprachkommunikation

Die Sprachkommunikation beansprucht derzeit einen Anteil von nahezu 50% der Arbeitszeit eines Fluglotsen [4]. Hier werden in Zukunft durch Einführung moderner Datenlinks zwischen Flugzeug und Zentrale operationelle Entlastungen erzielt.

3.4. Flugplandaten

Vor dem Start ist jeder Pilot verpflichtet, einen Flugplan mit Daten des Flugzeugs und der beabsichtigen Route abzugeben. Regelmäßige Flüge werden zentral gespeichert. Vor dem Flug werden die Flugpläne aktiviert und über das Aeronautical Telecommunication Network (AFTN) an die relevanten Flugsicherungstellen übermittelt [4]. Die Flugplaninformation dient den zivilen Zentralen zur Koordinierung der Luftraumnutzung sowie den militärischen Zentralen zur Identifizierung der Flugobjekte.

3.5 Sonstige Informationen

Eine Reihe bisher nicht aufgeführter Informationen, die aber nichtsdestoweniger im Air Space Management eine wesentliche Rolle spielen, sind z.B. Wetterdaten, Statusinformationen über Flugplätze, Flugzeuge etc.. Diese Daten werden über Weitverkehrsnetze ähnlich wie Flugplandaten an die zuständigen Stellen übertragen.

4. Einsatz der Informationstechnik beim Air Space Management

Viele der bisher genannten Informationselemente wären ohne Einsatz von Informationstechnik nicht oder nur stark eingeschränkt nutzbar. Allein in der EUROCONTROL-Zentrale in Maastricht werden pro Jahr z.Zt. 700.000 Flüge verarbeitet, so daß an die Leistungsfähigkeit künftiger Systeme hohe Anforderungen gestellt werden. Im folgenden sollen, ohne Anspruch auf Vollständigkeit, verschiedene Aspekte der Informationstechnik betrachtet werden, die die Qualität des Air Space Managements wesentlich beeinflussen. Die zunehmende Einführung von offenen Standards im industriellen Bereich lassen in Zukunft eine erhöhte Wirtschaftlichkeit und Operabilität der Flugsicherungs-Systeme erwarten.

4.1 Rechenleistung

Eine Erhöhung der Rechenleistung bewirkt quantitativ eine größere Anzahl kontrollierbarer Flugziele. Eine qualitative Verbesserung kann durch die statistische Kombination (Kalman Filter) der Daten mehrerer Radare (Multi Radar Tracking) erreicht werden [5]. Dies ermöglicht eine präzisere Lokalisierung der Flugobjekte sowie eine Erweiterung des Sichtbereiches einer Kontrollstation (Erhöhung der Flugsicherheit). Ebenso läßt sich ein höherer Automatisierungsgrad für den Operateur erzielen (z.B. Konflikterkennung und -auflösung), wobei bereits z.Zt. auch Ansätze in Richtung Expertensysteme untersucht werden.

4.2 Local Area Networks (LAN)

Lokale Netze stehen in direktem Zusammenhang mit der Rechenleistung, da sich durch die Verteilung der Rechenlast auf mehrere Computer die

Rechenkapazität wesentlich erhöhen läßt. Die LAN´s ersetzen dabei die bei älteren Systemen eingesetzten, meist proprietären Bus-Systeme der Systemhersteller. Hierdurch erhalten moderne Systeme eine neue Qualität im Hinblick auf Modularität, Erweiterbarkeit und Ausfallsicherheit.

4.3 Mensch-Maschine-Schnittstelle (MMS)

Die Verbesserungen der MMS haben für den operationellen Betrieb naturgemäß eine vordringliche Bedeutung. Durch die Einführung von Window-Systemen werden die komplexen mehrfach belegten Konsolen-Tastaturen in Zukunft obsolet werden [6]. Hinzu kommt die erhöhte Ergonomie von Großbildschirmen, die zudem durch entsprechende Farbdarstellung eine wesentlich größere Informationsdichte darstellen können. Heute werden bei neuen Anlagen in der Regel Auflösungen von 2048x2048 Pixeln gefordert [7].

4.4 Betriebssysteme

In der Radardatenverarbeitung werden Echtzeit-Anforderungen an die Systeme gestellt. Echtzeit-Betriebssysteme waren früher jedoch eher die Ausnahme und meist auf spezifische Hardware und Anwendungen abgestimmt. Insbesondere Realzeit-Unix-Varianten (z.B. RTU mit seinen Erweiterungen für die Interprozeßkommunikation und Synchronisation) werden sich in Zukunft wahrscheinlich durchsetzen, da sich auf diese Weise die Produktvielfalt und Standards der Unix-Welt nutzen lassen.

4.5 Programmiersprachen

Der Einsatz moderner Programmiersprachen und entsprechender Software-Tools nimmt Einfluß auf Ausfallsicherheit und Fehlerrate der Systeme. Gerade im Bereich der Flugsicherung werden hier hohe Anforderungen gestellt. Ebenfalls wird die Software-Wartung und -Pflege verbessert. Als Programmiersprachen werden z.Zt. in modernen Systemen vor allem Ada und C(++) eingesetzt.

5. Zusammenfassung und Ausblick

Aufgrund der (kostenbedingten) langen Nutzungszeiten (durchschnittlich ca. 15-20 Jahre) sind die derzeitigen Flugsicherungssysteme in der Regel technisch veraltet, so daß die EUROCONTROL und die BFS z. Zt. Planungen für zukünftige Systeme durchführen [8,9]. Hierbei wird klar, daß die eigentliche Radarmessung, mit der historisch gesehen begonnen wurde, immer stärker in den Hintergrund treten wird. So ist geplant, durch Einführung von Datenlinks zwischen Flugzeugen und Kontrollstationen den Informationsaustausch wesentlich zu erhöhen. Hierbei kann das bei vielen Fluggesellschaften bereits eingeführte Flight Management System (FMS) genutzt werden. Eine Positionsbestimmung kann das Flugzeug per GPS-Satellitennavigation durchführen [10], wobei Genauigkeiten zwischen 15-100 m erreicht werden. Dies ist wesentlich genauer als bei Radarmessungen.

Literaturverzeichnis:

[1] H.R.Hertz, Pogg. Ann. Physik u. Chem. (2) Bd. 34 (1888), S.197

[2] Ch. Hülsmeyer, mehrere internationale Patente

[3] Merill Skolnik, Radar Handbook, McGraw Hill Inc, New York, 1990

[4] H. Mensen, Moderne Flugsicherung, Springer-Verlag, Berlin, Heidelberg, New York 1989

[5] S.S. Blackman, Multiple-Target Tracking with Radar Applications, Artech House Inc., Norwood, 1986

[6] EUROCONTROL, Common Operational Performance Specifications for the Future Controller Operating Environment, Doc.Nr. GD-0046-03, 1991

[7] Dan Varon et al., 35th Annual Air Traffic Control Association Proceedings, 1990

[8] EUROCONTROL, ECAC Strategy for the 1990's

[9] BFS, Zukünftiges Betriebskonzept der Bundesrepublik Deutschland (CATMAC), BFS/Z-I6, 1990

[10] J. Porter et al., GPS/NAVSTAR System Overview, in: Global Civil Satellite Navigation Systems, Conference of the Royal Institute of Navigation, 1984

Lösung rechenaufwendiger Probleme

H. Ch. Zeidler
Universität der Bundeswehr,
Hamburg

K. Waldschmidt
J. W. Goethe Universität,
Frankfurt

Die Lösung rechenaufwendiger Probleme stellt einen der wesentlichen Berührungspunkte zwischen der Physik und der Informatik dar. Vordergründig erscheint diese Beziehung zunächst einseitig: Komplexe Fragestellungen der Physik sollen algorithmisch erfaßt und programmtechnisch bearbeitet werden, damit sie mit Hilfe eines Rechners effizient gelöst werden können. Andererseits aber benötigt auch die Informatik solche teilweise extremen Anforderungen als Initialzündung und Bewertungsmaßstab bei der Suche nach neuen leistungsfähigen Strukturen von Rechensystemen. Der weite Bereich der anwendungsbezogenen Architekturen und Systeme zeigt u.a. die fruchtbare Zusammenarbeit.

Viele Vorträge aus anderen Sitzungen stellen gleichzeitig auch einen Beitrag zum obigen Thema dar.

"Grand Challenges" - wie weit tragen die Antworten des Supercomputing?

F. Hoßfeld
Zentralinstitut für Angewandte Mathematik
Forschungszentrum Jülich (KFA)
Postfach 1913
5170 Jülich

1 "Grand Challenges" und "Computational Science"

Physik, Chemie, Biologie, Ökologie und Umweltforschung, Technik und Wirtschaftswissenschaften wie auch andere Bereiche außerhalb der exakten Naturwissenschaften werden mit Phänomenen der Dynamik komplexer nichtlinearer Systeme konfrontiert, deren Problembarrieren mit den heute verfügbaren theoretisch-analytischen Methoden und dem Wissen über das Verhalten und die Behandlung linearer Systeme nicht überwunden werden können. Die Untersuchung und Bewertung komplexer nichtlinearer Systeme ist deshalb darauf angewiesen, durch Ausschöpfen der Möglichkeiten der mathematischen Modellierung und der Computersimulation und den Ausbau ihrer Methoden erst das Gerüst an Wissen und Kenntnissen zu schaffen. Die Komplexität vieler Systeme (d.h. die Zahl der Freiheitsgrade, die Dimensionalität, vor allem die topologische un dynamische Struktur) überschreitet bei weitem das Potential konventioneller Verfahren und Rechenanlagen [1]. Diese Situation hat die Entwicklung einer dritten wissenschaftsmethodischen Kategorie erzwungen: die "Computational Science", die Theorie und Experiment qualitativ und methodisch (einer experimentellen Disziplin vergleichbar) ergänzt.

Computational Science ist synonym mit der Untersuchung komplexer Systeme; ihr Instrument ist der Supercomputer, ihre Methode die Simulation. Ihr umfassender strategischer Ansatz läßt sich daher nur schwer mit dem deutschen Begriff des wissenschaftlichen Rechnens einfangen; sie zielt vielmehr auf die großen ungelösten wissenschaftlichen Probleme, die in ihrer Wichtigkeit und Tiefe nicht nur die betreffende wissenschaftliche Disziplin herausfordern, sondern von außerordentlicher Bedeutung und Auswirkung für die Gesellschaft und ihre Bewältigung von Zukunft sind. Die Identifikation solcher Probleme hat in den USA zur Klassifizierung als "Grand Challenges to Computational Science" geführt, zu denen heute vor allem die Kernfragen aus den Gebieten der atmosphärischen Chemie (d.h. Umweltforschung), der Astrophysik, der Materialforschung, der Molekularbiologie, der Elementarteilchenphysik und der Aerodynamik gezählt werden [2].

Es besteht Übereinstimmung, daß diese Kernprobleme der Grand Challenges nur durch beständige Fortschritte auf dem Gebiet der Computational Science, und das heißt: in der Technologie, der Architektur, der Programmierung und der Algorithmik für innovative Rechner, einer Lösung näher gebracht werden können und schon von daher ein hohes Maß an interdisziplinärer Zusammenarbeit verlangen. Diese Erkenntnis hat in den USA seit Beginn der 80-er Jahre zu einer Kette, d.h. zu einer Folge von strategisch zusammenhängenden Initiativen geführt [3], die nicht zuletzt auch "die Mathematik als eine kritische Ressource für die Zukunft" ausweisen [4], und gerade jetzt appelliert die Association for Computing Machinery (ACM) für eine Revision des Selbstverständnisses der Computer Science/Informatik und für die verstärkte Hinwendung der Forschungsanstrengungen auf die Probleme der Computational Science ("..encourage more computing researchers to embrace computational science by joining in projects with physical scientists, bringing their expertise in algorithms and architectures." [5]).

Das Potential der Computational Science und damit des Supercomputing in der Forschung verstärkt verfügbar zu machen, ist eine dringende forschungsstrategische Antwort auf die Herausforderungen durch die ungelösten komplexen Probleme; dies erfordert die breitbandige Erschließung des Instrumentariums: Mathematik + Algorithmen + Computerfunktionen ("capability") + Computerleistung ("capacity"). Spät, aber vielleicht noch in der Zeit, wurde der Stellenwert der Computational Science und des Supercomputing und die Notwendigkeit für entsprechendes Handeln auch in Europa erkannt [6]. Es ist zu hoffen, daß die Umsetzung der Erkenntnis in Aktionspläne und Programme mit langfristig tragfähigen Rahmenbedingungen für kontinuierliche Forschungs- und Entwicklungsanstrengungen in allen Regionen Europas, in denen das Potential dafür vorhanden ist oder erfolgversprechend aktiviert werden kann, fruchtbar wird. Das verlangt sicherlich eine starke Kohärenz interdisziplinärer Forschungsanstrengungen und somit neue Wechselwirkung.

2 Vektorsupercomputer: Status und Perspektiven

Noch 1970 hat J.H. Wilkinson in seiner "Turing Award Lecture" beklagt, daß die numerischen Mathematiker versagt hätten, die Entwicklung der Rechner in dem für technisch-wissenschaftliche Anwendungen notwendigen Maße zu beeinflussen, und so gezwungen gewesen sein, das Beste aus Rechnern zu machen, die wegen des angeblich zu kleinen Marktes solcher

Anwendungen für andere Zwecke entworfen seien [7]. In den Jahren seitdem hat sich die Situation sowohl hinsichtlich des Spektrums der für technisch-wissenschaftliche Anwendungen entwickelten Höchstleistungsrechner als auch bezüglich des effektiven und zukünftigen Marktes geändert, wie die wachsende Zahl der weltweit installierten Supercomputer der Vektorrechnerarchitektur und der massiv-parallelen Rechnerarchitektur zeigt [8-12].

Das Supercomputing wird heute noch weitestgehend bestimmt von der Architektur der Vektorrechner. In diese Klasse gehören neben den Systemen des Marktführers Cray Research die Rechner der japanischen Hersteller Fujitsu, Hitachi und NEC; auch die mit Vektoreinheiten ausgestatteten Universalrechner von IBM sind - mit gewissen Einschränkungen - dazuzuzählen.Im Anhang ist die derzeitige Ausstattung in Deutschland mit Vektorsupercomputern tabellarisch ausgewiesen. Die Leistungsfähigkeit dieser Supercomputer liegt für wesentliche Anwendungen immerhin um eine bis zwei Größenordnungen über der Leistungsfähigkeit von Universalrechnern der höchsten Kapazitätsklasse; ihr Leistungsprinzip ist das Pipelining und seine auch die Software, die Programmierung und die Speicherorganisation betreffenden Ausprägungen. Aus der Auseinandersetzung mit der Vektorrechnerarchitektur hat sich ein großes Arsenal an technischem Wissen und algorithmischen Methoden entwickelt, die etwa mit den Begriffen BLAS und LAPACK verknüpfbar sind [13-15], so daß diese Vektorsupercomputer heute ganz wesentlich die "Arbeitspferde" der Computational Science sind [16].

Sehr früh schon (1982) haben die Cray-Vektorrechner - im Gegensatz zu den japanischen Linien bislang - den Schritt zu Mehrprozessorsystemen und damit zur Parallelverarbeitung vollzogen. Damit ging auch die rechtzeitige Entwicklung zu "offenen Systemen" einher, wie sie sich in dem Unix-basierten Betriebssystem UNICOS und der Unterstützung der TCP/IP-Kommunikation manifestiert. Es hat sich erwiesen, daß das Multitasking-Konzept, insbesondere in der heutigen Ausprägung des Autotasking, im Verbund mit sehr leistungsfähigen Compile-Systemen und Parallelisierungswerkzeugen die effiziente Implementierung paralleler Algorithmen gestattet und die Plattform für Gigaflops-Leistung auch bei großen Anwendungen der Grand-Challenge-Klasse gewährleistet [17-20]. Eine ausführliche Beschreibung der Vektorrechner von Cray, NEC und Fujitsu liefern die Veröffentlichungen [21, 22]. Während die neue Vektorrechnerfamilie von Fujitsu VP-S noch immer wesentlich bei der Einprozessorstruktur verharrt, vollzieht NEC mit dem System SX-3 mit jetzt vier CPUs ebenfalls den Übergang zu Mehrprozessorsystemen (und damit wohl auch zum Multitasking) wie Cray Research. Ohne Zweifel wird sich diese Architekturlinie kräftig weiterentwickeln; ausgehend von den vier bzw. acht Prozessoren der NEC SX-3 bzw. der CRAY Y-MP8 werden mit dem Ausbau des "schwachen" Parallelismus in den nächsten Jahren weitere Leistungssteigerungen erzielt werden. So wird Cray Research mit der soeben angekündigten Maschine CRAY Y-MP C-90 mit 16 Prozessoren und einer Peak-Leistung von 16 Gigaflops auf die Leistungsebene der SX-3 mit der Peak-Leistung von 24 Gigaflops nachrücken, wobei jeweils die Verkürzung der Zykluszeit auf 2,9 nsec bei NEC bzw. 4 nsec bei Cray ergänzt wird durch architektonische Verbesserungen. Systeme mit 64 Prozessoren erscheinen in Reichweite wie etwa bei dem Nachfolger der C-90 oder bei Steve Chen's Projekt bei SSI, die anstreben, die Grenze der 100 Gigaflops zu überschreiten. Wie die Entwicklungsaufwendungen für solche Systeme schon heute zeigen, stoßen sie nicht nur an technisch-technologische Grenzen, sondern auch an die Barrieren der Finanzierbarkeit seitens der Kunden und Anwender. Dabei ist mit der massiv-parallelen Rechnerarchitektur eine von der absoluten Leistungsfähigkeit wie vom Preis/Leistungsverhältnis her zukunftsträchtige Konkurrenz erwachsen. Mit dem Schlagwort "Killer-Micros" wurde vor wenigen Jahren die Bedrohung der Vektorrechner durch solche, auf Mikroprozessoren als Rechnerknoten aufbauende Parallelrechner mit Hunderten und Tausenden von parallel arbeitenden Prozessoren artikuliert [10, 23]. Es festigt sich aber mehr und mehr die Überzeugung, daß nach einer Phase der Koexistenz der beiden Architekturlinien die wechselseitige Integration der beiden Konzepte unter Ausnutzung der jeweiligen Vorzüge die sinnvolle, wenn nicht gar unabdingbare Entwicklung der Rechnerarchitektur sein wird, da nur dadurch die Lösung der großen Probleme der Computational Science näherrücken werde. Das von Cray Research mit Hochdruck verfolgte MPP-Projekt für ein massiv-paralleles Rechnersystem folgt daher konsequent dieser Entwicklungslinie [24].

3 Massiv-parallele Rechner

Die technologischen Fortschritte der Mikroelektronik, welche den Durchbruch der Supercomputer erst möglich gemacht haben, liefern neben der Expansion der Speichergröße die Grundlage für den jetzt bevorstehenden - oder mit den Neuankündigungen Connection Machine CM-5 von Thinking Machines mit neuem Architekturkonzept als Nachfolger von CM-2 bzw. CM-200 sowie Paragon von Intel als Nachfolger von iPSC/860 bzw. als "Spin-off" der Touchstone-Pilotsysteme Delta und Sigma bereits vollzogenen - Durchbruch der massiv-parallelen Rechnerarchitektur. Von der Leistungsfähigkeit wie vom Preis/Leistungsverhältnis her übertreffen diese Systeme schon jetzt die Vektorsupercomputer in ähnlicher Weise, wie die Mikroprozessoren, insbesondere die Prozessoren mit RISC-Struktur und die damit ausgestatteten Workstations die klassischen Universalrechner ("Mainframes") übertreffen werden (vgl. dazu [25], insbesondere Figure 1). Massiv-parallele Rechner werden so mit Sicherheit die zukünftigen Leistungsträger des Supercomputing stellen, zunächst eigenständig und zunehmend als integrale Systemkomponenten heterogen strukturierter Supercomputerkomplexe. Diese Entwicklung wird begleitet - oder bestimmt - werden durch weitere deutliche Verbesserungen der Architekturelemente [26].

Die neuen Systeme CM-5 sowie Paragon sind in der Prozessorzahl - und damit in der nominellen Maximalleistung ("Peak-Rate") - hoch skalierbar auf derzeit 1024 Prozessoren mit insgesamt 128 Gigaflops und 128 Gigabytes verteilten Hauptspeicher bzw. auf 2048 Prozessoren mit insgesamt 150 Gigaflops und 64 Gigabytes verteilten Hauptspeicher. Beide Systeme bieten daher gegenüber ihren Vorgängern CM-2 mit 14 Gigaflops (und der aufgerüsteten Zwischenversion CM-200) bzw. iPSC/860 mit 7,6 Gigaflops Peak-Rate beachtliche Leistungssprünge.

Die CM-5 hat dabei gleichzeitig den Schritt weg von der SIMD-Architektur [27] vollzogen, ohne jedoch das Konzept des extensiven Datenparallelismus aufzugeben, auf dem die bisherige effektive Leistung der CM-2 beruhte [28-30]; das neue System stützt sich dabei auf ein neues Prinzip für die Datenkommunikation (und die Kontrollkommunikation). Intel baut das MIMD-Konzept mit Message Passing [27] weiter aus, wobei als Knotenprozessor der Nachfolger des i860 verwendet wird; allerdings trägt Intel im Hinblick auf die Skalierung der Prozessorzahl über die bisher 128 Knoten hinaus der Limitierung der Hypercube-Struktur Rechnung: Das Verbindungsnetz für die Datenkommunikation im Paragon-System baut wie bereits die Touchstone-Prototypen Delta und Sigma [11, 12, 31] auf die Torus-Struktur des zyklisch-geschlossenen ebenen Gitters aus Nächste-Nachbarn-Verbindungen auf. Wie Thinking Machines, Intel, BBN, nCube, Meiko, Parsytec und Cray strebt auch Kendall Square Research mit dem neu angekündigten massiv-parallelen Rechner UCS ("Unlimited Computer System") die Teraflops-Grenze in der zweiten Hälfte der 90-er Jahre an; die UCS-Linie will das Konzept des virtuell gemeinsamen Hauptspeichers (d.h. real verteilten, aber über einen globalen Adreßraum virtuell als gemeinsamen Speicher zu verstehenden Hauptspeicher) mit einer hierarchisch aus Ringen aufgebauten Architektur verwirklichen. Dabei zielt Kendall Square Research mit UCS wohl auch (oder vor allem) auf komplexe transaktionsorientierte Anwendungen aus der kommerziellen Welt. Früher oder später dürften wohl alle wichtigen massiv-parallelen Rechner das Konzept des virtuell gemeinsamen Hauptspeichers adaptieren.

4 Algorithmen und Anwendungen

Es hat sich bisher gezeigt, daß die effektiv aus massiv-parallelen Rechnern zu gewinnenden Leistungsdaten realer Anwendungen im allgemeinen deutlich kontrastieren zu den schon jetzt beachtlichen Peak-Raten, ohne damit die durchaus bemerkenswerten Spitzenergebnisse auf CM-2, iPSC/860 oder anderen massiv-parallelen Systemen schmälern zu wollen. Diese Diskrepanzen haben ihre Ursache einerseits in der begrenzten Leistungsfähigkeit der bisherigen Datenkommunikation und in diesem Zusammenhang in der mangelnden Lokalität - oder Ausnützung der Lokalität - der parallelen Algorithmen; sie liegt anderseits aber auch in dem unter Umständen nicht beliebig skalierbaren Parallelismus der Algorithmen oder gar der Probleme selbst.

Solange die Probleme auf Verfahren der linearen Algebra geführt werden können, lassen sich entsprechend hohe Leistungszugewinne durch massiven Parallelismus noch gewährleisten; dafür gibt es ein sehr belastbares Arsenal von parallelen Verfahren [14, 32]; vgl. dazu auch die Literaturübersichten in [14] und [32]. Schwieriger stellt sich die Situation bei anderen Problemstrukturen, insbesondere natürlich bei Fragen, die auf Methoden der diskreten Mathematik, wie etwa Graphenalgorithmen und kombinatorische oder geometrische Probleme, führen [33-35]. Dennoch muß festgestellt werden, daß mit der faktischen Realisierung von Parallelverarbeitung und dem Eindringen der Konzepte in Lehre und Ausbildung [36] die Basis für durchschlagende Erfolge in der Forschung auf den Gebieten der Rechnerarchitektur, Software-Systeme, Algorithmen und Anwendungen deutlich verbessert worden ist [37], wenngleich die Informatik wie die Anwendungsdisziplinen - trotz aller Ungeduld - erst am Anfang der neuen Ära des massiven Parallelismus im Computing und in der Computational Science stehen.

Obwohl viele Fragestellungen der Computational Science von einer geradezu abschreckenden Komplexität sind, wie etwa durch den im Prinzip exponentiell mit der Problemgröße wachsenden Berechnungsaufwand bei Vielteilchensystemen und ihrer Wechselwirkungsdynamik (z.B. in Phasenübergängen und biologisch-chemischen Prozessen [38]), sind gerade in den letzten Jahren immer wieder zunächst konzeptionell, dann auch in den Implementierungen wirkungsvolle Ideen umgesetzt worden. Ein überaus bemerkenswerter Fortschritt ist so beispielsweise bei der Behandlung des N-Körper-Problems im Gravitationsfeld durch die Entwicklung des sog. Greengard-Algorithmus [39], der über Multipolentwicklungen und unter Anwendung der Quadtree-Datenstruktur von einem sequentiellen $O(N^2)$- zu einem sequentiellen O(NlogN)-Algorithmus führt und in seiner rekursiven Struktur Ähnlichkeit mit den bedeutenden Renormierungsgruppenmethoden sowie mit den effizienten Multi-Level-Methoden und schließlich mit der außerordentlich wirkungsvollen Entwurfsstrategie für parallele Algorithmen, dem - parallelen - Divide-and-Conquer-Prinzip, aufweist und sich von daher auf Parallelrechner übertragen läßt [40].

5 Ausblicke

Vor allem die Hersteller massiv-paralleler Systeme geben heute Prognosen ab, daß sie ab 1995 Systeme mit Peak-Raten oder gar gewährleisteter Dauer-Performance ("sustained performance") jenseits von 1 Teraflops auf den Markt bringen werden. Solche Prognosen erscheinen in Anbetracht der bislang erfahrenen Leistungsdaten kühn, sind doch neben großen Herausforderungen auf dem Gebiet der Betriebssysteme und der Datenkommunikation für massiv-parallele Systeme zumal schwierige Fragen der Technologie selbst erst zu beantworten, die die Größe, den Energieverbrauch und die Abwärme solcher Systeme bestimmt; insbesondere Energieverbrauch und Abwärme überschreiten derzeit wohl noch die Megawatt-Grenze. Dennoch besteht Grund zu Hoffnung, wenn sich die technologischen Fortschritte eine Zeitlang so extrapolieren lassen, wie es die Historie der letzten Jahrzehnte vorgibt [25]. Hält man sich aber die Anforderungen der "Grand Challenges" vor Augen [2, 37], so muß man feststellen - und das kann nur den Ignoranten überraschen -, daß das Supercomputing heute auf diese Herausforderungen nur Teilantworten geben kann, die zwar als solche Ausdruck ganz hervorragender technischer und wissenschaftlicher Leistungen auf vielen der interdisziplinär wechselwirkenden Gebiete sind und sicherlich zu neuen Leistungen anspornen werden, dennoch nur ein kleines Stück des Weges zur Lösung der großen Probleme tragen; dies wird auf manchen Gebieten - wie z.B. der Fluid- und Aerodynamik - deutlicher bewußt als auf anderen [41-47].

Aus der theoretischen Informatik kommen andererseits Konzepte hervor, die durchaus geeignet erscheinen, ihrerseits wesentliche Durchbrüche auf dem Gebiet der parallelen Systeme vorzubereiten, wenn nicht zu erzielen, indem sie neue paradigmatische Grundlagen schaffen, die sowohl für den Entwurf neuer Parallelrechner wie für die Kommunikation in solchen Systemen und schließlich für die Programmierung und den Entwurf paralleler Algorithmen in fruchtbares Neuland führen können [48, 49].

Literatur

[1] Hossfeld, F.; Nonlinear Dynamics: A Challenge on High-Speed Computation, Proceedings Parallel Computing '83, North-Holland 1984, S. 67

[2] Wilson, K.G., Grand Challenges to Computational Science, Future Generation Computer Systems 5 (1989), No. 2&3, 171-189

[3] Dixon, D.A., and Raveché, H.J., A National Computing Initiative: A Summary, Future Generation Computer Systems 5 (1989), No. 2&3, 339-345

[4] Board on Mathematical Sciences of the National Research Council (USA), The David II Report: Renewing U.S. Mathematics - A Plan for the 1990s, in: Notices of the American Mathematical Society, May/June 1990, 542-546; September 1990, 813-837; October 1990, 984-1004

[5] Denning, P., Computing, Applications, and Computational Science, in: An ACM Response: The Scope and Directions of Computer Science, Comm. ACM 34 (1991), No. 10, 121-131

[6] Commission of the European Communities, Report of the EEC Working Group on High-Performance Computing, February 1991

[7] Wilkinson, J.H., Some Comments from a Numerical Analyst, J. ACM 18 (1971), 137

[8] Hossfeld, F., Vector-supercomputers, Parallel Computing 7 (1988), 373-385

[9] Jarp, S., A Review of Japan and Japanes High-end Computers, Report CN/91/01, CERN/Geneva, January 1991

[10] Almasi, G.S., and Gottlieb, A., Highly Parallel Computing, Benjamin/Cummings 1989

[11] Fox, G.C., Parallel Computing Comes of Age: Supercomputer Level Parallel Computations at Caltech, Concurrency: Practice and Eperience 1 (1989), No. 1, 63-103

[12] Fox, G.C., et al., Solving Problems on Concurrent Processors - Volume 1 and 2, Prentice-Hall 1988 and 1991

[13] Dongarra, J.J., et al., Solving Linear Systems on Vector and Shared Memory Computers, SIAM Publications 1990

[14] Ortega, J.M., Introduction to Parallel and Vector Solution of Linear Systems, Plenum Press 1988

[15] Zima, H., and Chapman, B., Supercompilers for Parallel and Vector Computers, ACM Press 1990, Addison-Wesley

[16] Dongarra, J.J., Performance of Various Computers Using Standard Linear Equations Software, Report CS-89-85, ONRL, July 9, 1991

[17] Hossfeld, F., Knecht, R., and Nagel, W.E., Multitasking: Experiences with Applications on a CRAY X-MP, Parallel Computing 12 (1989), 259-283

[18] Nagel, W.E., Exploiting Autotasking on a CRAY Y-MP: An Improved Software Interface to Multitasking, Parallel Computing 13 (1990), 225-233

[19] Knecht, S., Laermann, E., and Nagel, W.E., Parallelizing QCD with Dynamical Fermions on a CRAY Multiprocessor System, Parallel Computing 15 (1990), 3-20

[20] Bailey, D.H., Experience with Parallel Computers at NASA Ames, Paper presented at ICS '91, Cologne, 17-21 June 1991 (Preprint)

[21] Bönninger, T. et al., Vergleichende Darstellung von Höchstleistungsrechern anhand eines Kenngrößenkataloges, Angewandte Informatik H. 2 (1988), 51-70

[22] Bönninger, T., Esser, R., und Krekel, D., CRAY Y-MP, NEC SX-3 und Siemens VP-S: Vergleichende Darstellung von Höchstleistungsrechnern, Wirtschaftsinformatik, 32. Jahrgang (1990), H. 3, 273-286

[23] Brooks, E.D., and Warren, K.H. (eds.), The 1991 MPCI Yearly Report: The Attack of the Killer Micros, Lawrence Livermore Nat.Lab., March 1991

[24] Trier, B., A CRAY by Any Other Name, The Spang Robinson Report on High Performance Computing 5, No. 9, (1991) 3-4

[25] Hennessy, J.L., and Jouppi, N.P., Computer Technology and Architecture: An Evolving Interaction, IEEE Computer 24 (1991), No. 9, 18-29

[26] Stone, H.S., and Cocke, J., Computer Architecture in the 1990s, IEEE Computer 24 (1991), No. 9, 30-38

[27] Hwang, K., and Briggs, F.A., Computer Architecture and Parallel Processing, McGraw-Hill 1984

[28] Hillis, W.D., The Connection Machine, MIT Press 1985

[29] Bailey, J., Implementing Fine-grained Scientific Algorithms on the Connection Machine Supercomputer, Thinking Machines Technical Report Series TR90-1, 1990

[30] Thinking Machines Corporation, The Connection Machine CM-5 Technical Summary, October 1991

[31] Intel Corporation/Supercomputer Systems Division, Toward TeraFlop Performance: An Update on the Intel/DARPA Touchstone Program, March 1991

[32] Gallivan, K.A., Plemmons, R.J., and Sameh, A.H., Parallel Algorithms for Dense Linear Algebra Computations, SIAM Review 32 (1990), No. 1, 54-135

[33] Shahookar, K., and Mazunder, P., VLSI Cell Placement Techniques, ACM Computing Surveys 23 (1991), No. 2, 143-220

[34] Aurenhammer, F., Voronoi Diagrams - A Survey of a Fundamental Geometric Data Structure, ACM Computing Surveys 23 (1991), No. 3, 345-405

[35] Mehlhorn, K., Data Structures and Algorithms, Vol. 2: Graph Algorithms and NP-Completeness; Vol. 3: Multidimensional Searching and Computational Geometry, Springer-Verlag 1984

[36] Akl, S.G., The Design and Analysis of Parallel Algorithms, Prentice-Hall 1989

[37] Johnson, G.M., Exploiting Parallelism in Computer Science, Future Generation Computer Systems 5 (1989), No. 283, 319-337

[38] Berendsen, H.J.C., et al., Molecular Dynamics on CRAY, CYBER and DAP, in: J. Kowalik (ed.), Proc. NATO Advanced Research Workshop on High-Speed Computation, Lectur Notes in Computer Science, Springer-Verlag 1984, S. 425

[39] Greengard, L.F., The Rapid Evaluation of Potential Fields in Particle Systems, MIT Press 1988

[40] Makino, J., and Hut, P., Gravitational N-Body Algorithms: A Comparison between Supercomputers and a Highly Parallel Computer, Computer Physics Reports 9 (1989), 199-246

[41] Birkhoff, G., Numerical Fluid Dynamics, SIAM Review 25 (1983), 1; Birkhoff, G. , Fluid Dynamics Reactor Computations and Surface Representation, in: S.G. Nash (ed.), History of Scientific Computing, ACM Press / Addison-Wesley 1990, p. 63-87

[42] Special Double Issue: Grand Challenges to Computational Science, Future Generation Computer Systems 5 (1989), No. 2&3

[43] Baillie, C.F., et al., Status and Prospects of the Computational Approach to High-Energy Physics, The Journal of Supercomputing 4 (1990), 277-300

[44] Decker, J.F., et al., Large-Scale Problems and Supercomputing in the Department of Energy, Proc. IEEE 77 (1989), No. 7, 1020-1037

[45] Halem, M., Scientific Computing Challenges Arising From Space-Borne Observations, Proc. IEEE 77 (1989), No. 7, 1061-1091

[46] Peterson, V.L., et al., Supercomputer-Requirements for Selected Disciplines Important to Aerospace, Proc. IEEE 77 (1989), No. 7, 1038-1055

[47] Clementi, E., et al., Supercomputing and Supercomputers for Science and Engineering in General and for Chemistry and Biosciences in Particular, Intern. J. Quantum Chemistry Vol. XXXV (1989), 3-89

[48] Valiant, L.G., General Purpose Parallel Architectures, in: J. van Leeuwen (ed.), Handbook of Theoretical Computer Science, North-Holland 1990

[49] Valiant, L.G., A Bridging Model for Parallel Computation, Comm. ACM 33 (1990), No. 8, 103-111

Anhang

Supercomputer in der Bundesrepublik Deutschland - Stand: Oktober 1991 -		
Institution	**Ort**	**Supercomputer**
MPI	Garching	CRAY X-MP/24
DLR	Oberpfaffenhofen	CRAY Y-MP2/232
KFA	Jülich	CRAY X-MP/416 CRAY Y-MP8/832
KfK	Karlsruhe	SIEMENS VP-50
Deutsches Klima-RZ	Hamburg	CRAY-2S/4-128 CRAY Y-MP4E/364
Konrad-Zuse-Zentrum	Berlin	CRAY X-MP/24 CRAY Y-MP2E/164
Leibniz-Rechenzentrum	München	CRAY Y-MP4/464
Universität	Stuttgart	CRAY-2D/4-256 CRAY Y-MP2E/116
Universität	Karlsruhe	SIEMENS VP-S600/20
Universität	Kiel	CRAY X-MP/216
Universität	Kaiserslautern	SIEMENS VP-100
Universität	Bochum	CDC CYBER 205
Universität	Köln	NEC SX-3/11
Universität	Frankfurt	SIEMENS VP-S200/10
Universität	Gießen	SIEMENS VP-S100/10
Universität	Kassel	SIEMENS VP-S100/10
Universität	Bremen	SIEMENS VP-100
Techn. Hochschule	Aachen	SIEMENS VP-S600/20
Techn. Hochschule	Darmstadt	SIEMENS VP-S400/40
Techn. Universität	Hannover	SIEMENS VP-S600/20
Techn. Universität	Dresden	SIEMENS VP-200EX
SIEMENS	München	SIEMENS VP-200 SIEMENS VP-100EX
Prakla Seismos	Hannover	CRAY Y-MP2/232
IABG	München	SIEMENS VP-200
Regierung	Bonn	CRAY ?
BMW	München	CRAY Y-MP8/364
Daimler-Benz	Stuttgart	CRAY Y-MP4/216
Opel	Rüsselsheim	CRAY X-MP/18
VW	Wolfsburg	CRAY Y-MP8/364
Continental	Hannover	CRAY X-MP/18se
Deutscher Wetterdienst	Offenbach	CRAY Y-MP4/432
Deutsches Museum	München	CRAY-1S

Analog Circuits Solving Problems from Linear Algebra

Knut Hüper, Steffen Paul and Josef A. Nossek
Technical University Munich,
Institute for Network Theory and Circuit Design,
D-8000 Munich 2, Arcisstr. 21, Germany,
e-mail: knhu@nws.e-technik.tu-muenchen.de
e-mail: stpa@nws.e-technik.tu-muenchen.de

Abstract

During the last years analog computing has gained increasing interest especially in the area of neural networks. In this field scientists and engineers hope to have more success using analog parallel computing than they have had using digital parallel computing.

In this paper a dynamical system is presented, which reveals the strong connections between classical mechanics and analog circuits on the one hand, and numerical linear algebra on the other hand.

The dynamical properties of the *finite nonperiodic Toda lattice*, a system of particles with exponential interaction between nearest neighbors, are well explored. Under a special change of variables Hamilton's equations of motion for such a mechanical system reduce to a *matrix-differential-equation*. In other words, the Toda lattice induces an isospectral flow on the space of symmetric tridiagonal matrices. In this manner a continuous generalization of the discrete *QR-algorithm*, an iterative process for diagonalizing symmetric tridiagonal matrices, is achieved.

Integrating the system of ordinary differential equations numerically on sequential digital computers, comparable results to other well-known discrete algorithms are obtained.

An *analog system* is presented, consisting of basic circuit elements like operational amplifiers or multipliers, which is able to diagonalize a symmetric tridiagonal matrix.

1 Introduction

The dynamical system to be considered comes from Hamiltonian mechanics: The finite nonperiodic Toda lattice, a system of unit mass particles on the line with exponential interaction force between nearest neighbors.

Fig. 1: Finite nonperiodic Toda lattice

With Hamiltonian

$$H = \frac{1}{2}\sum_{k=1}^{n} p_k^2 + \sum_{k=1}^{n-1} e^{(x_k - x_{k+1})}, \tag{1}$$

Hamilton's equations of motion become

$$
\begin{aligned}
\dot{x_k} &= \frac{\partial H}{\partial p_k} = p_k \qquad (k = 1, 2, \ldots, n),\\
\ddot{x_k} &= \dot{p_k} = -\frac{\partial H}{\partial x_k} = e^{(x_{k-1}-x_k)} - e^{(x_k - x_{k+1})} \qquad (k = 2, \ldots, n-1),\\
\dot{p_1} &= -\frac{\partial H}{\partial x_1} = -e^{(x_1 - x_2)},\\
\dot{p_n} &= -\frac{\partial H}{\partial x_n} = e^{(x_{n-1}-x_n)}.
\end{aligned}
\tag{2}
$$

The last two equations in (2) correspond to the formal boundary condition $x_0 = -\infty$ and $x_{n+1} = +\infty$. Applying Flaschka's nonlinear transform [Flas74a],[Flas74b]

$$
\begin{aligned}
a_k &= -\frac{1}{2} p_k,\\
b_k &= \frac{1}{2} e^{\left(\frac{x_k - x_{k+1}}{2}\right)},\\
b_0 &= b_n = 0,
\end{aligned}
\tag{3}
$$

the following system of nonlinear differential equations is obtained

$$
\begin{aligned}
\frac{da_k}{dt} &= \dot{a_k} = 2\left(b_k^2 - b_{k-1}^2\right),\\
\frac{db_k}{dt} &= \dot{b_k} = b_k(a_{k+1} - a_k),\\
b_0 &= b_n = 0.
\end{aligned}
\tag{4}
$$

Using the commutator form $[\mathbf{A}, \mathbf{B}] = \mathbf{AB} - \mathbf{BA}$, one gets the compact expression

$$
\frac{d\mathbf{A}(t)}{dt} = \dot{\mathbf{A}}(t) = [\mathbf{A}(t), \mathbf{B}(t)],
\tag{5}
$$

with

$$
\mathbf{A}(t) = \begin{bmatrix}
a_1(t) & b_1(t) & 0 & \cdots & 0\\
b_1(t) & a_2(t) & b_2(t) & \ddots & \vdots\\
0 & b_2(t) & \ddots & \ddots & 0\\
\vdots & \ddots & \ddots & \ddots & b_{n-1}(t)\\
0 & \cdots & 0 & b_{n-1}(t) & a_n(t)
\end{bmatrix}
\tag{6}
$$

and

$$
\mathbf{B}(t) = \begin{bmatrix}
0 & -b_1(t) & 0 & \cdots & 0\\
b_1(t) & \ddots & -b_2(t) & \ddots & \vdots\\
0 & b_2(t) & \ddots & \ddots & 0\\
\vdots & \ddots & \ddots & \ddots & -b_{n-1}(t)\\
0 & \cdots & 0 & b_{n-1}(t) & 0
\end{bmatrix}.
\tag{7}
$$

Further by substituting $b_k^2 = c_k$ and time scaling $2t \rightarrow t$ one arrives at

$$\begin{aligned} \frac{da_k}{dt} &= c_k - c_{k-1} \quad (k = 1, 2, \ldots, n), \\ \frac{dc_k}{dt} &= c_k(a_{k+1} - a_k) \quad (k = 1, 2, \ldots, n-1), \\ c_0 &= c_n = 0. \end{aligned} \tag{8}$$

The soliton-like solutions of (1) in the infinite case ($n = \infty$) were first found by Toda himself [Toda81]. An arbitrary ensemble of solitons will change into another *ordered* ensemble after waiting for a sufficient long time (ordered in the sense of increasing or decreasing amplitudes). Waiting for a sufficient long time is equivalent to solving an initial value problem: A time dependent symmetric tridiagonal matrix converges for $t \rightarrow \pm\infty$ to a diagonal matrix:

$$\mathbf{A}(t = 0) \longrightarrow \mathbf{A}(t \longrightarrow \pm\infty) = \mathrm{diag}(\lambda_1, \lambda_2, \ldots, \lambda_n) \quad (\lambda_1 \gtrless \lambda_2 \gtrless \ldots \gtrless \lambda_n). \tag{9}$$

Because the finite nonperiodic Toda lattice induces an isospectral flow on the space of symmetric tridiagonal matrices [ChNo88][Nand85], the entries of this diagonal matrix (9) are the ordered eigenvalues of the original $\mathbf{A}(t = 0)$. In Section 3 it will be shown, how to exploit such an ordering scheme.

Symes [Syme82] demonstrated the strong relation between the QR-algorithm, known from numerical linear algebra, and solving the Toda lattice equations. Let $\mathbf{A}(t)$ be a time dependent tridiagonal symmetric matrix. Initializing the Toda flow with $\mathbf{A}(0)$, i.e. solving an initial value problem by integrating (4) numerically, a sequence of isospectral tridiagonal matrices $\mathbf{A}(t_i)$ can be obtained, where the t_i's are the time instances for sampling the flow. This sequence corresponds to the QR-sequence in the following sense: Taking the exponential of the matrix $\mathbf{A}(t_i)$ of the Toda trajectory at integer times $t_i = k$ ($k = 1, 2, \ldots, \infty$), the same results as by applying k times the QR-algorithm to the matrix $\mathbf{A}_0 = e^{\mathbf{A}(0)}$ are obtained:

$$\mathbf{A}_k = e^{\mathbf{A}(k)}. \tag{10}$$

In [ChNo88] and [WaEl88] it is shown how to develop continuous generalizations of other iterative algorithms, known from numerical linear algebra, as for example singular value decomposition.

2 Application to Numerical Linear Algebra

As an alternative to iterative algorithms, the solution of the eigenvalue problem can be obtained by numerical integration of (8). This way is at least competitive to the most efficient *discrete* technique for computing eigenvalues on sequential computers, the QR-algorithm with Wilkinson-shift [HüPaWa91]. Several different techniques for integrating (8) numerically have been used. The fourth-order Runge-Kutta method [PrFlTeVe88] turned out to be the fastest technique. To achieve some predetermined accuracy in the solution with minimum computational effort, an adaptive step size control has been used. Taking for example

$$\mathbf{A}(0) = \begin{bmatrix} -1 & 1 & 0 & 0 & 0 \\ 1 & -10 & -1 & 0 & 0 \\ 0 & -1 & -7 & 1 & 0 \\ 0 & 0 & 1 & 15 & -1 \\ 0 & 0 & 0 & -1 & -1 \end{bmatrix} \tag{11}$$

as initial data, the matrix entries evolve in time as shown in Fig. 2. Whenever the trajectories of two diagonal entries cross, simultaneously a nondiagonal entry passes through a local maximum or minimum. Fixed strategies like step size doubling every $n/2$ iterations as proposed by Deift et.al.[DeNaTo83], turned out to be not accurate enough [HüPaWa91], since the initial step size depends on the matrix and has to be chosen carefully.

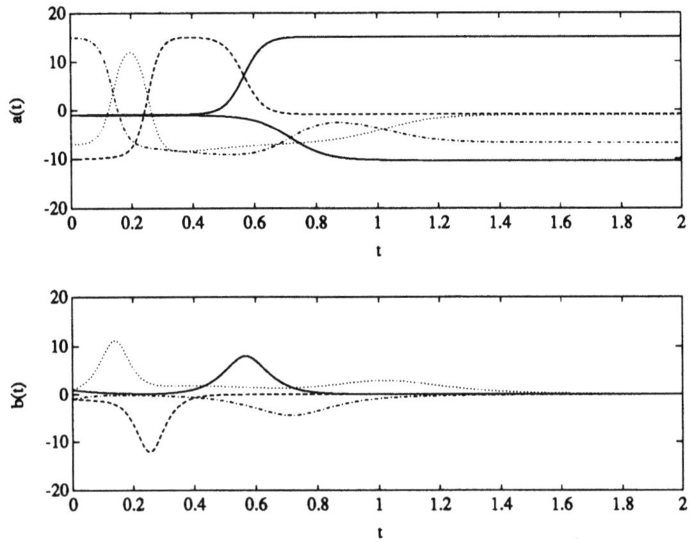

Fig. 2: Time evolution of matrix entries $a_k(t)$ and $b_k(t)$

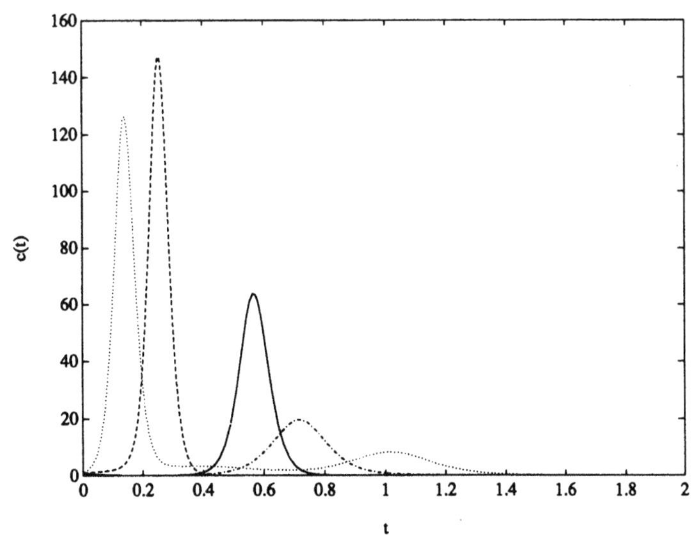

Fig. 3: Time evolution of $c_k(t)$

All computations were done on a Sparc II workstation. So far the algorithms were implemented without deflation, meaning that case, where a $(n \times n)$-matrix can be split into two matrices of smaller dimensions, i.e. into two uncoupled systems of ordinary differential equations. But deflation is intrinsic to the eigenvalue problem and higher order algorithms like the quadratic Toda flow [DeNaTo83] can only improve under deflation. Thus it is clear that the Toda method is not only suited to vector processing but also to parallel processing.

3 Analog Circuits for Computing Eigenvalues and Rank Filtering

The main advantages of analog circuits are simple implementation of nonlinearities, the high processing speed and the low power consumption at the cost of reduced accuracy. The class of regular arrays consisting of interconnected nonlinear analog processing cells finds increasing interest in various information processing applications. The signal flow graph of an analog circuit for diagonalizing symmetric tridiagonal $(n \times n)$-matrices is presented in Fig. 4. Eq. (4) as well as (8) is well suited for analog computing. There are only local connections between the state variables $a_k(t)$ and $c_k(t)$ $(b_k(t))$. The circuit only consists of basic elements like operational amplifiers and multipliers. Cascading these cells as shown in Fig. 4 allows solving eigenvalue problems of any dimension. The analog circuit works as follows: Initial data has to be stored as charges on the capacitors in the feedback paths of the integrators. A circuit for setting these initial conditions is presented in Fig. 5. There are two analog switches for loading and another one for starting computations. Simulating the analog system with realistic circuit models under SPICE demonstrated very good agreement with theoretic predictions. For an investigation from a more network theoretically motivated point of view see [PaHüNo92b].

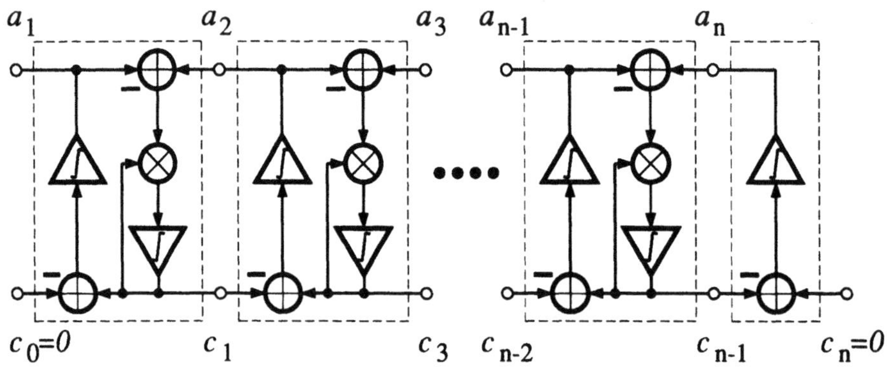

Fig. 4: Analog circuit for integrating Eq. (8)

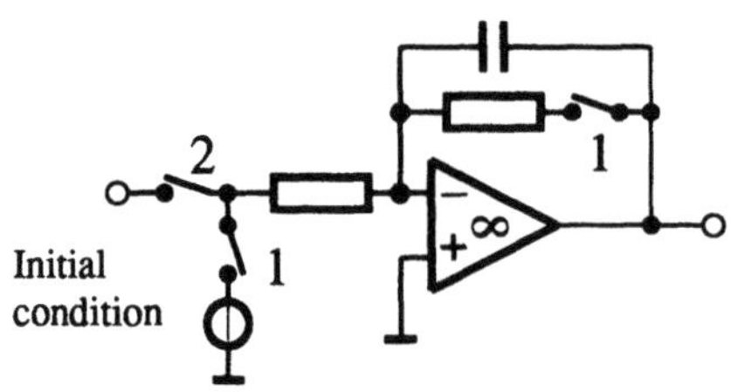

Fig. 5: Integrator circuit including initialization

The presented dynamical systems (4) and (8) have only one stable equilibrium point ([DeNaTo83]): For $t \to \pm\infty$ the eigenvalues are sorted in decreasing (increasing) order. To exploit the ordering characteristics the scalar rank filtering problem is embedded in a matrix-valued problem. From a given sequence of n elements to be sorted in decreasing (increasing) order the i-th value is selected as the filter output. The entries of an unsorted data vector **a** can be expressed in matrix form:

$$\mathbf{A}(0) = \begin{bmatrix} a_1 & 0 & \dots & 0 \\ 0 & \ddots & \ddots & \vdots \\ \vdots & \ddots & \ddots & 0 \\ 0 & \dots & 0 & a_n \end{bmatrix}. \tag{12}$$

Taking

$$\mathbf{A}'(0) = \mathbf{A}(0) + \delta \begin{bmatrix} 0 & 1 & 0 & 0 \\ 1 & \ddots & \ddots & 0 \\ 0 & \ddots & \ddots & 1 \\ 0 & 0 & 1 & 0 \end{bmatrix}, \delta << 1 \tag{13}$$

as initial data, the analog system is able to leave the unstable equilibrium point $\mathbf{A}(0)$ and to converge to the stable one with ordered data elements [PaHüNo91][PaHüNo92a].

4 Conclusion

In this paper a dynamical system was presented, which reveals the strong relations between classical mechanics and analog circuits on the one hand, and numerical linear algebra on the other hand. Solving problems from numerical linear algebra by integrating systems of ordinary differential equations gives not only an encouraging approach to parallel processing but also to nonlinear filtering.

Bibliography

[ChNo88] M. T. Chu and L. K. Norris. Isospectral flows and abstract matrix factorizations. *SIAM J. Num. Anal.*, 25:1383–1391, 1988.

[DeNaTo83] P. Deift, T. Nanda, and C. Tomei. Ordinary differential equations and the symmetric eigenvalue problem. *SIAM J. Num. Anal.*, 20:1–22, 1983.

[Flas74a] H. Flaschka. On the Toda Lattice I. *Phys. Rev. B*, 9:1924–1925, 1974.

[Flas74b] H. Flaschka. On the Toda Lattice II. *Prog. of Theor. Physics*, 51:703–716, 1974.

[HüPaWa91] K. Hüper, St. Paul, and A. Wachtel. Vergleich numerischer Algorithmen zur Lösung des symmetrischen Eigenwertproblems. Technical Report TUM-LNS-TR-91-20, Technische Universität München, November 1991.

[Nand85] T. Nanda. Differential equations and the QR algorithm. *SIAM J. Num. Anal.*, 22:310–321, 1985.

[PaHüNo91] St. Paul, K. Hüper, and J. A. Nossek. Analog rank filtering. Technical Report TUM-LNS-TR-91-21, Technical University Munich, November 1991.

[PaHüNo92a] St. Paul, K. Hüper, and J. A. Nossek. Analog median filtering. In *Proc. IEEE Int. Conf. on Acoust., Speech Signal Processing*, 1992.

[PaHüNo92b] St. Paul, K. Hüper, and J. A. Nossek. A class of nonlinear lossless dynamical systems. *submitted for publication Archiv für Elektronik und Übertragungstechnik*, 1992.

[PrFlTeVe88] W. H. Press, B. P. Flannery, S. A. Teukolsky, and W. T. Vetterling. *Numerical Recipes in C*. Cambridge, 1988.

[Syme82] W. W. Symes. The QR algorithm and scattering for the finite nonperiodic toda lattice. *Physica 4D*, pages 275–280, 1982.

[Toda81] M. Toda. *Theory of nonlinear lattices*. Springer, 1981.

[WaEl88] D. S. Watkins and L. Elsner. Self-similar flows. *Lin. Algebra & Applic.*, 110:213–242, 1988.

Numerical Simulation of Crystal Growth

Klaus Kassner

Institut für Festkörperforschung, Forschungszentrum Jülich, W-517 Jülich, Germany

The state of the art of numerical approaches to diffusion-limited crystal growth is reviewed. Recent numerical results on scaling laws and growth morphologies in dendritic growth and directional solidification, on secondary instabilities in lamellar eutectic growth, and on the transition to chaos in directional solidification are reported. Questions for future work and possible extensions or improvements of the current numerical methods are discussed.

1 Introduction

Crystallization is a nonlinear process. Depending on growth conditions, a large variety of crystal shapes and structures are formed (Chernov 1984, 1989). The abundance of crystal morphologies and microstructures is accompanied by a considerable amount of research activity, which has led to numerous models and theoretical results. Nevertheless, the microscopic details of most crystal growth processes are not understood well enough to allow numerical calculations starting from first principles. In many cases this is not even desirable, because computations on atomic length scales imply very limited system sizes.

Models of crystal growth are therefore based either on phenomenological equations (Burton et al. 1951, Müller-Krumbhaar et al. 1977, Langer 1980) or on simplified "microscopic" dynamics (Gilmer et al. 1974), where the building blocks of the crystal in the simulation need not be identical with those of the physical system (Saito and Ueta 1989, Shochet et al. 1991).

Owing to the huge extent of the field of crystal growth, any single contribution can cover a small fractional part of it at best. The model I would like to discuss here is suitable for the description of the growth of pure substances or alloys from their melt and of the ordering transition of liquid crystals. The common feature of these systems is that the growth dynamics are predominantly controlled by the diffusion of heat or material. Such a situation arises when the attachment kinetics of particles are fast on the time scale of the diffusion, because it is then the latter that limits the growth speed. The surface of the growing crystal is microscopically rough and macroscopically rounded, i.e., the crystal does not have facets. Kinetic effects can be taken into account (Classen et al. 1991), but this is a fairly new development, which has not yet been extended to the case of nonlinear kinetics.

The equation of motion in the liquid and solid phases is very simple within this model, being just the linear diffusion equation. It is the boundary conditions at the liquid-solid interface (from thermodynamics and local heat or mass conservation) that render the problem nonlinear. Solving it means to determine the motion of the interface, which in turn enters the equations – the problem is of the *moving-boundary* type.

For physicists, the model is interesting because it allows to describe and understand various nonequilibrium systems exhibiting intriguing properties of pattern formation and mode selection. Complex patterns arise because of the competition between the destabilizing effects of diffusion and the stabilizing counteraction of surface tension or interface kinetics (Müller-Krumbhaar and Kurz 1991).

One well-studied example is the growth of *dendritic crystals*. The question, how a needle shaped crystal selects, for given undercooling, its unique tip radius and velocity, has recently found a satisfactory answer (Ben Amar and Pomeau 1986, Caroli et al. 1986, Barbieri et al. 1987). Numerical simulations (Saito et al. 1987, 1988) confirmed the quality of the selection criterion following from the analytic calculations even in the case of a sidebranching dendrite; they furthermore provided "experimental" verification of analytic scaling relations.

Another process that is encompassed by the general model, is crystallization in an external temperature gradient, called *directional solidification*. Again, instabilities are the key to pattern formation. At a certain critical growth rate, a planar interface, stable at smaller velocities, goes unstable and turns into an array of cellular shapes, which in some systems give way, at even higher velocities, to needlelike and then dendritic structures (Saito et al. 1989). Other systems, in which diffusion in the solid is comparable to that in the liquid, never even get to the needle stage. New aspects of spatiotemporal pattern formation have been found by numerical simulation for systems of this type (Kassner et al. 1991).

The microstructures of *eutectic alloys* are, next to dendritic ones, the second large class of patterns seen in diffusion-controlled crystal growth. For thin samples, lamellar structures with an alternating arrangement of the two solid phases constitute the normal growth mode. Experiments have shown that these fairly regular structures can, via a series of "secondary" instabilities turn into tilted (Faivre et al. 1989) or oscillatory states and even become chaotic (Faivre et al. 1991). At least one of these instabilities has been understood in more detail in the framework of the model to be discussed here (Kassner and Misbah 1990, 1991a-c).

This article is organized as follows. Section 2 gives the model equations for different physical systems. In Sect. 3, the numerical methods for addressing the main questions of the dynamic evolution of the system and of the structure of its stationary states are described. A few recent results obtained with these methods are discussed in Sect. 4 together with some unsolved questions. Interesting directions for future work and ideas how to go beyond current numerical approaches in order to transcend present limits on tractability are outlined in Sect. 5.

2 Model Equations

2.1 Free Growth

Consider a solid of a pure substance that is in contact with an infinite reservoir of its supercooled melt. Clearly, it will be thermodymically favourable for the solid to crystallize, if it exceeds a certain critical size. An increase in volume decreases its free energy by the released latent heat. This effect is counteracted by an increase in surface free energy due to the growing liquid-solid interface. If the crystal is large enough, the volume effect will always win. During the growth process, the generated latent heat will heat up the solid at the interface to (roughly) its melting temperature. Further growth is impeded to a certain extent by this excess heat, which has to be transported away from the interface by thermal diffusion into the liquid. The driving force of the crystallization process is the temperature gradient pointing from the liquid into the solid.

If the material under consideration contains impurities, i.e. is a mixture, spontaneous growth is possible even in an isothermal situation, i.e., when the temperatures of the liquid and the solid are equal. To see this consider the phase diagram of a binary alloy given in Fig. 1. Suppose we cool the liquid down below its equilibrium temperature along the dotted equiconcentration line (describing a dilute alloy). Doing this carefully, we can avoid crystallization until the system is in the metastable state between the liquidus and solidus lines. Crystallization may then be started by providing a nucleation center. Assume that thermal diffusion is fast in comparison with the chemical diffusion of impurities (which is a good approximation for most materials, in particular for metals) and that the experimental setup is such that latent heat is easily withdrawn from the system, i.e., heat production can be neglected and the temperature remains constant. The growing solid will have the concentration of the solidus line at the given temperature, while the concentration of the liquid is below its equilibrium value, or stated differently, the *temperature* of the liquid in the reservoir is below the equilibrium value for the given concentration. This situation has been termed *constitutional supercooling* (Langer 1980). At the interface, the concentration of the liquid is increased by the impurities that cannot be incorporated into the crystal until the value of the liquidus line is reached. This concentration excess diffuses into the liquid, the impurities are rejected by the solid; the driving force is now the concentration gradient.

The first (purely thermal) and second (chemical) cases can be treated within the same mathematical framework by introduction of a dimensionless diffusion field u,

$$u = \begin{cases} (T - T_\infty)/(Lc_p^{-1}) & \text{(thermal model)} \\ (\mu - \mu_\infty)/(\Delta c\, \partial\mu/\partial c) & \text{(chemical model)} \end{cases} \tag{1}$$

In the thermal model, u is the difference between the local temperature T and the temperature T_∞ in the liquid far from the interface, normalized by the ratio of latent heat L and specific heat c_p. In the chemical model, u is the difference between the local chemical potential μ and its value μ_∞ far out in the liquid, normalized by the chemical potential increase corresponding to the miscibility gap Δc (at the reference temperature).

The general geometry of the system is depicted in Fig. 2. In the frame of reference moving with the interface, the diffusion equation takes the form

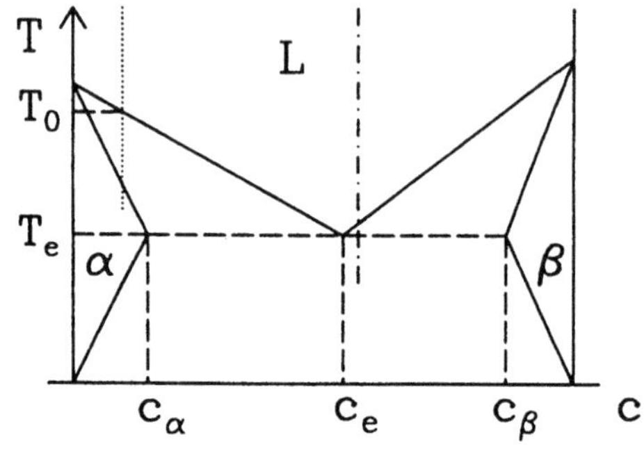

Fig. 1. Simplified generic phase diagram of an alloy. The regions L, α, β correspond to one-phase equilibrium states of the liquid, the solid α and the solid β phases, respectively. The L region is bordered from the bottom by the two liquidus lines, the α and β regions from the top by the solidus lines. Along these lines, the liquid coexists with a single solid phase, in two-phase equilibrium. The equilibrium concentrations of the liquid and the two solid phases at the triple or eutectic point are denoted by c_e, c_α, and c_β. Experiments on *dilute* alloys operate in a concentration range indicated by the dotted line. One can then focus on the left part of the diagram; the miscibility gap Δc is in this case defined as the concentration difference between liquid and solid, when they are in equilibrium at the reference temperature T_0. If the experimental concentration is close to c_e (dash-dotted line), one has to consider two-phase solid structures; the miscibility gap $\Delta c = c_\beta - c_\alpha$ then refers to the eutectic temperature T_e.

$$\begin{aligned} \frac{1}{D}\frac{\partial u}{\partial t} &= \nabla^2 u + \frac{2}{l}\frac{\partial u}{\partial z} \quad , \\ \frac{1}{D'}\frac{\partial u}{\partial t} &= \nabla^2 u + \frac{2}{l'}\frac{\partial u}{\partial z} \quad . \end{aligned} \tag{2}$$

Herein, D and D' are the diffusion constants in liquid and solid, respectively. l and l' are the *diffusion lengths*, given by $l = 2D/V$ and $l' = 2D'/V$, where V is the (constant) velocity, in the z direction, of the moving interface.

In the chemical case, one can often neglect diffusion in the solid ($D' = 0$), we then have the so-called *one-sided model*. In the thermal case, sometimes the *symmetric model* is considered ($D' = D$).

I will restrict the discussion to the two-dimensional case ($u = u(x, z, t)$), since this is the only case for which quantitative numerical results have been obtained to date.

Next we have to examine the boundary conditions to the equations of motion (2). At infinity we require the field to vanish both in the liquid and solid phases

$$u(x,\infty) = u(x,-\infty) = 0 \quad , \quad u(\pm\infty, z) = 0 \quad . \tag{3}$$

More interesting is the boundary condition at the liquid-solid interface, where the assumption of (infinitely) fast attachment kinetics implies local thermodymamic equilibrium, meaning that the liquid and solid phases are at their equilibrium temperature or concentrations, respectively. However, due to the Gibbs-Thomson effect these quantities

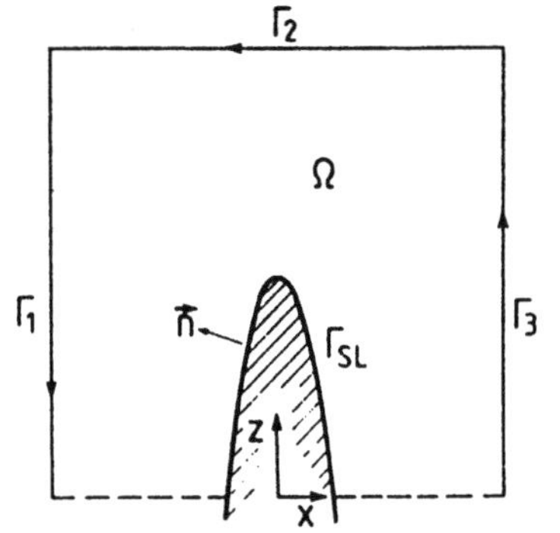

Fig. 2. Geometrical configuration of a crystal (here needle-shaped) growing into the (supercooled) liquid Ω, enclosed by a solidification front Γ_{sl} and three boundaries Γ_1 ($x = -\infty$), Γ_2 ($z = \infty$), and Γ_3 ($x = \infty$).

do not have their bulk equilibrium values; they contain a surface term which makes them curvature dependent (Langer 1980, Kassner 1989b):

$$\begin{aligned} u\Big|_{\text{interface}} &\equiv u_i = \Delta - d\kappa \quad \text{(liquid)} \quad , \\ u\Big|_{\text{interface}} &= k\,(u_i - 1) \qquad \text{(solid)} \quad . \end{aligned} \tag{4}$$

Here Δ is the normalized undercooling

$$\Delta = \begin{cases} (T_m - T_\infty)/(Lc_p^{-1}) & \text{(thermal model)} \\ (\mu_{eq} - \mu_\infty)/(\Delta c\,\partial\mu/\partial c) & \text{(chemical model)} \end{cases} \quad , \tag{5}$$

where T_m is the equilibrium melting temperature of the bulk and μ_{eq} correspondingly the bulk equilibrium chemical potential. In the thermal model, $k = 1$, while in the chemical model, k is the partition coefficient, here the ratio of the (bulk) equilibrium concentrations of the liquid and solid (Fig. 1). κ is the local curvature of the interface $\zeta(x,t)$,

$$\kappa = -\frac{\partial^2\zeta/\partial x^2}{\left(1 + (\partial\zeta/\partial x)^2\right)^{3/2}} \quad , \tag{6}$$

which is positive for a convex solid, and d is the capillary length, whose anisotropy is described in terms of the angle ϑ between the normal vector $\mathbf{n}$ on the interface (pointing from the solid into the liquid) and the z direction:

$$d = [\gamma(\vartheta) + \gamma''(\vartheta)] \begin{cases} T_m c_p L^{-2} & \text{(thermal model)} \\ (\Delta c)^{-2}\,(\partial\mu/\partial c)^{-1} & \text{(chemical model)} \end{cases} \quad , \tag{7}$$

γ and γ'' are the anisotropic surface tension and its second angular derivative. A simple model expression for the surface tension in the case of m-fold crystalline anisotropy is

$$\gamma = \gamma_0[1 + \tilde{\epsilon}_m \cos(m\vartheta)] \quad , \tag{8}$$

which for the capillary length implies

$$d = d_0[1 - \epsilon_m \cos(m\vartheta)] \tag{9}$$

with

$$\epsilon_m = \tilde{\epsilon}_m(m^2 - 1) \quad . \tag{10}$$

Typical capillary lengths are on the order of a few Å.

From local mass conservation at the interface one obtains a continuity equation

$$\left(u\Big|_{\text{liquid}} - u\Big|_{\text{solid}} \right) v_n = -D\frac{\partial u}{\partial n}\Big|_{\text{liquid}} + D'\frac{\partial u}{\partial n}\Big|_{\text{solid}} \quad , \tag{11}$$

where $v_n = \mathbf{v}\mathbf{n}$ is the normal velocity of the interface. Denoting by n_z the z component of the normal vector, we obviously have

$$v_n = (V + \dot{\zeta})n_z \quad . \tag{12}$$

The left hand side of (11) can be rewritten using (4), which results in

$$[u_i(1 - k) + k]\, v_n = -D\frac{\partial u}{\partial n}\Big|_{\text{liquid}} + D'\frac{\partial u}{\partial n}\Big|_{\text{solid}} \quad . \tag{13}$$

Equations (1) – (13) constitute a complete mathematical description of the problem. Given initial conditions for u and an initial interface $\zeta(x, t_0)$, a solution strategy consists in the following steps. First calculate u on the boundary from (4). Solve the diffusion equation for a small timestep to obtain u in the bulk. Determine the normal derivatives of u at the interface from the known bulk and interface values of u and calculate the normal velocity using (11). During a sufficiently small time step, move the interface according to this velocity. Repeat the process.

2.2 Directional Solidification

In directional solidification, the sample (usually a thin strip of transparent material, which is best suited for observation through a microscope) is pulled (or pushed) at a controlled velocity between two stationary hot and cold contacts, whose temperatures are above and below the melting point, respectively (see Fig. 3). Essentially the same assumptions are made that have been discussed above for isothermal solidification, namely that heat diffusion is fast enough to decouple from chemical diffusion and that the thermal conductivities of the solid and liquid phases are (approximately) equal. Then a constant temperature gradient G will be established throughout the interesting part of the sample. Let us first look at a *dilute* alloy.

The dimensionless field quantity to be considered is now

$$u = (c - c_\infty)/\Delta c \quad , \tag{14}$$

and its motion is again governed by (2). However, the growth rate V is now imposed externally – in stationary growth its magnitude must be equal to the pulling velocity.

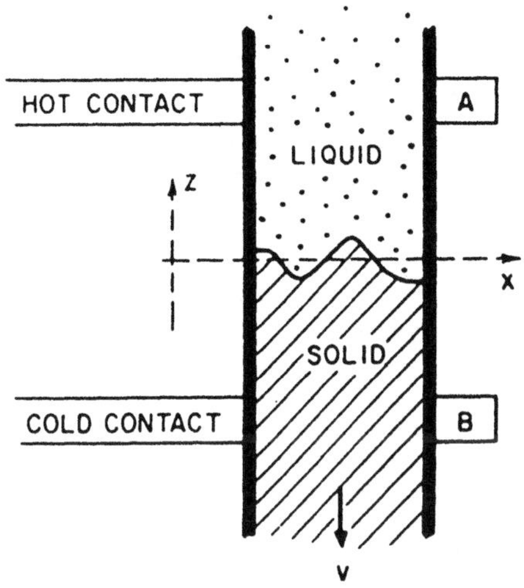

Fig. 3. Schematic setup of a directional solidification experiment.

(If it were smaller, the interface would be pulled into colder regions, and crystallization would accelerate, if it were larger, the ensuing advancement of the interface into the hot region would favour melting, leading to a deceleration.)

As to the boundary conditions, the first equation from (3) remains unchanged, the second however is modified; instead of structures with a needle-like backbone (see Fig. 2) we now consider (guided by experiment) periodic ones. Denoting the wavelength of the pattern by λ we therefore have

$$u(x,\infty) = u(x,-\infty) = 0 \quad , \quad u(x+\lambda,z) = u(x,z) \quad . \tag{15}$$

Furthermore, the Gibbs-Thomson condition, i.e. the first equation in (4), takes a slightly different form, while the second equation is not altered. Due to the external thermal gradient the interface temperature must become position dependent:

$$u_i = 1 - \frac{\zeta}{l_T} - d\kappa \quad . \tag{16}$$

Herein

$$l_T = \frac{m\Delta c}{G} \tag{17}$$

is the so-called *thermal length*. m is the absolute value of the slope of the liquidus line describing coexistence of the liquid and the solid (see Fig. 1). Typical thermal lengths are on the order of a few cm. The isotropic prefactor of the capillary length becomes

$$d_0 = \frac{\gamma T_m}{Lm\Delta c} \quad . \tag{18}$$

Instead of Δ, there is a 1 in (16), which means that we have the equivalent of unit undercooling. This has to do with the fact that the motion of an interface with a planar envelope cannot be stationary for other undercoolings (unless kinetics are taken into account (Misbah et al. 1991)) – this is different for needle-shaped structures.

The continuity equation is, of course, identical to the one for free growth, i.e., it is given by (13).

All these equations do not only apply to liquid-solid phase transitions, but also to other first-order transitions such as the isotrope-nematic phase transition of a liquid crystal (Simon et al. 1988). The experiment is then called directional ordering.

2.3 Lamellar Eutectic Growth

Lamellar eutectic structures are abundant in the casting of alloys. Their controlled experimental investigation usually proceeds in the directional solidification setup described above. However, the equations of motion are a little more complex due to the presence of a second solid phase. Also the growth mechanism is somewhat different from that of dendritic growth or directional solidification of dilute alloys, where excess concentration has to be rejected to infinity. Here, it can diffuse laterally to the companion phase whose growth it even favours. That such a *cooperative* growth mode is possible is one reason for the abundance of lamellar microstructures in alloys.

For simplicity, our consideration will be restricted to the one-sided model which normally is a good approximation for eutectics.

There is a natural reference point for concentration measurements in eutectic systems; it is the equilibrium concentration c_e of the liquid at the triple or eutectic point. The most convenient definition of the diffusion field is therefore

$$u = (c - c_e)/\Delta c \quad , \tag{19}$$

where $\Delta c = c_\beta - c_\alpha$ refers to the eutectic temperature T_e (see Fig. 1). The concentration far out in the liquid, c_∞, is generally not much different from that of the eutectic point (see again Fig. 1, the dash-dotted line).

In the laboratory frame, where the growth front is stationary, the equation of motion is again given by (2). But whenever laterally traveling states occur, the pattern will not be at rest in the laboratory. For reasons that will become clear below, the most useful coordinate system is the one that is attached to the interface. Transformed to this frame of reference, the diffusion equation reads

$$\frac{1}{D}\frac{\partial u}{\partial t} = \nabla^2 u + \frac{2}{l}\left(\frac{\partial u}{\partial z} + \tan\phi\,\frac{\partial u}{\partial x}\right) \quad , \tag{20}$$

where ϕ is the *tilt angle* of the lamellar pattern and the lateral velocity is related to V by $V_x = \tan\phi\, V$.

The boundary condition (15) changes into

$$u(x,\infty) = u(x,-\infty) = u_\infty \quad , \quad u(x+\lambda, z) = u(x,z) \quad , \tag{21}$$

and the two-phase nature of the solid first shows up in the Gibbs-Thomson equation

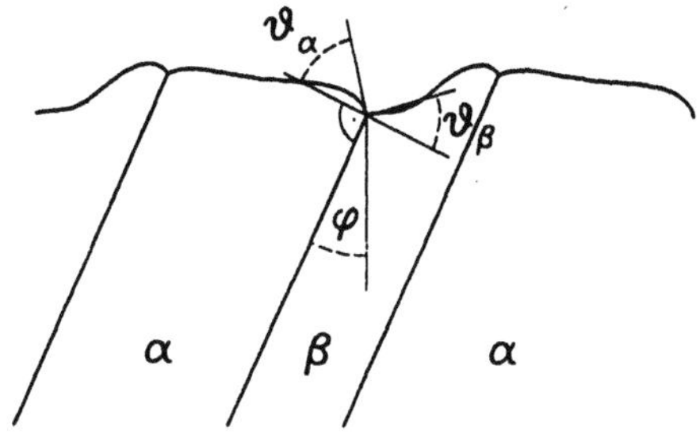

Fig. 4. Definition of the contact angles ϑ_α, ϑ_β, and the tilt angle ϕ. Note that ϕ is counted positive for a tilt to the right, while ϑ_α and ϑ_β are always positive.

$$u = \begin{cases} -\zeta/l_T^\alpha - d_0^\alpha \kappa & \alpha \text{ phase} \\ \zeta/l_T^\beta + d_0^\beta \kappa & \beta \text{ phase} \end{cases} . \tag{22}$$

Here another simplification has been made. The capillary lengths,

$$d_0^\alpha = \frac{\gamma_{\alpha l} T_e}{L_\alpha m_\alpha \Delta c} \quad , \quad d_0^\beta = \frac{\gamma_{\beta l} T_e}{L_\beta m_\beta \Delta c} \quad , \tag{23}$$

of the two liquid-solid interfaces have been taken isotropic: anisotropy is much less important in eutectic than in dendritic growth. $\gamma_{\alpha l}$ and $\gamma_{\beta l}$ are the interface tensions between the liquid and the solid α and β phases, respectively; the L_i ($i = \alpha, \beta$) are efective latent heats per unit volume (Kassner et al. 1991b) and the m_i are the slopes of the liquidus lines (taken positive) in the eutectic phase diagram (Fig. 1). There are also two thermal lengths in the problem now:

$$l_T^\alpha = \frac{m_\alpha \Delta c}{G} \quad , \quad l_T^\beta = \frac{m_\beta \Delta c}{G} \quad . \tag{24}$$

Moreover, the two-phase structure appears in the continuity equation:

$$-D\,\frac{\partial u}{\partial n} = \begin{cases} [(1-k_\alpha)u + \delta]\, v_n & \alpha \text{ phase} \\ [(1-k_\beta)u + \delta - 1]\, v_n & \beta \text{ phase} \end{cases} . \tag{25}$$

In this equation, k_α and k_β are the partition coefficients of the two phases, given by the ratios of the slopes of the liquidus and solidus lines in our linearized phase diagram (Fig. 1); $\delta = (c_e - c_\alpha)/\Delta c$ is the reduced miscibility gap of the α phase (and $1 - \delta$ that of the β phase). It should be mentioned that in (25) a capillary term has been neglected. This term vanishes for dilute alloys (Sect. 2.2) and is small for eutectics with small temperature gaps $m_i \Delta c$ (Kassner and Misbah 1991b).

An additional relation that must hold in eutectic systems is the mechanical equilibrium condition for the surface tensions of the three interfaces meeting at a triple point. The contact angles ϑ_α and ϑ_β, whose definition can be inferred from Fig. 4 are determined by this constraint (on a set of 2D vectors) which explicitly reads

$$\gamma_{\alpha l} \sin \vartheta_\alpha + \gamma_{\beta l} \sin \vartheta_\beta = \gamma_{\alpha\beta} \quad ,$$
$$\gamma_{\alpha l} \cos \vartheta_\alpha - \gamma_{\beta l} \cos \vartheta_\beta = 0 \quad . \tag{26}$$

This completes the set of equations defining the eutectic problem.

2.4 Liquid Crystals

Directional ordering of liquid crystals is described, as has been explained above, by the same equations as directional solidification. However, it has not yet been possible to attack the full dynamical problem numerically. An approximation had to be introduced (it will be discussed below) that is expected to miss certain interesting aspects of the dynamics in the high velocity regime. Due to the smallness of the diffusion length, describing the range of nonlocal retardation effects, the dynamics are quasilocal at very high velocities, which are nevertheless accessible in liquid crystal experiments. When the diffusion length ($\propto 1/V$) becomes smaller than the capillary length, the planar interface restabilizes. Close to the threshold of restabilization, one can derive analytically, from the set of equations for the full two-dimensional problem, a *local* equation of motion for the one-dimensional interface (by a method going back to Sivashinsky (1977)). This has been done (Kassner et al. 1991) for the symmetric model with partition coefficient 1, and the resulting partial differential equation which is second order in time and fourth order in space reads

$$3\zeta_{xxxx} - 4\zeta_{xxt} + \zeta_{tt} + 8\zeta_{xx} + 8l_T^{-1}\zeta = 4\zeta_x\zeta_{xt} + 2\zeta_{xx}\zeta_t - 4(\zeta_x^2)_{xx} - 2(\zeta_x^3)_x + 2\zeta_{xx}^2 \quad . \tag{27}$$

x and t are renormalized *slow* variables here; likewise, l_T^{-1}, the only remaining parameter of the equation, is a renormalized thermal gradient.

Equation (27) should be solved with periodic boundary conditions in a range of wavelengths λ

$$\zeta(x + \lambda, t) = \zeta(x, t) \quad . \tag{28}$$

The numerical technique for its solution and the results will be described in the following sections.

3 Numerical Methods

3.1 Quasistationary Approximation

We are interested in the motion of the liquid-solid interface and the patterns arising from it, including its stationary states. We are less interested in the values of the diffusion field in the bulk. Moreover, any discretization of the bulk would impose the anisotropy of the chosen grid on the discretized problem. For free growth, anisotropy has a strong influence on the types of possible patterns – there are no dendrites without anisotropy. Therefore, it should be introduced in a controlled manner and not by accident.

The physics of the problem allows to achieve this by a numerical method, in which bulk discretization is avoided, i.e., only the interface itself is dicretized. For not too large growth rates one can assume that the diffusion field follows the interface adiabatically, which means that the profile is stationary on the time scale of diffusion, hence the name *quasistationary approximation*. Even in the moving frame, this does *not* mean that all the time dependence is lost, because $\dot{\zeta}$ in (12) is retained as a time dependent quantity. It is clear that the quasistationary approximation becomes exact for true stationary states. In addition, it correctly identifies instabilities of stationary patterns, if the bifurcation is not of Hopf type, i.e., if the bifurcating state does not oscillate at a finite frequency at the onset of the instability. Of course, traveling states, whose velocity is constant, still qualify as stationary in the comoving coordinate system and can therefore also be found exactly (for this reason we use (20) to describe eutectics).

Formally, one sets the diffusion constants D and D' equal to infinity on the left hand side of (2) (but keeps their ratio and hence the ratio of l and l' constant), which means that the diffusion equation turns into a homogeneous elliptic equation. The latter can be reformulated, using its Green's function, into an integral equation, where the domain of integration is just a contour, namely the boundary of the liquid or solid subsystem, respectively. In order to simplify the discussion, we restrict ourselves to the one-sided model, and specific formulas will be written for the eutectic case. The pertinent boundary integral equation is

$$\int_{\Gamma_{sl}} d\Gamma' \, g(\mathbf{r},\mathbf{r}') \frac{\partial u}{\partial n'} = \int_{\Gamma_{sl}} d\Gamma' \, h(\mathbf{r},\mathbf{r}';\mathbf{n}') \left(u(\mathbf{r}') - u_\infty\right) \quad , \tag{29}$$

where ($\Delta x = x - x', \Delta z = z - z'$)

$$g(\mathbf{r},\mathbf{r}') = \frac{1}{2\pi} \, \mathrm{e}^{-(\Delta z + \Delta x \tan\phi)/l} \, K_0\left(\frac{|\mathbf{r}-\mathbf{r}'|}{l\cos\phi}\right) \tag{30}$$

is the mentioned Green's function (vanishing at infinity) for a half-infinite slab. K_0 is the modified Bessel functon of zeroth order (Macdonald's function). Furthermore,

$$\begin{aligned} h(\mathbf{r},\mathbf{r}';\mathbf{n}') = \frac{1}{2\pi l} \, \mathrm{e}^{-(\Delta z + \Delta x \tan\phi)/l} \Bigg\{ & -(n'_z + n'_x \tan\phi) K_0\left(\frac{|\mathbf{r}-\mathbf{r}'|}{l\cos\phi}\right) \\ & - \frac{\mathbf{n}'(\mathbf{r}'-\mathbf{r})}{\rho\cos\phi} K_1\left(\frac{|\mathbf{r}-\mathbf{r}'|}{l\cos\phi}\right)\Bigg\} - c_\mathbf{r}\delta(\mathbf{r}-\mathbf{r}') \quad , \end{aligned} \tag{31}$$

where K_1 is the modified Bessel function of first order. The δ function is one dimensional, i.e., it is defined via a contour integral, not (as usual) a (2D) volume integral. The prefactor $c_\mathbf{r}$ of the δ function is $\frac{1}{2}$ if the interface is smooth at $\mathbf{r}$; for more details see Saito et al. (1988) and Brebbia (1978). Due to global mass conservation, $h(\mathbf{r},\mathbf{r}';\mathbf{n}')$ obeys a sum rule

$$\int_{\Gamma_{sl}} d\Gamma' \, h(\mathbf{r},\mathbf{r}';\mathbf{n}') = -1 \quad , \tag{32}$$

which has been derived for the general case of a tilted interface ($\phi \neq 0$) in Kassner and Misbah (1991c). The integration contour has been denoted by Γ_{sl}, because the originally closed contour Γ for the boundary integral formulation (the sum $\Gamma_{sl} + \Gamma_3 + \Gamma_2 + \Gamma_1$ in Fig. 2) can be reduced to the solid-liquid interface, since $u - u_\infty$ as well as $\partial u/\partial n'$ vanish for $z \to \infty$ and both g and h decay fast enough when $|x| \to \infty$. For free growth or directional solidification, one simply sets $u_\infty = 0$ and $\phi = 0$ in (29)–(31).

The discretized version of (29) is the basis of most numerical approaches that I will discuss.

Even before discretization, the first step in the numerical treatment of (29) is to reduce the number of length scales of the problem (by one) via introduction of dimensionless ratios. There are three length scales in the free growth problem – the capillary length, the diffusion length and the tip radius R of the evolving dendrite; in directional solidification, the latter is replaced by the wavelength λ of the pattern, and the thermal length provides a fourth length scale; finally, the existence of *two* capillary and thermal lengths increases the number of length scales to six in the eutectic problem. One of the dimensionless ratios that are formed from these lengths is theoretically important enough to have a name of its own; it is called Péclet number and given by $P = R/l$ in the dendritic case and by $P = \lambda/l$ otherwise.

The second step is to exploit symmetries of the problem to reduce the integration contour. In the case of dendritic growth the only symmetry is the reflection symmetry with respect to the axis of the dendrite. For directional solidification, periodicity, i.e. invariance with respect to a discrete set of lateral translations, is an additional symmetry. In eutectics, the axisymmetry gets lost, as soon as tilted patterns appear. For this case, the integral equation becomes (lengths are measured in units of λ)

$$\begin{aligned} &\int_{x'=0}^{1} d\Gamma' \sum_{m=-\infty}^{\infty} g(x,\zeta;x'+m,\zeta')\frac{\partial}{\partial n'}u(x',\zeta') \\ &= \int_{x'=0}^{1} d\Gamma' \sum_{m=-\infty}^{\infty} h(x,\zeta;x'+m,\zeta')(u(x',\zeta')-u_\infty) \quad . \end{aligned} \tag{33}$$

Discretization is performed according to a boundary element scheme (Brebbia 1978), in which the diffusion field as well as its derivatives are approximated by functions, not just constant values, on the discretization intervals. The boundary curve itself is replaced with a polygon. The position of a point $\mathbf{r}$ on the polygon edge between the corners $\mathbf{r}_j$ and $\mathbf{r}_{j+1}$ is described by a parameter ξ between -1 and 1 as

$$\mathbf{r} = \Phi_1(\xi)\mathbf{r}_j + \Phi_2(\xi)\mathbf{r}_{j+1} \quad , \tag{34}$$

where

$$\Phi_1(\xi) = (1-\xi)/2\,, \quad \Phi_2(\xi) = (1+\xi)/2 \tag{35}$$

The diffusion field and its derivative, whose values at the discretization points we denote by

$$u_j = u(\mathbf{r}_j) \quad , \quad q_j = \left.\frac{\partial u}{\partial n'}\right|_{\mathbf{r}_j} \quad , \tag{36}$$

are obtained by linear interpolation on the boundary elements:

$$\begin{aligned} u(\xi) &= \Phi_1(\xi)u_j + \Phi_2(\xi)u_{j+1} \quad , \\ q(\xi) &= \Phi_1(\xi)q_j + \Phi_2(\xi)q_{j+1} \quad . \end{aligned} \tag{37}$$

Defining

$$\mathbf{s}_j = \mathbf{r}_{j+1} - \mathbf{r}_j = \begin{pmatrix} s_{xj} \\ s_{zj} \end{pmatrix} \quad , \quad s_j = |\mathbf{s}_j| \quad , \tag{38}$$

we can write the integration on Γ' as follows

$$\int_0^1 d\Gamma' \ldots = \sum_j \int_{\mathbf{r}_j}^{\mathbf{r}_{j+1}} d\Gamma_j \ldots = \sum_j \frac{s_j}{2} \int_{-1}^{1} d\xi \ldots \quad , \tag{39}$$

and obtain for the discretized form of (29)

$$\sum_j G_{ij} q_j = \sum_j H_{ij} (u_j - u_\infty) \quad , \tag{40}$$

where G_{ij} and H_{ij} are defined by

$$\begin{aligned} G_{ij} &= \sum_m \frac{s_j}{2} \int_{-1}^{1} d\xi \Phi_1(\xi) g(x_i, \zeta_i; x_j + \Phi_2(\xi) s_{xj} + m, \zeta_j + \Phi_2(\xi) s_{zj}) \\ &+ \frac{s_{j-1}}{2} \int_{-1}^{1} d\xi \Phi_2(\xi) g(x_i, \zeta_i; x_j - \Phi_1(\xi) s_{x,j-1} + m, \zeta_j - \Phi_1(\xi) s_{z,j-1}) \quad , \end{aligned} \tag{41}$$

$$\begin{aligned} H_{ij} &= \sum_m \frac{s_j}{2} \int_{-1}^{1} d\xi \Phi_1(\xi) h(x_i, \zeta_i; x_j + \Phi_2(\xi) s_{xj} + m, \zeta_j + \Phi_2(\xi) s_{zj}) \\ &+ \frac{s_{j-1}}{2} \int_{-1}^{1} d\xi \Phi_2(\xi) h(x_i, \zeta_i; x_j - \Phi_1(\xi) s_{x,j-1} + m, \zeta_j - \Phi_1(\xi) s_{z,j-1}) \quad . \end{aligned} \tag{42}$$

The slightly more complicated formulae for the symmetric case can be found in Kassner and Misbah (1991b). In the dendritic case there is no spatial periodicity that could be exploited; furthermore, the tail of the dendrite has to be treated separately (Saito et al. 1988). For $i \neq j$ and $i \neq j \pm 1$, G_{ij} and H_{ij} are evaluated by the four-point Gauss-Legendre integration rule. The other cases need special treatment, because $g(\mathbf{r}, \mathbf{r}')$ diverges logarithmically for $\mathbf{r} \to \mathbf{r}'$ and $h(\mathbf{r}, \mathbf{r}'; \mathbf{n}')$ contains a δ function. The logarithmic divergence is handled by using a Gauss integration rule for integrands exhibiting a logarithmic singularity (Abramowitz and Stegun 1972, pp. 877ff.). The δ function singularity plays a role only in the calculation of the diagonal elements of the matrix H. These are therefore not evaluated directly but via the sum rule (32), which in matrix notation reads

$$H_{ii} = -1 - \sum_{j \neq i} H_{ij} \quad . \tag{43}$$

By this we gain the additional advantage of ensuring global mass conservation.

For eutectics, the discretized integral equation (40) is imposed at all discretization points except the triple points. There we require instead that (26) is fulfilled, i.e. that the *pinning* angles (measured from the x axis and to be distinguished from the *contact* angles, Fig. 4) take on the values $\vartheta_\beta + \phi$, $\vartheta_\alpha - \phi$, if the triple point lies to the left of the α phase, and $\vartheta_\alpha + \phi$, $\vartheta_\beta - \phi$, if it lies on its right side. These equations are most easily discretized within a hybrid representation of the boundary by the arclength s, the normal angle ϑ on the interface curve, and the coordinates of one triple point (Kassner and Misbah 1991b).

Given the discretization, there are two ways to deal with interface motion. The first, used to describe free growth or the dynamics of directional solidification, leads to a dynamical code, allowing to follow the temporal evolution of the interface (Saito et al. 1987,1988), while the second directly solves for the stationary shape of the interface and has been applied to directional solidification and eutectics (Kessler and Levine 1989, Levine and Rappel 1990, Kassner and Misbah 1990, 1991a-c).

Dynamical Code. This is basically the strategy mentioned in Sect. 2.1. Starting from an initial interface profile, one calculates the matrix elements G_{ij} and H_{ij}, which are only shape dependent, and u_j, from the Gibbs-Thomson condition. Then (40) is solved for the q_j, i.e., the matrix G_{ij} has to be inverted, which is a problem of linear algebra. Given the q_j, one calculates the velocities of the discretization points from the continuity equation and steps the interface forward in time, taking care by an appropriate choice of the time step that no large curvature changes occur. After each time step, the number of discretization points is adapted, by interpolation and addition or by removal of points (followed by an equilibration of their distances), to accommodate length changes of the interface without losing accuracy. In free growth, the velocity of the coordinate system has to be chosen as close as possible to that of the actual dendrite (in order to maintain the quality of the quasistationary approximation), therefore an additional kinetic equation is introduced to have it relax to the tip velocity of the dendrite

$$\frac{dV}{dt} = -\frac{1}{\tau}\left[V(t) - v_{\text{tip}}(t)\right] \quad , \tag{44}$$

where the relaxation time τ is taken small in comparison with all physical time scales.

The matrix G_{ij} is fully occupied, so its inversion takes $O(N^3)$ elementary operations (multiplications) if its dimensions are $N \times N$, i.e., when N is the number of discretization points. In order to compare this expenditure with that of methods discretizing the bulk, one should consider what number of discretization points N_b would be needed there. A rough estimate is $N_b \approx N^2$ and hence our method is $O(N_b^{3/2})$ per time step, i.e. per solution of the quasistationary diffusion equation. This is in principle comparable with simultaneous overrelaxation algorithms (Press et al. 1986). However, the method avoids the artificial anisotropy introduced by most discretizations of the bulk. In addition, it has the aforementioned adaptive feature to maintain the chosen accuracy independent of the (increasing) interface length, which would be hard to implement in any bulk method.

As an aside, it may be mentioned, that for system sizes that could be achieved up to now ($N \lesssim 500$), the time consumption is still dominated by the calculation of the modified Bessel functions (after Abramowitz and Stegun (1972), p. 379), so that the method is actually $O(N_b)$.

Stationary Code. Newton Method. In the case of eutectics, interest has up to now focused on stationary patterns. These could be obtained, in principle, by running the dynamical code until the pattern does not change in time any more. However, it is possible to calculate them directly.

For given q_j, u_j, G_{ij}, and H_{ij}, (40) constitutes a set of implicit *nonlinear* equations determining the interface points $\mathbf{r}_j$. Now there are N equations and the $\mathbf{r}_j$ correspond to $2N$ Cartesian coordinates, but if one requires the discretization points to be equidistant along the α and β-interfaces, respectively, i.e. the arclength increments s_j to be constant, each of the discretization intervals can be represented by a single variable ϑ_j. Furthermore, to establish the mechanical equilibrium condition (26), the integral equation is imposed only at the inner points of the α and β-phase boundaries. At each of the (four) endpoints, the pinning angle conditions described above are used, adding up to four equations. Together with the $N-4$ equations for the inner points we have N equations for N hybrid variables, the coordinates x_e, z_e of one eutectic point and the normal angles ϑ_j. Hence, the problem is well-defined.

When the profiles are symmetric ($\phi = 0$), this is the whole story, because the symmetry guarantees continuity of the periodic continuation of a half-cell solution. For asymmetric profiles, we have an additional variable, the tilt angle ϕ, but also periodicity must be required as an additional constraint, we need to set

$$\zeta_1(\phi) = \zeta_N(\phi) \quad . \tag{45}$$

Hence we have $N+1$ equations and $N+1$ variables, including ϕ, and one can expect a discrete set of solutions for ϕ only.

The described set of nonlinear equations is solved iteratively by a standard Newton algorithm (NAG routine).

3.2 Local Equation

Stationary solutions to the local equation (27) are found pretty much the same way as explained in the last subsection, except that the problem is much simpler because it is one-dimensional to begin with. Of course, no tilt angle can be defined, and so the additional variable that has to be introduced in a search for asymmetric traveling solutions is the transverse velocity V_x.

However, the local equation was not derived in order to calculate steady-state solutions but to study the full time dependence of the dynamical problem. Note that as with the quasistationary approximation, tractability is achieved here through an effective reduction of the dimension of the mathematical problem.

For not too large systems, a partial differential equation describing a time-dependent one-dimensional system, can be numerically integrated with standard methods. We have used Gear's backward difference method (Gear 1971) to simulate the long-time behaviour of the dynamics. It may be noteworthy that the spatial derivatives in (27) were evaluated by first calculating the Fourier transform of the interface, multiplying it with that power of the wavevector (times the imaginary unit) that corresponds to the order of the desired derivative, and transforming back to real space. The accuracy of this "infinite" order method compares impressively with a simple $O(h^2)$ scheme in long-time runs. This gain in accuracy precludes, however, the investigation of very large systems, because the matrix to be inverted in an implicit integration scheme is not sparse.

4 Numerical Results and Open Questions

In this section, I will describe a few representative results obtained in numerical simulations of the model. An attempt at completeness is not intended; more results can be found in the given references. Here, some emphasis will be put on the necessary computational expense and on the arising limitations for numerical investigations of open questions. The next section will then sketch alternative present and possible future approaches.

4.1 Free Growth

Figure 5 shows the temporal evolution of a numerically grown dendritic crystal (Saito et al. 1987). The left hand side of the picture shows the tip region only, whereas on the right hand side a transition region needed to connect the tip region to a parabolic tail is also displayed (therefore, the dendrite does not appear symmetric). The height of the tip region in this example is 25 in units where the capillary length is 10^{-3} and within this region the discretization intervals are not allowed to become longer than $s_{max} = 0.1$. Hence there are about 400 discretization points in one side of the tip region. The transition region is as finely discretized as the tip, while in the tail, which extends to at least five diffusion lengths from the tip (to keep the influence of the cutoff at the dendrite's bottom small) the discretization is coarser. Both the transition and tail regions are taken into account when the set of linear equations (40) is solved, but their dynamics are not governed by the solutions. The shown dendrite grew in roughly 20 h CPU time on a CRAY XMP 2/2.

Dendritic structures could only be obtained for anisotropic surface tension ($\epsilon_4 \neq 0$) as anticipated by theory. When ϵ_4 is set equal to zero, the initial needle crystal (a parabola) is destroyed by tip splitting as time evolves.

The velocity of simulated dendrites converges to a unique value for given undercooling Δ and anisotropy ϵ_4 and so does the tip radius. Hence the program exhibits velocity and tip radius selection and provides numerical values for the selected quantities.

The dependence of these quantities on the undercooling and anisotropy was found to follow the scaling laws

$$V = \frac{2D}{d_0}\,\sigma(\epsilon_4)\,p(\Delta)^2 \quad , \tag{46}$$

$$R = d_0\,\rho(\epsilon_4)/\sigma^2(\epsilon_4)\,p(\Delta) \quad , \tag{47}$$

where σ is a dimensionless growth rate that depends on anisotropy only. $p(\Delta)$ is a special Péclet number, which I will explain shortly. In the limit of vanishing surface tension, i.e. for $d \equiv 0$, the growth equations (1)–(13) are analytically solvable. The solutions are parabolas moving at a constant velocity (Ivantsov 1947). However, for given undercooling Δ there is not just one so-called Ivantsov parabola but a continuous infinite family of solutions with different velocities V and tip radii R_0 but fixed product of these quantities. This product is essentially the Péclet number $p = R_0/l$; it is determined uniquely by the undercooling via the equation (in two dimensions)

$$\Delta = \sqrt{\pi p}\,e^{p}\,\mathrm{erfc}(\sqrt{p}) \quad , \tag{48}$$

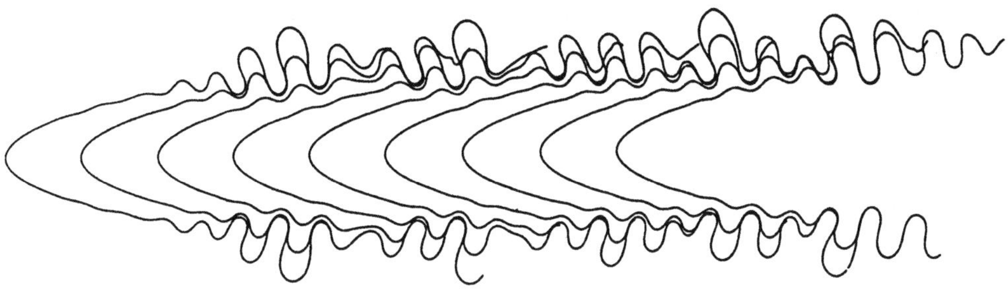

Fig. 5. Stroboscopic plot of the evolution of a dendritic crystal moving from bottom to top (Saito et al. 1987,1988). The time interval between two successive shapes is constant. The undercooling is $\Delta = 0.25$, the anisotropy of the capillary length, $\epsilon_4 = 0.10$. In the steady state, the diffusion length has converged to $l = 2D/V = 34$, and the tip radius is $R = 0.885$. Length units are referring to a capillary length of $d_0 = 0.001$.

(For $d \neq 0$, the infinite family of solutions disappears, and it was one of the successes of the theory to show that the presence of anisotropy allows a *discrete* solution family to appear.) The parameter $\rho(\epsilon_4)$ is the ratio of the true Péclet number P defined in Sect. 3.1 and the Ivantsov one, $p(\Delta)$, and as the notation suggests this ratio depends only on the anisotropy.

The nontrivial and remarkable fact about the numerical confirmation of the scaling relations (46) and (47) is that they do not only hold for for the needle crystal (for which they were derived analytically) but also for a dendrite with sidebranches, thus establishing the relatively passive role of the sidebranches.

In addition, the numerics provided scaling results for the wavelength of the sidebranches which were consistent with earlier estimates (Barber et al. 1987, Caroli et al. 1987).

Finally, it is worth mentioning that to date this is the only numerical simulation of dendritic growth which did not just produce nice pictures but also numbers for a quantitative comparison.

Open Questions. In microcuts of alloys one often sees *arrays* of dendrites. How the primary spacing of these dendrites is selected, is not known. From the given discussion, it should be clear that even within the quasistationary approximation the treatment of such a system is presently beyond our capabilities.

Another open question concerns the limit of vanishing anisotropy. Since dendrites become unstable against tip splitting, a different growth morphology must take over. For this structure which is characterized through a cascade of tip-splitting processes the name *dense-branching morphology* has been coined (Ben-Jacob et al. 1986). A picture conveying an impression of this morphology is given in Fig. 6, which has been produced by a program of the simplified microscopic dynamics type (Shochet et al. 1991). Anisotropy is visible in the preferential 45-degree growth directions, and it cannot be avoided in that program, because solid particles are sitting on a square lattice. Most interesting is the recent development of an analytic scaling theory (Brener et al. 1991) making predictions about the average velocity and branch width of this morphology, also called *seaweed* structure, *in the absence of anisotropy*. In order to numerically check this theory, one would have to solve the free-growth model for a morphology which is stationary only on average, therefore the quasistationary approximation should be given up, and for very large systems, as one can estimate from the scaling theory.

Although numerical results compare favourably with experimental data (which are obtained in three-dimensional systems), dendritic growth in *three dimensions* belongs into the class of unsolved problems from a theoretical point of view. For three-dimensional dendritic growth with real anisotropy in the azimuthal direction there has neither been any satisfactory analytic treatment (for an unsatisfactory one see Kessler and Levine 1987) nor are there at present quantitative numerical simulations. (By contrast, the selection problem of *axisymmetric* three-dimensional dendritic growth has been solved (Caroli et al. 1986).)

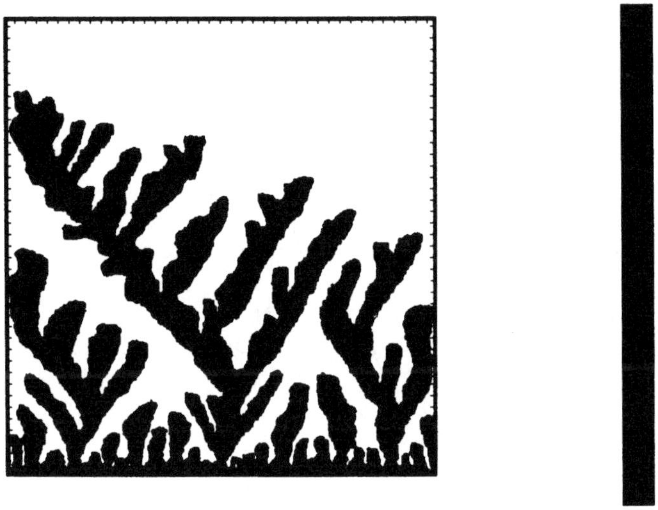

Fig. 6. An example for the dense-branching morphology, grown on a lattice.

4.2 Directional Solidification

Numerical investigations are also needed in directional solidification, because the range of validity of analytical approximations such as amplitude equations is extremely small, as is known from general considerations concerning the flatness of the so-called neutral stability curve of the model and has been shown explicitly in numerical work by Brattkus and Misbah (1990).

Simulations of directional solidification (Saito et al. 1989, Misbah et al. 1990) have led to the identification and characterization of a transition from cellular patterns to dendritic ones, which appear when the diffusion length l becomes smaller than the cell spacing λ ($P > 1$), in agreement with experiments (Kurowski et al. 1990).

Furthermore, it was found that at high velocities there is a wavelength selection mechanism via a *tail instability*, a result which appears unattainable analytically. A pattern that develops this tail instability to reduce its cell spacing after a velocity jump is shown in Fig. 7.

Scaling relations for small wavelengths of the pattern were found to agree quantitatively with analytic results for the Saffman-Taylor viscous fingering problem (Pelcé and Pumir 1985). At large wavelengths and velocities they agree with results for freely growing dendrites (see (46,47)).

Computing time for one parameter set until convergence to a steady state is reached, varied between 10 min and 20 h on a CRAY XMP, with a fully vectorized code, increasing from small to large growth rates and anisotropies.

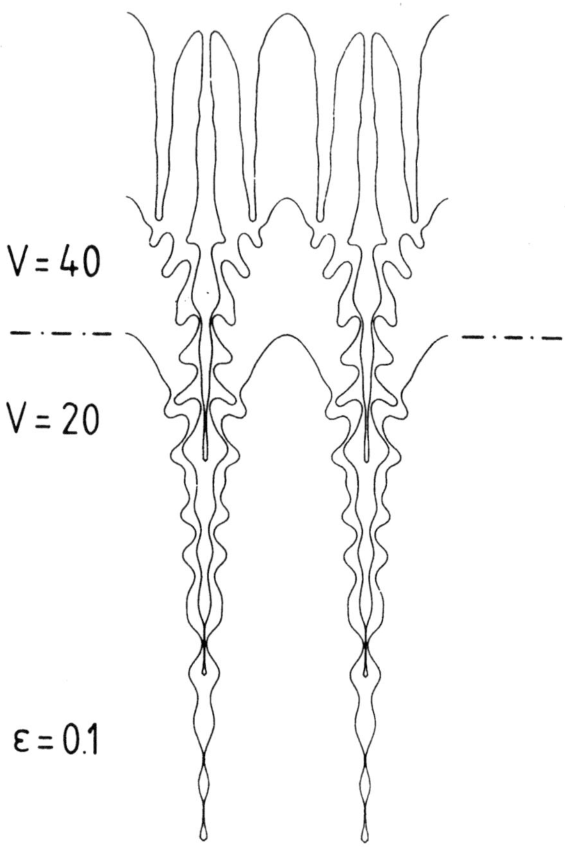

Fig. 7. Stroboscopic plot of a sidebranching structure moving from bottom to top (Saito et al. 1989). The bottom dendrite growing at velocity $V = 20$ has reached a stationary sidebranching state. Then V is suddenly increased to 40. The middle and top profiles show the evolution of a tail instability, leading to the formation of a new primary cell (plus its mirror image) out of a sidebranch catching up with the tip from a tail position.

Open Questions. No numerical evidence was found for *oscillatory instabilities* which are seen in experiments (Flesselles et al. 1991). This may be due to the quasistationary approximation. Little is known about the ability of a code using this approximation to detect Hopf-type bifurcations.

The problem of *wavelength selection* is unsolved in directional solidification. (In the dendritic regime, this is again the problem of a dendritic array.) Above the stability threshold of the planar front, the system numerically finds a stationary state for ***any*** prescribed wavelength within a certain range. The only hard evidence to date for selection of a unique wavelength comes from a recent analysis of a pattern of small amplitude cells under the influence of a spatially dependent ramp-like thermal gradient (Misbah 1989, Misbah et al. 1990).

4.3 Eutectic Growth

Two major recent results in the theory of eutectic growth are the numerical demonstration of the existence of extended tilted patterns for isotropic surface tension (Fig. 8), which includes the quantitative determination of the tilt angle as a function of the pulling velocity (Fig. 9), and the numerical evaluation of a scaling function relating the wavelength of the pattern to its velocity (Fig. 10).

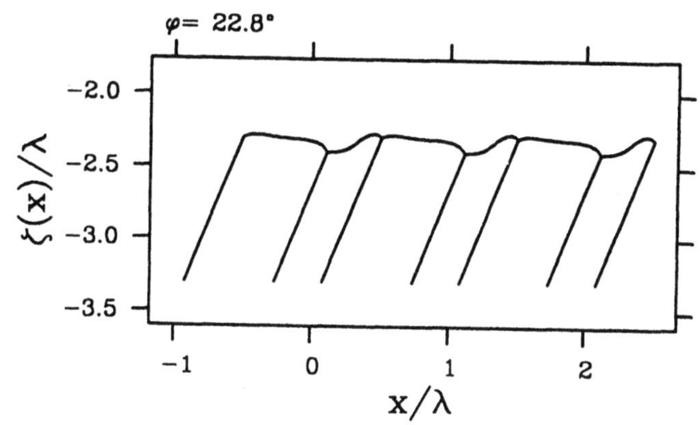

Fig. 8. A tilted pattern close above the critical velocity determining the onset of the tilt instability (Kassner and Misbah 1991c).

It was a surprise that the bifurcation from a symmetric untilted pattern to an asymmetric (and traveling) tilted one is *supercritical*. Before this was demonstrated (Kassner and Misbah 1990/91), the belief was generally held that it should be subcritical (Coullet et al. 1989), since in experiments tilted domains appeared only as localized inclusions escorted by domains of symmetric lamellae (Faivre et al. 1989). This made it plausible that both tilted and untilted states were stable in a certain parameter range and hence the transition from one to the other was first-order like, i.e. subcritical. Similar experimental results had been found in the directional solidification of liquid crystals (Simon et al. 1988) and there the bifurcation has also been shown to be supercritical (Levine and Rappel 1990). Following a suggestion by ourselves, Faivre et al. (1991) could meanwhile produce extended homogeneously tilted states, thus providing experimental support for the supercriticality of the bifurcation, and there are theoretical developments explaining the existence of localized inclusions even in this picture (Caroli et al. 1991).

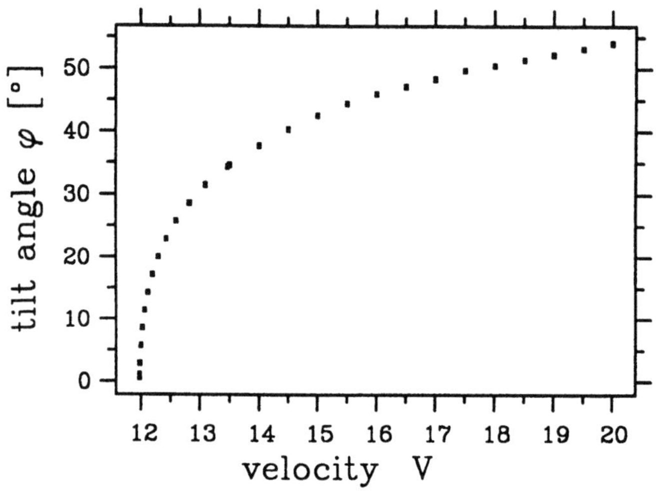

Fig. 9. Tilt angle ϕ as a function of the velocity, showing that the bifurcation from the state with $\phi = 0$ is supercritical (Kassner and Misbah 1991c).

From an analytically derived similarity equation (Kassner and Misbah 1991a), we concluded that at small Péclet numbers $P = \lambda/l$ (which are typically realized in experiments) the selected wavelength of any lamellar pattern (be it symmetric or asymmetric) must behave as

$$\lambda \sim V^{-1/2} f\left(\frac{l}{l_T}\right) \quad . \tag{49}$$

Apart from the interesting prediction, which has not yet been tested experimentally, that for constant ratio of diffusion and thermal lengths, i.e. for constant G/V, the selected wavelength scales as $V^{-1/2}$ exactly, this calls for the determination of the scaling function f, which can only be done numerically. We have calculated f under the assumption, put forward already by Jackson and Hunt (1966), that the selected wavelength of symmetric patterns is determined by the criterion that the average undercooling

$$\langle \Delta T \rangle = -G \left[\int_{\alpha\ \mathrm{phase}} \zeta(x) dx + \int_{\beta\ \mathrm{phase}} \zeta(x) dx \right] \tag{50}$$

be minimum (in comparison with undercoolings at other wavelengths and the same pulling velocity). The result of this calculation is presented in Fig. 10. It agrees in tendency with experiments described by Lesoult (1980), which gave evidence for a scaling behaviour of the type $\lambda \sim V^{-b}$, with an eponent b that is smaller than $\frac{1}{2}$ at low velocitites. Values of b down to 0.3 were reported.

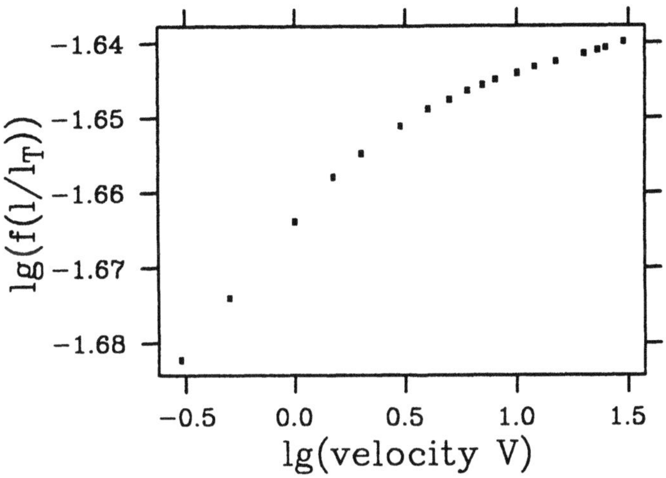

Fig. 10. Scaling function $f(l/l_T) = \lambda V^{1/2}$ as a function of the velocity in double logarithmic representation (Kassner and Misbah 1991b). A constant value corresponds to $\lambda \propto V^{-1/2}$, a slope a to $\lambda \propto V^{a-1/2}$. At low velocities, $a \approx 0.033$, in the high velocity regime, $a \approx 0.008$.

The numerical calculation of a single pattern at not too low velocities is not extremely costly. With a number of discretization points on the order of 60, convergence is, starting from a good initial guess, usually obtained within between one and twenty minutes, increasing from large to small velocities. The tendency is opposite to that in the dynamic case (where high velocities are more expensive, see the last subsection); in this sense the stationary code is complementary to the dynamic one. The program is fully vectorized and averages – as does the dynamical code – a little over 100 MFlops on a CRAY XMP 4/16. Nevertheless, the calculation of the data for Fig. 10 required on the order of 100 h CPU time. One reason is that one must go to very small velocities to see the deviation from $\lambda \sim V^{-1/2}$, and that at these velocities ($V \lesssim 1.0$) the calculation of a single profile takes well above an hour CPU time. What is worse, at each velocity many ($\gtrsim 10$) profiles have to be calculated, for different wavelengths, in order to determine the wavelength providing the minimum average undercooling.

Open Questions. Oscillatory states are also found in eutectics (Faivre et al. 1991); to investigate them numerically, it will be necessary to compute the dynamical evolution of lamellar eutectics, a description which must include dynamic equations for the triple points. Furthermore, it seems likely that the quasistationary approximation will be insufficient.

This holds to an even higher degree for any quantitative characterization of *chaotic* states that have been seen experimentally in eutectics. While a numerical treatment capturing the onset of oscillatory instabilities seems quite feasible at present, it is not clear whether one can get as easily into the chaotic regime.

4.4 Liquid Crystals

Basically the same phenomena of asymmetric cells, oscillating and chaotic states are found, as in eutectics, in experiments on liquid crystals (Simon et al. 1988, Flesselles et al. 1990), and it is believed that they have a common underlying mechanism.

For liquid crystals, however, the system parameters are such (large capillary length (μm), slow impurity diffusion) that experiments can be performed close to the restabilization threshold, where the aforementioned quasilocal description should be a good approximation; many of the interesting phenomena occur in that region, fortunately.

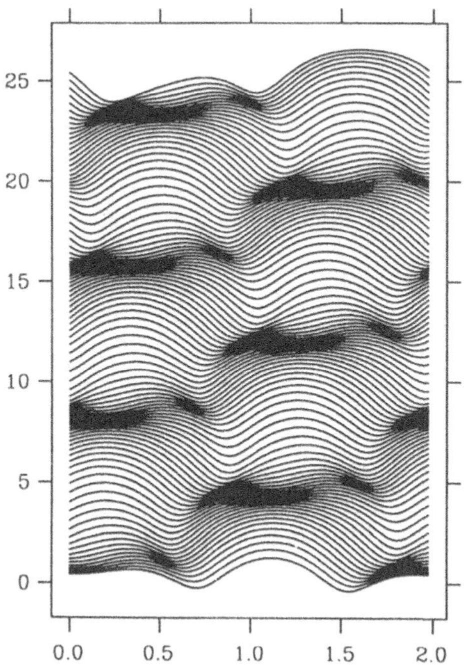

Fig. 11. Interface dynamics of a VB mode (Kassner et al. 1991). The vertical coordinate represents time.

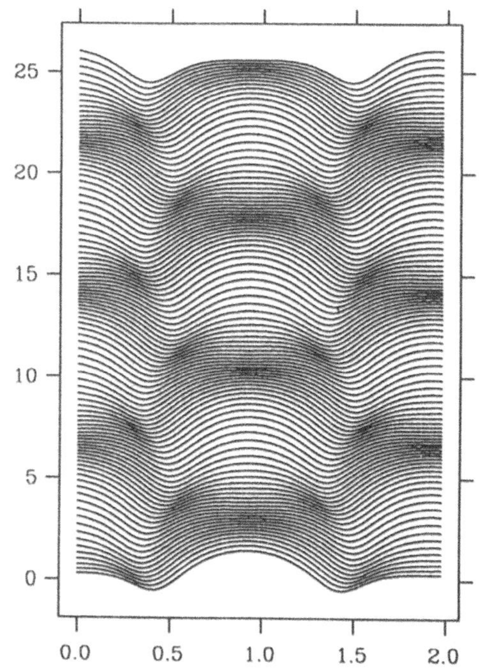

Fig. 12. Interface dynamics of a quasiperiodic mode (Kassner et al. 1991).

It is very easy to numerically obtain stationary symmetric or traveling asymmetric solutions to the simplified system (27), but the emphasis of our investigation (Kassner et al. 1991) was laid of course on the detection of more complicated dynamical behaviour. A full linear stability analysis of stationary solutions to (27) is readily performed (by contrast, it has not yet been done for the eutectic system), and it allows to identify the points in parameter space where oscillatory instabilities occur. This space is here spanned by just two variables – the (renormalized) inverse thermal length l_T^{-1} and the wavenumber q of the pattern.

Figures 11–13 contain stroboscopic plots of the interface for fixed wavenumber q but varying l_T^{-1}.

The first picture shows the temporal evolution for $q = 0.9$ and $l_T^{-1} = 0.29$, after transients have decayed. As expected from the stability analysis, the profile exhibits an oscillatory instability resulting in spatial period doubling. The width of a given cell oscillates in phase opposition with that of its neighbours, which leads to a vacillating motion of the minima of the interface. The position of the maximum of each cell remains stationary but the top of the cell oscillates, a motion that reminds of breathing. Therefore, we adopted the name *vacillating-breathing* (VB) mode for this type of interface motion that has been observed in various systems (Flesselles et al. 1991, Rabaud et al. 1990, de Cheveigné and Guthmann 1991). Note that this mode of motion has a symmetry axis, i.e. the interface has a well defined parity under reflection on this axis.

When l_T^{-1} is decreased, the oscillatory mode goes unstable with respect to *parity breaking* at $l_T^{-1} \approx 0.8$. As a consequence, a lateral drift is superimposed on the VB mode (Fig. 12). Now a given point of the interface "sees" two frequencies, one from the basic oscillation and one that is produced by shifting the interface across that point and whose value is qV_x. The Fourier spectrum of the time evolution of the interface amplitude, at fixed abscissa, clearly shows these two frequencies. Because they do not bear any relation to each other, their ratio is most likely irrational and hence the motion is *quasiperiodic*. This is confirmed by the nondiscrete nature of an appropriate Poincaré map of the system dynamics.

On further decrease of l_T^{-1}, the motion first remains quasiperiodic without any appearance of mode locking. Eventually, it becomes *chaotic*, and at $l_T^{-1} = 0.24$, the Poincaré map reveals a fully developed strange attractor.

In Fig. 13, we give a very short piece of the history of an interface in the chaotic regime ($l_T^{-1} = 0.23$). At the bottom of the cell, the motion corresponds essentially to a VB mode, and there are two cells fitting into the considered periodicity unit. As time increases, one of the cells becomes wider, until it reaches a critical local wavelength, at which it is destroyed by tip splitting. At the top, we then have *three* cells in a periodicity unit. This looks like an attempt of the system to select a more convenient wavelength. Of course, this new wavelength does not correspond to a stable state either for the given system parameters. In its further evolution the interface erratically runs through all kinds of different states, including drift motion to the left and to the right as well as intermittent almost steady states, taking turns with oscillatory modes.

In a nutshell, we have identified a quasiperiodic scenario for the transition to chaos in directional solidification, and we believe that this scenario is generic for systems where oscillatory and parity-breaking modes coexist.

In Figs. 11–13, the aspect ratio of the system is just two, i.e., the basic periodicity unit corresponds to two wavelengths of an underlying stationary solution. For these small aspect ratios and a discretization of the interface into 81 points, CPU-time considerations essentially did not play any role. Runs taking an hour CPU-time were usually sufficient to explore the relevant features of long-time dynamics. On the other hand, for an aspect

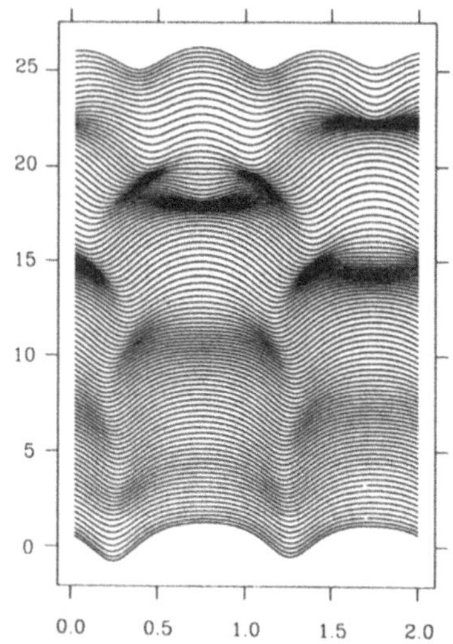

Fig. 13. Chaotic interface motion. Note the increase in cell number from two to three by a tip splitting process at time $t \approx 18$.

ratio of five (with 201 discretization points), a run for 10 h would sometimes be too short to decide, into what state of motion the slow part of the system dynamics might eventually settle. This is of course a consequence of the fact that the method is based on the inversion of a nonsparse matrix and hence $O(N^3)$.

Open Questions. Since the investigation of chaos in directional solidification is a relatively new field, the range of validity of certain scaling laws in the Fourier spectrum that have been predicted to hold close to the onset of chaos (Feigenbaum et al. 1982, Ostlund et al. 1983) has not yet been examined.

Furthermore, in all the systems considered, spatial coherence of the interface is preserved. There are however, experimental patterns that have been interpreted as specimens exhibiting *spatiotemporal* chaos, the analog of turbulence in hydrodynamics. The exploration of this kind of motion would be very interesting, and it will clearly require the study of systems with large aspect ratios (exceeding 10).

5 Alternative Approaches and Possible Future Developments

As we have seen throughout this contribution, the basic model defined in Sect. 2 is mathematically complex enough that it became numerically tractable only after some simplification was introduced. The simplification was achieved either by use of the quasistationary approximation, by restricting the consideration to stationary states, or by reducing the model to its quasilocal essence.

All alternative approaches in the literature that I know of employ comparable simplifications.

Karma (1987) investigated the *dynamic* evolution of eutectics using a random walker method which was originally developed by Kadanoff (1985). However, apart from simplifying the phase diagram, he solved the Laplace equation instead of the diffusion equation. This is akin to the quasistationary approximation in replacing the diffusion equation with a homogeneous elliptic one.

Even back in 1984, a *finite element* scheme had been constructed that produced numerical solutions to the *full* model equations for directional solidification (Ungar and Brown 1984a, 1984b, 1985, Ungar et al. 1985). However, the method works only for interfaces that are single-valued functions of the lateral coordinate, which restricts its applicability to cellular interface structures. Dendrites such as the ones in Figs. 5 and 7 cannot be grown. Nevertheless, nice results on the bifurcation diagram close above the stability threshold of the planar interface have been found. In particular, a Hopf bifurcation to an oscillatory state could be detected (Bennett and Brown 1989), which is probably out of the reach of the quasistationary approximation. A similar scheme might be applicable in search of oscillatory instabilities in eutectics.

A very different way to deal with the *physics* contained in the model has been proposed by the protagonists of *phase-field models* (Caginalp and Fife 1986). The phase field, a constitutive element of these models besides the temperature or concentration field, is a means to describe to which phase a point in space belongs, e.g. a point in the liquid would have a phase field value of 1, while one in the solid would have -1. Points with intermediate values can be considered to be part of the interface. The motion of the phase field is coupled to that of the physical field by phenomenological equations. Mathematically, the model of section 2, a so-called *sharp-boundary model*, is obtained from an appropriate phase field model by taking a singular limit, in which the interface thickness goes to zero. An advantage of phase field models in numerical simulations is that no explicit interface treatment is necessary. There are just two coupled fields in the bulk, and if a calculation of the interface position is desired, one simply determines the zeros of the phase field (e.g. by a program plotting contour lines). On the other hand, these models are unphysical in that they have (at least) one parameter too many and the mapping to physical quantities is somewhat obscure. There are no quantitative numerical results from this approach so far, even though the growth of a *three-dimensional dendrite* (with a symmetry axis) has as yet only been simulated in the context of a phase-field model (Kobayashi 1991).

Returning to the sharp-boundary model it seems fair to say that existing numerical approaches are based on simplifications which in most cases can be characterized as reducing the effective dimension of the mathematical problem by 1. Even these reduced problems may become very time-consuming. Often this happens in physically interesting situations where the reduced description itself becomes doubtful (Sects. 4.1,2). The quasilocal description of Sect. 4.4 is an exception; however, it is easy to see in this case that that there are no unsurmountable obstacles in the way to a more efficient solution of the one-dimensional equation (27). Basically, one has to go back one step and give up the spectral method for high-accuracy calculation of the spatial derivatives. Representing them by simple 5-point molecules, one will need more discretization points to obtain the same accuracy, but this is more than compensated for by the gain that the system evolution matrix becomes sparse (having a bandwidth of 5) and its inversion takes only $O(N)$ operations. This should allow numerical work with several hundred cells.

Therefore, I would like to consider, from now on, possible future approaches to the numerical solution of the *full* (two-dimensional) model. Throughout the discussion, I will keep in mind as the most difficult (and most interesting) test problem the simulation of the temporal evolution of the dense-branching or seaweed morphology without anisotropy. Any numerical code that can successfully handle that problem, will also be able to cope with the other open questions discussed above.

Since we reintroduce the time-dependence in (2) or (20), it is not a priori clear that Green's function methods are favourable. Let us first consider various ways to solve the diffusion equation in the bulk, requiring the minimization of anisotropy effects and the ability to deal with a complicated moving boundary.

A very attractive candidate at first sight are boundary-fitted coordinate systems (Thompson et al. 1982). Since within this approach the boundary is a coordinate line, the coordinate system changes with the motion and one can be pretty sure that orientational correlations as well as translational order will disappear, at least in the vicinity of the interface, as soon as the latter becomes sufficiently complex. Also normal derivatives will be discretized with high accuracy due to the adaptation of coordinate lines to the interface shape, which is important if growth directions resulting from physical instabilities are to be represented correctly. On the other hand, the generation of a boundary-fitted grid is a nontrivial task for the complex interface structure to be expected in our model problem. In principle, it can be performed via the solution of a Laplace or Poisson equation for each interface configuration; this requires roughly the same expenditure as *solving* the problem in quasistationary approximation. Furthermore, if the discretization points are approximately equidistant on the boundary curve, they will have very different distances on the rectangle to which they are mapped; if instead the discretization is equidistant on the rectangle, troughs in the pattern might not get discretization points at all.

Discretization of the bulk into *finite elements* would also avoid the anisotropy problem, and we have seen above that a finite element method has already been applied successfully in restricted geometries (Ungar and Brown 1984). However, for the system sizes we expect to have to deal with (well above 100.000 discretization points), a fully implicit scheme with a nonbanded evolution matrix is expected to be hopelessly slow. This type of approach as well as boundary-fitted grids might however prove useful in the eutectic problem below the chaotic regime.

Looking for a general solver, we considered the following less conventional approach a few years ago. Initially, the bulk is discretized via a random distribution of points, obtained from a Poisson-distributed set by an equilibration procedure that reduces the variance of nearest-neighbour distances. Neighbourhood relations are established most naturally by constructing the associated Voronoi tesselation of the geometry (Fig. 14). The Voronoi cell corresponding to a given point A is the set of all points of the plane (in

3D: space) sitting not closer to any other point of the set than to A. The *dual* graph to the Voronoi tesselation defines a triangulation of the plane, where each line connects a pair of nearest neighbours.

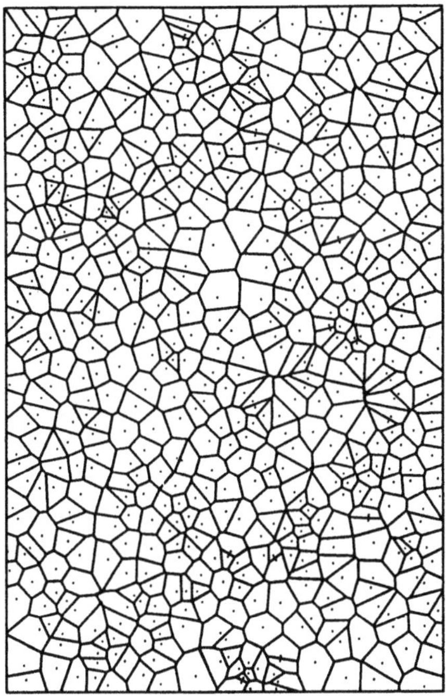

Fig. 14. A set of random points in the plane and the corresponding Voronoi tesselation.

Clearly, the simulation should not be attempted on a *static* random grid, which might exhibit accidental local anisotropies which are even stronger than in a regular grid. The idea of the scheme was to have a repulsive interaction between grid points forcing them to diffuse into the bulk ahead of the moving interface, which would be described by a distinguished subset of grid points. Since the total area of the bulk does not change much during one time step, the main effect of this diffusion would be that the arrangement of points in front of the interface adapts itself to the interface shape, thus creating a boundary-fitting grid structure.

In principle, one could discretize the Laplacian on such a grid by expanding the field variable about each grid point in a power series, truncating the series at second order, and evaluating the coefficients from the field values of the nearest neighbour points via solution of a (small) system of linear equations. This is possible, if the number of nearest neighbours is at least five (otherwise next-nearest neighbours have to be taken into account), but there are problems with negative weights and a nonnegligible fraction of singular matrices to be inverted in the solution of the linear equations (necessitating singular value decomposition which is slow). Fortunately, a direct representation of the

Laplacian on random lattices has been derived in a field-theoretic context (Itzykson 1983)

$$(\nabla^2\phi)_i = \frac{1}{\sigma_i}\sum_{j(i)}\sigma_{ij}\frac{\phi_j - \phi_i}{\ell_{ij}} \quad . \tag{51}$$

Herein, the subscript i denotes the Voronoi cell or grid point, $j(i)$ its neighbours; σ_i is the area of the cell, σ_{ij} the length of the edge it shares with cell j, and ℓ_{ij} the distance between the neighbour points i and j (which is the length of an element of the dual graph). One easily checks that this formula reduces to the ordinary 5-point formula for the Laplacian on a square lattice.

Computing the Voronoi cells for an arbitrary assembly of points can be done by a divide and conquer algorithm, which is $O(N \ln N)$, and this is optimal (Shamos 1978). As soon as the local neighbourhoods of the points are known and changes are only gradual, one can nonetheless reduce the expenditure to $O(N)$, if one forces the number of nearest neighbours to be smaller than a fixed value, say 20 (this is not a severe restriction, since for a Poissonian random lattice only a fraction on the order of 10^{-10} of the cells has larger edge numbers (Itzykson 1983)). In spite of this apparently favourable situation, it turns out that the *construction alone* of the Voronoi cells sets severe limitations to the performance of the algorithm. The reason is that this calculation does not vectorize, basically because it is not known in advance for a given point, how many edges the corresponding Voronoi cell will have. On a CRAY XMP 4/16, the average CPU time to construct a single cell is on the order of 0.3 ms. If we assume that not all cells have to be recalculated at each time step but only those close to the interface, let us say some 20.000 cells, this would set a lower limit of 6 s CPU time per internal time step of the program. Since the method would have to be an explicit one, a time step could not be large. All this would not be so bad, were it not for another disadvantage: in the random case, the discretization (51) of the Laplacian is only $O(h)$ where $h = \langle \ell ij \rangle$. By comparison, the ordinary 5-point molecule on a square lattice is $O(h^2)$. Hence to obtain the same accuracy as with a square grid, one would need a much finer discretization, and there is no guarantee that the estimate given above for the number of cells that need to be recalculated after each step forward of the interface is sufficient.

Why do I give such a lengthy account of a method that seems so bad? Because it might become useful on a different computer architecture. The construction of a Voronoi cell is completely independent of that of its neighbour cells, hence it is a process well suited for *parallelization*. If there are enough processors so that each of them has to calculate but a few cells, grid generation would not be a cost factor any more. Then the advantages that the method offers in avoiding large scale anisotropy effects might compensate for the necessity of a finer discretization and make it competitive.

A very recent idea how to deal with the anisotropy problem in large systems is to solve the diffusion equation on a set of (at least) two *incommensurable* grids. Figure 15 gives an impression of the components of the system: the bulk is covered by two square grids, the second of which is rotated by 45° with respect to the first and shifted by $\frac{1}{4}$ lattice unit in both the x and y directions; the interface lives on neither grid but has its own (coarser) discretization. Since the diagonal length of a square is an irrational multiple of its side length, these two lattices have no points in common and their union neither possesses translational symmetry nor an inversion center (it is quasiperiodic). Of course, there are still orientational correlations, but these correspond to an 8-fold symmetry and it is expected from experience with simulations of diffusion-limited aggregation (Meakin 1986) that for m-fold symmetry with $m \geq 6$ anisotropy effects will not become visible before extremely large crystal sizes have been reached (Family and Hentschel 1987, Kassner 1989a).

Fig. 15. Bulk discretization by two square grids. The interface is discretized independently.

The scheme of the simulation will then be to calculate the intersection points of the interface with both lattices in order to obtain the boundary conditions for bulk diffusion, to perform a diffusion step on both lattices, to obtain the ensuing velocities for the interface from both solutions and to move it according to their average. For starters, a simple alternating direction implicit method (Press et al. 1986) can be used to realize bulk diffusion – this is already $O(N \ln N)$ per time step, with not too small time steps due to the implicit nature of the approach.

If the method keeps its promise of allowing the study of isotropic systems (which can be tested by considering the growth of a circle), more sophisticated solution methods may be used for the bulk equations, including multigrid algorithms (Stüben and Trottenberg 1982). This will necessitate intensified communication between physicists and mathematicians, because the available black-box multigrid solvers are not suitable for the boundary geometries encountered in this kind of problem and physicists usually are not inclined to spend a large fraction of their time dealing with complexity that is of purely algorithmic nature. On the other hand, lack of sophistication in the boundary treatment may lead to poor performance and even divergence of multigrid routines (Joppich 1987).

An approach that bears some similarity to the one just discussed was suggested by Ben-Jacob (1991). There one would have only a single square grid but rotate it through a random angle after each time step. This would definitely destroy orientational correlations. The question is whether the book-keeping, i.e. the transformations from one lattice to the next, a procedure that does not vectorize well, will not slow down the method too much.

As a final possible approach to the full problem I would like to discuss a boundary element method which has not yet been studied in detail. If instead of the quasistationary diffusion equation the full equation (2) is transformed to an integral formulation, one obtains:

$$\int_0^{t^+} dt' \int_{\Gamma(t')} d\Gamma' \, g(\mathbf{r},t;\mathbf{r}',t') \frac{\partial u(\mathbf{r}',t')}{\partial n'} = \frac{1}{D} \int_{\Omega(0)} d\tau'^2 \, g(\mathbf{r},t;\mathbf{r}',t') u(\mathbf{r}',0)$$
$$+ \int_0^{t^+} dt' \int_{\Gamma(t')} d\Gamma' \left(\frac{\partial g(\mathbf{r},t;\mathbf{r}',t')}{\partial n'} - \frac{v_n}{D} g(\mathbf{r},t;\mathbf{r}',t') - \frac{1}{2}\delta(\mathbf{r}-\mathbf{r}')\delta(t-t') \right) u(\mathbf{r}',t') \quad , \tag{52}$$

where t^+ is a time that is larger than t by an infinitesimal amount and

$$g(\mathbf{r},t;\mathbf{r}',t') = \frac{1}{\sqrt{4\pi D(t-t')}} e^{-V(t-t')/2l} e^{-(z-z')/l} \exp\left(\frac{-(\mathbf{r}-\mathbf{r}')^2}{4D(t-t')} \right) \Theta(t-t') \quad . \tag{53}$$

$\Theta(t-t')$ is the Heaviside function.

When this equation is discretized in the same fashion as this was done in the quasistationary case, one arrives at a set of linear equations similar to (40). At first sight it looks as if one had to invert much larger matrices now to solve for the q_j, since each of the subscripts i and j corresponds to the discretization of a *pair* of spatial and temporal coordinates, i.e. to a 2D manifold in contrast with the previous effective 1D problem. However, at any given time, all the q_j on the left hand side corresponding to *preceding* times should be known already (directly or by interpolation from old solutions). Therefore, it would seem feasible to reduce the computation to the solution of a linear system of equations that is *not larger* than in the quasistationary case, because the unknowns are only the q_j at the present time step.

A major drawback of Green's function methods is, however, that they become inefficient in higher dimensions. Even for the quasistationary model the method would become $O(N_b^2)$ per time step in three dimensions, because then $N_b \propto N^{3/2}$.

Hence it seems reasonable to rather pursue one of the outlined bulk methods, whose generalization to three dimensions would not suffer from limitations of this kind.

References

Abramowitz, M., Stegun, I. (eds.) (1972): Handbook of Mathematical Functions. Dover, New York

Barber, M.N., Barbieri, A., Langer, J.S. (1987): Dynamics of dendritic sidebranching in the two-dimensional symmetric model of solidification. Phys. Rev. A **36**, 3340–3349

Barbieri, A., Hong, D.C., Langer, J.S. (1987): Velocity selection in the symmetric model of dendritic crystal growth. Phys. Rev. A **35**, 1802–1808

Ben Amar, M., Pomeau, Y. (1986): Theory of dendritic growth in a weakly undercooled melt. Europhys. Lett. **2**, 307–314

Ben-Jacob, E. (1991): Private communication

Ben-Jacob, E., Deutscher, G., Garik, P., Goldenfeld, N., Lareah, Y. (1986): Formation of dense branching morphology in interfacial growth. Phys. Rev. Lett. **57**, 1903–1906

Bennett, M.J., Brown, R.A. (1989): Cellular dynamics during directional solidification. Interaction of multiple cells. Phys. Rev. B **39**, 11705–11723

Brattkus, K., Misbah, C. (1990): Phase dynamics in directional solidification. Phys. Rev. Lett. **64**, 1935–1938

Brebbia, C.A. (1978): The Boundary Element Method for Engineers. Penetech, London

Brener, E., Müller-Krumbhaar, H., Temkin, D. (1991): Kinetic phase diagram and scaling relations for stationary diffusional growth. Submitted to Europhys. Lett.

Burton, W. K., Cabrera, N., Frank, W.C. (1951): The growth of crystals and the equilibrium structure of their surfaces. Phil. Trans. Roy. Soc. (London) **243**, 299–358

Caginalp, G., Fife, P.C. (1986): Phase-field methods for interfacial boundaries. Phys. Rev. B **33**, 7792–7794

Caroli, B., Caroli, C., Misbah, C., Roulet, B. (1986): Solvability condition for 3-D axisiymmetric needle crystals at large undercooling. J. Phys. France **47**, 1623–1631

Caroli, B., Caroli, C., Fauve, S. (1991): On the phenomenology of tilted domains in lamellar eutectic growth. Preprint

Caroli, B., Caroli, C., Roulet, B. (1987): On the linear stability of needle crystals: evolution of a Zel'dovich localized front deformation. J. Phys. France **48**, 1423–1437

Caroli, B., Caroli, C., Roulet, B., Langer, J.S. (1986): Solvability condition for needle crystals at large undercooling in a nonlocal model of solidification. Phys. Rev. A **33**, 442–452

Chernov, A.A. (1984): Modern Crystallography III: Crystal Growth. Springer-Verlag, Berlin

Chernov, A.A. (1989): Formation of crystals in solutions. Contemporary Physics **30**, 251–276

Classen, A., Misbah, C., Müller-Krumbhaar, H., Saito, Y. (1991): Directional solidification with interface dissipation. Phys. Rev. A **43**, 6920–6933

Coullet, P., Goldstein, R.E., Gunaratne, G.H. (1989): Parity-breaking transitions of modulated patterns in hydrodynamic systems. Phys. Rev. Lett. **63**, 1954–1957

de Cheveigné, S., Guthmann, C. (1991): Interface dynamics and anisotropy effects in directional solidification. To be published

Faivre, C., de Cheveigné, S., Guthmann, C., Kurowski, P. (1989): Solitary tilt waves in thin lamellar eutectics. Europhys. Lett. **9**, 779–784

Faivre, C., Guthmann, C., Mergy, J. (1991): Private communication

Family, F., Hentschel, H.G.E. (1987): Asymptotic structure of diffusion-limited aggregation clusters in two dimensions. Farad. Discuss. Chem. Soc. **83**, 139–144

Feigenbaum, M., Kadanoff, L.P., Shenker, S.J. (1982): Quasiperiodicity in dissipative systems: A renormalization group analysis. Physica **5D**, 370–386

Flesselles, J.-M., Simon, A.J., Libchaber, A.J. (1991): Dynamics of one-dimensional interfaces: An experimentalists view. Adv. Phys. **40**, 1–52

Gear, C.W. (1971): Numerical Initial-Value Problems in Ordinary Differential Equations. Prentice-Hall, Englewood Cliffs, NJ

Gilmer, G.H., Leamy, H.J., Jackson, K.A., Reiss, H. (1974): Pair approximation for interface kinetics. J. Cryst. Growth **24/25**, 495–498

Itzykson, C. (1983): Fields on a random lattice. In: T'Hoft (ed.) Progress in Gauge Fields Theory. Cargése summer school. Plenum Press, New York

Ivantsov, G.P. (1947): Dokl. Akad. Naut. SSSR **58**, 567

Jackson, K.A., Hunt, J.D. (1966): Lamellar and rod eutectic growth. Trans. Metall. Soc. AIME **236**, 1129–1142

Joppich, W. (1987): Ein Mehrgitteransatz zur Lösung einer nichtlinearen Diffusionsgleichung auf einem zeitabhängigen Gebiet unter Verwendung kartesischer Koordinaten. GMD-Studie Nr. **122**, Gesellschaft für Mathematik und Datenverarbeitung, Sankt Augustin

Kadanoff, L.P. (1985): Simulating hydrodynamics: a pedestrian model. J. Stat. Phys. **39**, 267–283

Karma, A. (1987): Beyond steady-state lamellar eutectic growth. Phys. Rev. Lett. **59**, 71–74

Kassner, K. (1989a): Scaling behaviour of generalized diffusion-limited aggregation: The correct form of the m-spoke model. Phys. Rev. A **39**, 4797-4800

Kassner, K. (1989b): Stefan Problem: Dendritische Kristallisation. In: 20. IFF-Ferienkurs: Computersimulation in der Physik. KFA Jülich GmbH, ZB, Jülich, 35.1–46

Kassner, K., Misbah, C. (1990/91): Parity breaking in eutectic growth. Phys. Rev. Lett. **65**, 1458–1461; Phys. Rev. Lett. **66**, 522(E)

Kassner, K., Misbah, C. (1991a): Similarity laws in eutectic growth. Phys. Rev. Lett. **66**, 445–448

Kassner, K., Misbah, C. (1991b): Growth of lamellar eutectic structures. The axisymmetric state. Phys. Rev. A **44**, 6513-6532

Kassner, K., Misbah, C. (1991c): Spontaneous parity-breaking transition in directional growth of lamellar eutectic structures. Phys. Rev. A **44**, 6533-6543

Kassner, K., Misbah, C., Müller-Krumbhaar, H. (1991): Transition to chaos in directional solidification. Phys. Rev. Lett. **67**, 1551–1554

Kessler, D.A., Levine, H. (1987): Pattern selection in three dimensional dendritic growth. Acta Metall. **36**, 2693–2706

Kessler, D.A., Levine, H. (1989): Steady-state cellular growth during directional solidification. Phys. Rev. A **39**, 3041–3052

Kobayashi, R. (1991): Private communication

Kurowski, P., Guthmann, C., de Cheveigné, S. (1990): Shapes, wavelength-selection, and the cellular-dendritic "transition" in directional solidification. Phys. Rev. A **42**, 7368–7376

Langer, J.S. (1980): Instabilities and pattern formation in crystal growth. Rev. Mod. Phys. **52**, 1-28

Levine, H., Rappel, W. (1990): Numerical study for traveling waves in directional solidification. Phys. Rev. A **42**, 7475–7478

Lesoult, G. (1980): La solidification orientée des systèmes eutectiques binaires: Aspects fondamentaux. Ann. Chim. Fr. **5**, 154–185

Meakin, P. (1986): Universality, nonuniversality, and the effects of anisotropy on diffusion-limited aggregation. Phys. Rev. A **33**, 3371–3382

Misbah, C. (1989): Wavelength selection in rotating solidification of binary mixtures. J. Phys. France **50**, 971–994

Misbah, C., Müller-Krumbhaar, H., Saito, Y. (1990): Dendritic growth and directional solidification. J. Cryst. Growth **99**, 156–160

Misbah, C., Müller-Krumbhaar, H., Temkin, D.E. (1991): Interface structure at large supercooling. J. Phys. I, 585–601

Müller-Krumbhaar, H., Burkhardt, T.W., Kroll, D. (1977): A generalized kinetic equation for crystal growth. J. Cryst. Growth **38**, 13–22

Müller-Krumbhaar, H., Kurz, W. (1991): Solidification. In: Cahn, R.W., Haasen, P., Kramer, E.J. (eds.) Materials Science and Technology. VCH Verlagsgesellschaft, Weinheim, pp. 553–632

Ostlund, S., Rand, D., Sethna, J., Siggia, E. (1983): Universal properties of the transition from quasiperiodicity to chaos in dissipative systems. Physica **8D**, 303–342

Pelcé, P., Pumir, A. (1985): Cell shape in directional solidification in the small Péclet number limit. J. Cryst. Growth **73**, 337-342

Press, W.H., Flannery, B.P., Teukolsky, S.A., Vetterling, W.T. (1986): Numerical Recipes. Cambridge University Press, Cambridge

Rabaud, M., Michalland, S., Couder, Y. (1990): Dynamical regimes of directional viscous fingering: Spatiotemporal chaos and wave propagation. Phys. Rev. Lett. **64**, 184–187

Saito, Y., Goldbeck-Wood, G., Müller-Krumbhaar, H. (1987): Dendritic crystallization: Numerical study of the one-sided model. Phys. Rev. Lett. **58**, 1541–1543

Saito, Y., Goldbeck-Wood, G., Müller-Krumbhaar, H. (1988): Numerical simulation of dendritic growth. Phys. Rev. A **38**, 2148–2157

Saito, Y., Misbah, C., Müller-Krumbhaar, H. (1989): Directional solidification: Transition from cells to dendrites. Phys. Rev. Lett. **63**, 2377–2380

Saito, Y., Ueta, T. (1989): Monte Carlo studies of equilibrium and growth shapes of a crystal. Phys. Rev. A **40**, 3408–3419

Shamos, M.I. (1978): Computational Geometry. PHD thesis. Yale University

Shochet, O., Kassner, K., Ben-Jacob, E., Lipson, S.G., Müller-Krumbhaar, H. (1991): Morphology transitions during non-equilibrium growth: I. Study of equilibrium shapes and properties. Physica A, in press.

Simon, A.J., Bechhofer, J., Libchaber, A. (1988): Solitary modes and the Eckhaus instability in directional solidification. Phys. Rev. Lett. **63**, 2574–2577

Sivashinsky, G.I. (1977): Nonlinear analysis of hydrodynamic instability in laminar flames. - I. Derivation of basic equations. Acta Astronautica **4**, 1177-1206

Stüben, K., Trottenberg, U. (1982): Multigrid methods: Fundamental algorithms, model problem analysis and applications. In: Hackbusch, W., Trottenberg, U. (eds.) Multigrid Methods. Lecture Notes in Mathematics, Vol. 960. Springer, Berlin, pp. 1–176

Thompson, J.F., Warsi, Z.U.A., Mastin, C.W. (1982): Boundary-fitted coordinate systems for numerical solution of partial differential equations – a review. J. Comp. Phys. **47**, 1–108

Ungar, L.H., Brown, R.A. (1984a): Cellular interface morphologies in directional solidification. The one-sided model. Phys. Rev. B **29**, 1367–1380

Ungar, L.H., Brown, R.A. (1984b): Cellular interface morphologies in directional solidification. II. The effect of grain boundaries. Phys. Rev. B **30**, 3993–3999

Ungar, L.H., Bennett, M.J., Brown, R.A. (1985): Cellular interface morphologies in directional solidification. III. The effects of heat transfer and solid diffusivity. Phys. Rev. B **31**, 5923–5930

Ungar, L.H., Brown, R.A. (1985): Cellular interface morphologies in directional solidification. IV. The formation of deep cells. Phys. Rev. B **31**, 5931–5940

Computer Algebra

Mit Buchstaben auf dem Computer rechnen:

Über die Anwendung der Computeralgebra in Mathematik, Naturwissenschaft und Technik

von

Friedrich W. Hehl und Hartmut Meyer

Institut für Theoretische Physik, Universität zu Köln, D(W)-5000 Köln 41

Zusammenfassung

Traditionellerweise rechnet man auf dem Computer mit Zahlen, d.h. numerisch. Inzwischen sind aber Software-Pakete entwickelt worden (z.B. Axiom, Derive, Macsyma, Maple, Mathematica, Reduce), die es erlauben, mathematische Formeln auf dem Computer zu manipulieren. Man kann etwa Polynome vereinfachen, nicht–lineare algebraische Gleichungen analytisch lösen, Funktionen differenzieren und integrieren, Matrizen invertieren usw. Anhand einfacher Beispiele wird gezeigt, wie leistungsfähig die heutigen Systeme sind und wie man sie in Mathematik, Naturwissenschaft und Technik typischerweise anwenden kann.

1 Einführung

„Computeralgebra ist der Teil der Informatik, der algebraische Algorithmen entwirft, analysiert, implementiert und anwendet.“[1]

Im Gegensatz zur Numerik, die sich parallel zu der Entwicklung von Computern in den letzten Jahrzehnten einen festen Platz in Wissenschaft und Technik mühelos erobern konnte, befindet sich die Computeralgebra (CA) noch immer im Dornröschenschlaf. Dies hat mehrere Gründe:

- Bis vor etwa 10 Jahren boten Computer eine Hardwareumgebung, die für universell anwendbare CA–Systeme nicht genügend leistungsfähig war. Dadurch waren die Möglichkeiten von vornherein begrenzt.
- Die verwendeten Algorithmen sind i.a. nicht trivial. Ihre Entwicklung und Implementation ist noch nicht abgeschlossen. Heute kann man deshalb mit CA Probleme angehen, für die vor ein paar Jahren noch keine Algorithmen existierten. Genauso kann man davon ausgehen, daß sich die Potenz der CA–Systeme in diesem Sinne auch in Zukunft noch weiter steigern wird.

[1] Dies ist die Definition von R. Loos in dem einführenden Kapitel der Monografie von B. Buchberger et al. [1] über Computer Algebra.

- Die meisten potentiellen Anwender wissen entweder überhaupt nicht, daß es CA gibt, oder haben keine genaue Vorstellung davon und unterschätzen deshalb die Möglichkeiten, die sich auch ihnen bieten.
- CA-Systeme werden erst in jüngster Zeit ernsthaft kommerziell vertrieben.

2 Einfache Beispiele

Im folgenden Abschnitt soll dem Neuling ein erster Eindruck von CA vermittelt werden. Die gegebenen Beispiele sind willkürlich und hängen i.a. nicht zusammen. Das verwendete CA-System heißt REDUCE. Wir demonstrieren hier die neueste Version 3.4 (Juli 1991). Reduce unterscheidet sich jedoch nicht wesentlich von seinen weiter unten beschriebenen Konkurrenten.

Man kann mit Reduce interaktiv arbeiten, indem man einen Befehl eingibt, auf die Antwort wartet, den nächsten Befehl eingibt usw. Um direkt mit einem Beispiel zu beginnen, das die Möglichkeiten der Numerik übertrifft, weisen wir Reduce mit

```
int(tan(x)^3, x);
```

an, das unbestimmte Integral $\int \tan^3 x \, dx$ zu berechnen. Reduce antwortet mit

```
                     2           2
      - LOG(TAN(X)  + 1) + TAN(X)
--> -------------------------------
                    2
```

Jeder wird einsehen, daß dieses Problem mit Fortran oder C quasi unlösbar wäre. Als Beweis für die Richtigkeit der Lösung mag in diesem Fall gelten, daß die Ableitung des Ergebnisses

```
df(ws,x);

           3
--> TAN(X)
```

wieder die ursprüngliche Funktion $\tan^3 x$ ergibt. Mit `WS` („workspace" oder Arbeitsspeicher) wurde hierbei auf den ersten Output zurückgegriffen.

Alle universell anwendbaren CA-Systeme (general purpose systems) können mit Matrizen rechnen. Die Determinante und den Rang einer Matrix erhält man z.B. mit

```
m:=mat((-2*a, 0, pi),
       (   0, a,  5),
       ( a*b, 5,  b));

det(m);

--> A*( - A*B*PI - 2*A*B + 50)

rank(m);

--> 3
```

Auf ähnlich einfache Weise können auch die Eigenwerte und Eigenvektoren von Matrizen bestimmt werden. Für Vektoralgebra stehen Zusatzpakete bereit, die komfortables Arbeiten mit Vektoren erlauben:

```
load_package orthovec;

a:=svec(ax,ay,az);                    % A und B werden
b:=svec(bx,by,bz);                    % als Vektoren definiert

a >< b;                               % Kreuzprodukt von A und B

-->  [AY*BZ - AZ*BY,AZ*BX - AX*BZ,AX*BY - AY*BX]

v:=svec(3*x^2+5*y,-12*y*z,2*x*y*z^2);
dlineint( v, svec(s,s,0), s, 0, 1);   % definite line integral

      7
-->  ---
      2
```

Mit Hilfe von DLINEINT wurde dabei für den Vektor $\mathbf{V} = (3x^2 + 5y, -12yz, 2xyz^2)$ das bestimmte Linienintegral

$$\int_{\mathbf{r}_1}^{\mathbf{r}_2} \mathbf{V} \cdot dr$$

über den Weg von $\mathbf{r}_1 = (0,0,0)$ nach $\mathbf{r}_2 = (1,1,0)$ berechnet. Neben den normalen kartesischen, kann der Anwender auch zylindrische oder sphärische Koordinaten wählen.

Bei der Berechnung mit dimensionsbehafteten Größen wird oft dazu geraten, das Ergebnis durch eine Dimensionsbetrachtung zu überprüfen. Wenn Sie mit einem CA-System arbeiten, können Sie die dimensionsbehafteten Größen direkt in die Rechnung eingeben. Im folgenden Beispiel wird der Thomson-Absorptionsquerschnitt $\sigma_0 = e^4/(6\pi\epsilon_0{}^2 m_e{}^2 c^4)$ des Elektrons aus Fundamentalkonstanten berechnet:

```
on rounded;

c    := 299792458          *meter/sekunde;
el   := 1.602189*(10^-19)  *coulomb;
eps0 := 8.8542*(10^-12)    *ampere*sekunde/(volt*meter);
m    := 9.10953*(10^-31)   *kg;

volt   := newton*meter/coulomb;
ampere := coulomb/sekunde;
newton := kg*meter/sekunde^2;

sigma0 := el^4/(6*pi*eps0^2*m^2*c^4);

                               2
-->  6.65243212159E-29*METER
```

Flüchtigkeitsfehler, wie das Einbringen einer Größe in den Nenner statt den Zähler, können damit leicht erkannt werden.

3 Einfache Programmierung

Selbstverständlich kann auch ein hochentwickeltes System wie Reduce nicht alles bieten, was sich der einzelne Anwender im speziellen wünschen mag. Deshalb ist es besonders wichtig, daß die dem System „angeborenen" Fähigkeiten vom Anwender bei Bedarf erweitert werden können. Reduce umfaßt eine einfach zu erlernende Programmiersprache, mit der kleine und große Probleme umgesetzt werden können. Mit einer elementaren **FOR**-Schleife werden etwa im folgenden Beispiel alle 19 Primzahlen zwischen 7000 und 7200 ausgegeben:

```
for k:=7000:7200 do
   if primep(k) then write k;
```

Dabei testet jeweils das Prädikat **PRIMEP(K)**, ob **K** eine Primzahl ist. Dem Zweifler sei empfohlen, dies entweder mit Papier und Bleistift oder auch mit einem Fortran-Programm zu überprüfen.

Die l'Hospitalsche Regel kann man mit Reduce in rekursiver Weise formulieren:

```
procedure lhospital(funct,var,lim);
if sub(var=lim,den(funct))=0 then
  if sub(var=lim,num(funct))=0 then
    lhospital(df(num(funct),var) / df(den(funct),var),var,lim)
  else
    rederr "infinity"
else
  sub(var=lim,funct);
```

Zur Erklärung muß hier nur noch gesagt werden, daß mit **NUM()** und **DEN()** der Zähler bzw. Nenner eines Ausdrucks ermittelt wird. **SUB()** ersetzt lokal eine oder mehrere Variablen eines Ausdrucks. Eine so definierte Prozedur läßt sich im weiteren genauso benutzen wie alle vordefinierten Prozeduren:

```
lhospital( (e^(a*x)-e^(-b*x))/log(1+x), x, 0);

-->  A + B

lhospital( ((1+x)^n-(1-x)^n)/(e^(1+x)-e^(1-x)), x, 0);

       N
-->   ---
       E
```

Als ein weiteres Beispiel für die leichte Umsetzbarkeit von algebraischen Ideen soll die Taylorsche Reihenentwicklung herhalten (die es allerdings schon als Zusatzpaket gibt):

```
procedure taylor(funct,x,x0,n);
for k:=0:n sum
      sub(x=x0,df(funct,x,k)) * (x-x0)^k / factorial(k);

taylor(x^x, x, 1, 5);

       5    4      2
      X  - X  + 8*X  - 5*X + 9
-->  --------------------------
                 12
```

Neben der Programmierung im Stil der „klassischen“ Programmiersprachen, also mit Kontrollstrukturen wie IF- und FOR-Anweisungen, gibt es in CA-Systemen die Möglichkeit Regeln zu definieren, nach denen bestimmte Ausdrücke durch andere Ausdrücke ersetzt werden. So kann man z.B. Auswertungsregeln für Operatoren festlegen:

```
operator binomial;             % Operator fuer Binomialkoeffizienten

let { binomial(~n,~p) => factorial(n)/(factorial(p)*factorial(n-p)) };

binomial(140,17);

--> 3115999030593616524000
```

Die wesentlichen Elemente eines typischen CA-Systems sind damit genannt. Unterschiede zwischen Reduce und den anderen universellen CA-Systemen bestehen hauptsächlich in der Umsetzung dieser Elemente in die Syntax des jeweiligen Systems.

4 Welche CA-Systeme gibt es?

Reduce ist eines der am weitest verbreiteten CA-Systeme. Es läuft vom kleinen PC bis hin zu den größten Computern, etwa einer Cray im Konrad-Zuse-Zentrum für Informationstechnik in Berlin. Andere bekannte universell anwendbare CA-Systeme sind Axiom, Macsyma, Maple und Mathematica, womit die „Großen Fünf“ genannt sind. Derive ist ein universelles System für MS-DOS Rechner. Für den Benutzer sehen diese verschiedenen CA-Systeme ziemlich ähnlich aus. Sobald Sie mit *einem* System umgehen können, sollten Sie mit Hilfe des entsprechenden Handbuches auch in der Lage sein, ein anderes System zu handhaben. Die „Konkurrenten“ im Überblick:

KREISE IM KORNFELD-WAR ES EIN UFO ODER WAR ES REDUCE ?

Axiom (früher Scratchpad II) ist eine relativ neue Entwicklung von IBM, ein speicherintensives System, das sehr vielversprechend scheint, da die Datenstrukturen von Axiom flexibler sind als die der älteren Systeme. Axiom ist das erste objektorientierte CA-System. Axiom ist in Lisp geschrieben. Sein Nachteil ist, daß es derzeit im wesentlichen nur auf IBM-Workstations vom Typ RS/6000 und auf *großen* IBM-Rechnern läuft und zudem schlecht erhältlich ist. Wohl eher für Forschungszwecke anwendbar.

Derive (Nachfolger von μ-Math), ist ein sehr kleines CA-System und läuft, wie der Name des Vorläufers μ-Math verrät, auf „Micros“, beispielsweise auf IBM-kompatiblen PC's. Trotz seiner Kompaktheit ist Derive recht leistungsfähig, also etwa für Schüler und Studenten als privat verfügbares System höchst geeignet. Derive ist in Lisp programmiert.

Macsyma wurde ursprünglich am MIT (Massachusetts Institute of Technology) entwickelt und ist, wie Axiom und Reduce, in der Sprache Lisp geschrieben. Es läuft auf Lisp-Machinen der

Firma Symbolics und auf DEC Rechnern, sowie in einer eingeschränkten Version, auf vielen Unix-Workstations. Macsyma ist ein großes und gutausgebautes speicherintensives System, das seit neuestem wieder gepflegt wird.

Maple stammt von der Universität Waterloo/Kanada und ist in der Sprache C geschrieben. Es ist auf zahlreichen Computern unter dem (ebenfalls in C geschriebenen) Betriebssystem Unix verfügbar. Maple besteht aus einem relativ kleinen Kern von ca. 20 000 Zeilen C-Code (Zum Vergleich: Der Kern von Mathematica besteht aus ca. 180 000 Zeilen C-Code). Der Rest des Systems ist in Maple-Code geschrieben und wird, je nach Bedarf, zugeladen. Dank dieses Konzepts ist Maple zum einen gut portierbar und zum anderen auch problemlos auf Multi-User-Systemen zu benutzen (etwa zu Unterichtszwecken), was bei anderen CA-Systemen schon wegen ihres Bedarfs an Arbeitsspeicher kritisch ist.

Mathematica (eine Weiterenticklung von SMP) ist neueren Datums. Neben CA und Numerik bietet es die Möglichkeit, Ergebnisse in 2D- oder 3D-Grafiken darzustellen. Auch Mathematica ist in C geschrieben und ist unter anderem für den Apple Macintosh, den NeXT, sowie für IBM-kompatible PC's und Sun-Workstations erhältlich. Mathematica ist das erste CA-System, das konsequent (man könnte auch sagen „aggressiv") vermarktet wurde, mit dem Erfolg, daß innerhalb eines Jahres ungefähr 50 000 Systeme verkauft wurden.

Nun sollen noch zwei bekanntere spezielle CA-Systeme erwähnt werden, wobei sich die Liste noch leicht verlängern ließe:

SAC-2 ist ein nichtkommerzielles System, das für wissenschaftliche Zwecke, hauptsächlich zur Algorithmen-Entwicklung gedacht ist. Der SAC-2 Quell-Code ist ursprünglich in der Spezialsprache Aldes geschrieben. Inzwischen gibt es aber auch Implementierungen in C, Fortran, Lisp und Modula-2. Ähnlich wie bei Reduce kann derjenige, der in der Implementierungssprache des jeweiligen CA-Systems bewandert ist, aus dem Quelltext des Systems viel über die in CA-Systemen verwendeten Algorithmen erfahren.

Schoonschip ist ein älteres System, das hauptsächlich für Anwendungen in der Elementarteilchenphysik in Frage kommt (Berechnung von Feynman-Diagrammen). Es wurde ursprünglich für CDC-Computer entwickelt, läuft jetzt aber auch auf PC's (z.B. auf dem Atari ST). Dank der Tatsache, daß Schoonschip in Assembler geschrieben ist, ist es schnell und kompakt.

5 Welche Probleme wurden beispielsweise erfolgreich mit Computeralgebra gelöst?

Wir selbst arbeiten auf dem Gebiete der Allgemeinen Relativitätstheorie und der *Eichtheorie der Gravitation.* Wir entwickelten eichtheoretische Modelle der Gravitation, deren klassische Feldgleichungen sich als Verallgemeinerung der Einsteinschen Feldgleichung ergaben. Es sind 34 inhomogene quasilineare partielle Differentialgleichungen 2.Ordnung in 34 Variablen. Für dieses System von partiellen Differentialgleichungen (PDG's) exakte (oder auch genäherte) Lösungen zu finden, ist nicht einfach. Es wurde uns schnell klar, daß ohne CA keine Aussicht auf Erfolg besteht, da wir uns mit numerisch approximierten Lösungen nicht zufriedengeben wollten.

Typischerweise geht man bei solchen Problemen so vor, daß man eine bestimmte räumliche Symmetrie, etwa die Axialsymmetrie ansetzt, damit in die PDG's eingeht und dann das homogene System löst. Nun gibt es heutzutage noch kein CA-System, mit dem man solche PDG's direkt „knacken" kann. Was man vielmehr macht ist, daß man mit Fingerspitzengefühl gewisse Lösungsansätze macht, diese einsetzt und nachschaut, unter welchen Bedingungen sie die PDG's erfüllen.

Diese Hau-Ruck-Methode mag das Herz des Algorithmen-Forschers zum Bluten bringen, uns brachte sie aber nach etwa 4-jähriger Arbeit in einem etwa 4-köpfigen Team einige schöne axialsymmetrische exakte Vakuum-Lösungen bestimmter eichtheoretischer Modelle der Gravitation [12]. Wir benutzten Reduce und insbesondere das Paket EXCALC, das von Schrüfer [13] für Rechnungen im äußeren Differential-Kalkül geschaffen wurde. Manches Mal stellt man dann allerdings am Ende fest, daß man das Problem auch ganz ohne CA hätte lösen können, wenn man nur schlauer gewesen wäre... Trotzdem (oder gerade deswegen) beabsichtigen wir auch in Zukunft uns mit der besten verfügbaren CA-Software einzudecken.

CA-Programme für die *Allgemeine Relativitätstheorie* sind bereits gut entwickelt [14]. Hier erweist es sich, daß beim Testen schon bekannter Lösungen und bei deren Klassifikation CA bereits ein unentbehrliches Hilfsmittel geworden ist. Auch für supersymmetrische Theorien, insbesondere für Stringtheorien, liegen Spezialpakete hauptsächlich zur Auswertung von Kommutatoren vor [15].

In der *Mechanik* hat sich die CA ebenfalls bei komplexen Problemen bewährt, etwa bei der Berechnung der Eigenwerte anharmonischer Oszillatoren [16] oder der dynamischen Matrix von Oberflächen-Schallwellen in der Kristallgitter-Physik [17]. Das letzte Problem ist von großer praktischer Bedeutung. Weitere Anwendungen in der Ingenieurtechnik finden sich in [6].

Im Umfeld der numerischen Integration der Navier-Stokes-Gleichungen ist man zuerst an deren Diskretisierung interessiert. Die Aufstellung entsprechender Differenzenschemata und deren Stabilitätsanalyse kann mit Hilfe von CA erheblich erleichtert werden [18,19]. Solche Verfahren für hyperbolische und parabolische Differentialgleichungen haben potentiell ein breites, für die Praxis bedeutungsvolles Anwendungsgebiet.

Die Anwendungen der CA erstrecken sich von der Mathematik (u.a. Zahlen- und Gruppentheorie, Differentialgeometrie) über die Physik (u.a. Plasma- und Elementarteilchenphysik, Relativitätstheorie) und Chemie (u.a. Reaktionskinetik, Quantenchemie) bis hin zur Biologie (u.a. Genetik, Populationsforschung) und Technik (u.a. Statik, elektrische Netzwerktheorie), wie man etwa in Crandall [9], der Mathematica benutzt, nachlesen mag.

Über interessante Fortschritte in der Algorithmen-Entwicklung zur Integration gewöhnlicher und partieller Differentialgleichungen berichtet Wolf [20,21], der auch ein Reduce-Paket CRACK für solche Zwecke vorhält.

Schließlich wollen wir auf ein schon klassisches Problem hinweisen, das mit Hilfe von CA gelöst wurde und Eingang in die Lehrbuch-Literatur gefunden hat: Die Berechnung der Lamb-Shift in zweiter Ordnung in der Feinstrukturkonstanten [22], also die Auswertung multi-dimensionaler Feynman-Integrale unter Einsatz von CA, die zu instabil sind, um numerisch berechnet werden zu können. Auswertung von Feynman-Integralen haben Hearn zur Kreation von Reduce und Veltman zu Schoonschip geführt.

6 Ausblick

In der Bundesrepublik hatte sich vor einigen Jahren eine Fachgruppe „Computeralgebra" der GI (Gesellschaft für Informatik), der DMV (Deutsche Mathematiker-Vereinigung) und der GAMM (Gesellschaft für Angewandte Mathematik und Mechanik) gebildet. Ihr derzeitiger Sprecher[2] bereitet eine Dokumentation über CA-Anwender und Arbeitsgruppen in Deutschland vor. In einem etwa zweimal jährlich erscheinenden „Computeralgebra-Rundbrief" trägt die Fachgruppe zur besseren Kommunikation zwischen den CA-Anwendern bei.

Wie wir gesehen haben, vermag man mit CA Rechnungen durchzuführen, die sonst aus Zeitgründen einfach nicht machbar wären. Als Beispiel wird gerne Delaunay erwähnt, der für seine Berechnungen zur Mondbewegung im Rahmen der Newtonschen Mechanik 20 Jahre brauchte, während man

[2]Prof. Dr. V.Weisspfenning, Lehrstuhl für Mathematik, Universität, W-8390 Passau.

mit einem CA-System (es war Macsyma) für entsprechende Rechnungen auf einem Computer nur wenige CPU-Minuten benötigte. Zur gerechten Beurteilung sollte man aber auch sagen, daß in die Entwicklung von Macsyma vielleicht 50 Mannjahre gesteckt wurden und das Schreiben und Testen des Programms sicher nicht nur Minuten gekostet hatte. Es kann aber keinem Zweifel unterliegen, daß heute mit CA Probleme gelöst werden, die noch vor 10 Jahren als nicht handhabbar abgetan wurden.

Bei der analytischen Integration hat ein CA-System auch die Funktion einer Integraltafel, auf die schnellstens zugegriffen werden kann. Integraltafeln haben nicht selten Fehler, womit aber nicht gesagt sein soll, daß dies CA-Systeme nicht auch hätten. Man kann jedoch mit CA zur Kontrolle (bei unbestimmten Integralen) sehr schnell differenzieren, was bei Integraltafeln, falls mit Hand ausgeführt, mehr Arbeit machte.

Die Exaktheit der Resultate bei der Anwendung von CA — im offenkundigen Gegensatz zu den bei der Numerik sich ergenbenden Näherungslösungen — wird schon immer als ein Hauptvorzug der CA betrachtet. Dies gilt allerdings nur, solange man nicht etwa in Reduce durch `ON ROUNDED` Fließkomma-Arithmetik erzwingt. Das Ideal der Exaktheit wird nicht immer durchzuhalten sein. Man wird oft nicht umhin können, irgendwann eine Näherung einzuführen. Man sollte aber solange analytisch bleiben, wie dies möglich ist.

Eine der Zununftsperspektiven bei der Anwendung der CA ist dementsprechend, daß die Verzahnung zwischen zwischen CA auf der einen Seite und etwa Fortran, Pascal und C auf der anderen Seite enger wird. Man kann in einem CA-System eine Rechnung analytisch durchführen, dann aber in die Numerik derart überleiten, daß der Output des CA-Systems beispielsweise als Fortran-Quelltext ausgegeben wird, so daß man sofort ein Fortran-Programm ausführen lassen kann. Diese Schnittstellen zwischen CA und Numerik werden immer wichtiger werden. Viel Entwicklungsarbeit wird bereits heute darauf geleistet, und es wird für die Anwender nützlich sein, sich dieser Verfahren kundig zu machen.

Auch die Ausgabe von Daten in grafischer Form, wie dies in schöner Weise etwa bei Mathematica realisiert ist, ist zum intuitiven Erfassen von Resultaten wichtig. So zeichnet sich für den Anwender in Mathematik, Naturwissenschaft und Technik folgender „Rechenschieber" der Zukunft ab: eine Workstation, auf der *Computer-Algebra*, *Numerik*, etwa realisiert durch Fortran, C und Pascal, und *Grafik*, 3-dimensional und farbig, in enger Verbindung implementiert sind und etwa bequem aus einer integrierten Benutzeroberfläche heraus aufgerufen werden können. Ein Apple Macintosh mit Mathematica (für etwa 6 kDM) könnte als eine Vorstufe eines solchen zukünftigen System angesehen werden.

7 Literatur

Grundlegendes über CA-Systeme und ihre Algorithmen:

[1] B. Buchberger, G.E. Collins and R. Loos (eds.): „Computer Algebra. Symbolic and Algebraic Computation." *Computing Supplementum 4*. Springer, Wien (1982).

[2] J.H. Davenport, Y. Siret and E. Tournier: „Computer Algebra. Systems and algorithms for algebraic computation." Transl. from the French. Academic Press, London (1988).

Vergleich verschiedener CA-Systeme:

[3] B. Fuchssteiner et al. (Herausg.): „mathPAD". Sonderausgabe zur Tagung der Deutschen Mathematiker-Vereinigung in Bielefeld. Vol.1, Heft 3, Universität Paderborn, September 1991.

[4] D. Harper, C. Wooff and D. Hodgkinson: „A Guide to Computer Algebra Systems." Wiley, Chichester (1991).

[5] G.H. Gonnet und D.W. Gruntz: „Algebraic Manipulation: Systems." In *The Encyclopedia of Computer Science and Engineering*. A. Ralston et al. (eds.). To be published (1991/92).

Anwendung von CA:

[6] B. Buchberger und B. Kutzler: „Computer-Algebra für den Ingenieur". In *Rechnerorientierte Verfahren* von B. Buchberger et al. Teubner, Stuttgart (1986).

[7] V. Winkelmann and F.W. Hehl: „Reduce for Beginners. Six Lectures on the Application of Computer-Algebra." In *Computer Simulation and Computer Algebra*, D. Stauffer et al. (eds.), 2nd ed. Springer, Berlin (1989). Deutsche Version, Köln (1991); wird demnächst veröffentlicht.

[8] M. MacCallum and F. Wright: „Algebraic Computing with Reduce." In *First Brazilian School on Computer Algebra, Vol.1*, M. J. Rebouças and W. L. Roque (eds.). Clarendon Press, Oxford (1991).

[9] R. E. Crandall: „Mathematica for the Sciences." Addison-Wesley, Redwood City (1991).

Jetziger Stand der CA und zukünftige Entwicklung:

[10] S.M. Watt, ed.: „ISSAC'91 — Proceedings of the 1991 International Symposium on Symbolic and Algebraic Computation. Bonn, July 1991." ACM Press, New York (1991).

[11] A.C. Hearn: „Future Directions for Research in Symbolic Computation." Report of a Workshop on Symbolic and Algebraic Computation. A. Boyle and B.F. Caviness (eds.). SIAM, Philadelphia (1990).

Verschiedene ernsthafte Anwendungen der CA:

[12] P. Baekler, M. Gürses, F. W. Hehl and J. D. McCrea: „The exterior gravitational field of a charged spinning source in the Poincaré gauge theory...". Phys. Lett. **128A** (1988) 245.

[13] E. Schrüfer, F. W. Hehl and J. D. McCrea: „Exterior calculus on the computer: The REDUCE-Package EXCALC Applied to General Relativity and to the Poincaré Gauge Theory". GRG Journal **19** (1985) 437.

[14] J. D. McCrea: „REDUCE in General Relativity and in Poincaré Gauge Theory". In *First Brazilian School on Computer Algebra, Vol.2*, M. J. Rebouças and W. L. Roque (eds.). Clarendon Press, Oxford (1991/92), in press.

[15] W. M. Seiler: „SUPERCALC - a REDUCE package for commutator calculations". Comp. Phys. Comm. **66** (1991) 363.

[16] Chen Zhijiang et al.: „A Reduce approach to one-dimensional polynomial anharmonic oscillators". Preprint IC/90/202, International Centre for Theoretical Physics, Trieste (1990).

[17] R. Berndt, A. Lock, G. Witte and Ch. Wöll: „Application of CA to surface lattice dynamics". In ISSAC'91 [10], p.433.

[18] V. G. Ganzha, B. Yu. Scobelev and E. V. Vorozhtsov: „Stability analysis of difference schemes by the catastrophe theory methods and by means of CA." In ISSAC'91 [10], p.427.

[19] V. G. Ganzha, E. V. Vorozhtsov und C. Zenger: „Stabilitätsuntersuchung von Differenzenverfahren mit Hilfe der Resultantenalgebra unter Verwendung des REDUCE-Systems". Vorabdruck TUM-I9120, Institut für Informatik, TU München (1991).

[20] T. Wolf: „The Integration of Under-determined Linear ODEs". Preprint School of Math. Sciences, Queen Mary and Westfield College, London (1991).

[21] T. Wolf: „The Symbolic Integration of Exact PDEs". Preprint School of Math. Sciences, Queen Mary and Westfield College, London (1991).

[22] T. Appelquist and S.J. Brodsky: „Order α^2 electrodynamic corrections to the Lamb shift". Phys. Rev. Lett. **24** (1979) 562.

Karikaturen:

[23] Peter Scherer, Niehler Str. 351, W-5000 Köln 60.

Computer Algebra and Neurodynamics

Dongming Wang
Research Institute for Symbolic Computation
Johannes Kepler University, A-4040 Linz, Austria

Bernd Schürmann
Corporate Research and Development
ZFE IS INF 24, Siemens AG, 8000 Munich 83, Germany

Abstract

We review the capabilities of computer algebra and the characteristics of neurodynamics. Possible applications of computer algebra to the study of neurodynamics are then discussed. We have investigated some of the applications, in particular, to the computer aided analysis and derivation of neurodynamics by using and developing tools in computer algebra systems. Part of our own work in this direction is summarized.

1. The Capabilities of Computer Algebra

In this section we briefly explain the terms *computer algebra* (CA) and *computer algebra systems*, indicate the research activities in the field and review the capabilities of CA and CA systems. For more details the reader may confer to the cited references [1, 8, 9, 10, 14] and references therein.

1.1. Computer Algebra

According to R. Loos [9, pp. 1], computer algebra is a part of computer science which designs, analyzes, implements and applies algebraic algorithms. The term *computer algebra* has long hesitated between and/or used along with *symbolic and algebraic computation*, *symbolic and algebraic manipulation*. In addition to algebra, computer algebra has relations to many other, if not all, branches of mathematics including analysis, geometry and numerical analysis, as mathematical problems are often given in an algebraic form or can be algebraized in one way or in the other. Therefore, in a broader sense computer algebra is also considered as computer mathematics, computerization of mathematical research, education and applications. One feature of computer algebra is its emphasis on the development of software systems and its application in science and engineering. The field of CA has existed and expanded since the early 1960's and its research activities have been coordinated by SIGSAM (the Special Interest Group on Symbolic and Algebraic Manipulation) of ACM, which publishes the SIGSAM Bulletin reporting all events and developments in the field, and by independent local research groups. There are periodic conferences and symposia including the now annual conference ISSAC (which was formerly called SYMSAC, EUROSAM, EUROCAM, etc.) in the field. Two major journals, the *Journal of Symbolic Computation* and the journal *Applicable Algebra in Engineering, Communication and Computing* were established in 1985 and 1990, respectively.

1.2. Computer Algebra Systems

Computer algebra systems are pieces of software developed for performing symbolic and algebraic computations on the basis of theoretical research results in computer algebra. There have been a large number of CA systems available for research and/or as commercial products. They can be roughly divided into two major families: special purpose systems and general purpose systems. Systems in the first family are designed to solve specific problems in various fields and generally can operate with relatively high speed. Such systems include CAMAL, SCHOONSCHIP, ALTRAN, SAC-1, SAC-2, CAYLEY and CoCoA. Systems in the second family provide the user with many mathematical capabilities and enjoy wide availability and accessibility. Such systems include MACSYMA, REDUCE, AXIOM (SCRATCHPAD II), MAPLE, MATHEMATICA and DERIVE which have been used in various institutions world-wide. CA systems run on different types of machines ranging in power from the IBM PC to Cray X-MP and in memory from one mega-byte to several mega-bytes. They are capable of performing many kinds of numerical, symbolic and graphical computations that incorporate high level programming languages and have been powerful tools for research and applications in mathematics, physics, computer science, engineering and many other fields.

1.3. Representation of Mathematical Objects

In CA systems, mathematical objects are generally represented in two forms: one is an internal form used for the internal computation of the systems and the other is an external form directed to the user for input and output. The former depends on the design and implementation philosophy of the systems and the features of the chosen implementing languages and it considers the memory storage and practical efficiency. The latter adopts some conventional notations (in accordance with the language's syntax) in linearized one- or two-dimensional form which are usually close to the natural mathematical language. For example, the unary polynomial $2x^3 + x - 5$ may be internally represented by an array of its coefficients: $[-5, 1, 0, 2]$, or by storing the exponent and the coefficient for all the non-zero terms: $((3, 2), (1, 1), (0, -5))$, whereas at the user level this polynomial is entered, e.g., in MAPLE, as `2*x**3+x-5` and its output may be given as `2*x**3+x-5` or as

```
     3
  2 x  + x - 5.
```

Within the current (general purpose) CA systems, the user may represent and define objects like sets, lists, integers and rationals with arbitrary precision, polynomials, rational functions, elementary functions, matrices of definite dimension, sums, products, series, groups, tensors, differential and integral equations, unspecified functions, indexed variables and functions. However, these systems are not capable of representing uncertain objects like the Kronecker symbol, piecewise defined functions and indefinite sums and products with non-interval domains, matrices of an indefinite dimension, indefinite geometric objects. Such capabilities could be improved for special purposes (see Section 4 for the case of indefinite sums and products).

1.4. Basic Operations and Manipulations

The basic operations and manipulations of elementary mathematical objects such as the multiplication of two integers, the inversion of a matrix and the transition of a rational number from its floating point representation to a fractional representation are all well-known and quite trivial in mathematics. However, from the viewpoint of computer algebra such operations and manipulations are crucial for the performance of the whole system. A great amount of work in the field has been devoted to improve the practical efficiency of basic operations and manipulations on elementary objects. Current CA systems have built-in efficient algorithms for operations on integers and rational numbers with arbitrary precision (addition, subtraction, multiplication, division), floating point numbers with high precision, sets and lists (union, difference, intersection, membership test), matrices with numerical and/or symbolic elements (multiplication, inversion, determinant computation, eigenvalue computation etc.), polynomial and rational functions (addition, subtraction, multiplication, division, resultant), special functions and constants, differentiation and integration (of various expressions), limit calculation, series expansion, manipulation of formulae (substitution, selection of parts, numerical evaluation, pattern recognition), transformation between different number domains and between different representations.

1.5. Advanced Computations

Persistent research in computer algebra has contributed many efficient algorithms for performing advanced algebraic computations which constitute the power of current CA systems. Such research touches many branches of mathematics as well as computer science such as commutative and non-commutative algebra, polynomial ideal theory, mathematical analysis, differential equations, group and number theory, constructive algebraic geometry, combinatorics, statistics, numerical mathematics, mathematical logic, computational complexity, artificial intelligence and computer graphics. Current CA systems have implemented most recent and efficient algorithms and provide advanced computations on greatest common divisors (of integers and of polynomials), (multivariate) polynomial factorization (over finite fields, integers, algebraic number fields), solving systems of algebraic equations, isolation of the real roots, proving geometric theorems, decision problem for real closed fields, groups, tensors, summation in finite terms, formal integration and solving differential equations.

1.6. Simplification and Deduction

A mathematical expression or formula can often be represented in several different forms. For example, $(x+1)(x-1)$ may be written as $(1+x)(x-1)$, $-1+x^2$ or x^2-1. It is therefore necessary to know whether an expression is simpler than and/or equivalent to the other. This is the problem of (automated) simplification and deduction. The simplification in computer algebra consists of two kinds: One is the so-called canonical simplification which has been more extensively studied and for which there are practical techniques available [9]. The second kind of simplification is to find *simpler* but *equivalent* expressions, for which we must give a specification to the term *simplicity* in order to measure what we call *simpler*. This is strongly problem-dependent. The current CA systems have integrated simplification mechanisms and are able to verify the equivalence of some different expressions, and some of them are able to heuristically simplify various expressions according to some prescribed rules and simplicity measures and by using techniques of term-rewriting and pattern matching.

1.7. Programming Facilities

Modern CA systems have incorporated high level programming languages and provide easy-to-learn programming facilities: functional and procedural programming, "generic" programming and "rewrite rules" style programming using pattern matching.

1.8. User Interface and Interface with Other Utilities

CA systems are user-friendly and highly interactive with powerful devices for input and output on many mathematical objects. They perform both numerical and symbolic computations and some of them provide advanced 2D and 3D graphics and animation. The terminal display of expressions are close to natural mathematical language and text-like. There is interface between different CA systems and between CA systems and textprocessing utilities. Packages have been developed for generating Fortran codes, C codes, Pascal codes from symbolic expressions produced by CA systems and for linking the CA systems to routines such as the NAG Fortran Library for numeric computation.

2. The Characteristics of Neurodynamics

Neural networks can be characterized by their architecture/topology and dynamics. For a fixed architecture, the behaviour of networks is then reflected completely by their dynamics. Neurodynamics may be roughly divided into three categories as shown below.

2.1. Activation Dynamics

The activation dynamics of a neural network describes mathematically the activation rule for combining the inputs impinging on a unit with the current state of that unit to produce a new level of activation for the unit, i.e., the change of the state of activations with respect to the time t. It takes the form

$$x_i(t+1) - x_i(t) = F_i(\boldsymbol{x}(t), \boldsymbol{w}(t), \boldsymbol{K}), \quad i = 1, \ldots, N,$$

in discrete time, or

$$\dot{x}_i = F_i(\boldsymbol{x}(t), \boldsymbol{w}(t), \boldsymbol{K}), \quad i = 1, \ldots, N,$$

in continuous time, where the overdot denotes the time derivative as usual, $\boldsymbol{x}(t) = (x_1(t), \ldots, x_N(t))$ is the activation state vector, $\boldsymbol{w}(t) = (w_{ij}(t))_{N\times N}$ is the weight matrix, $\boldsymbol{K}$ is an external pattern vector and N is the number of neurons. Typically, the function F_i has the form

$$F_i(\boldsymbol{x}, \boldsymbol{w}, \boldsymbol{K}) = f_i(x_i)[g_i(x_i, \boldsymbol{K}_i) - \sum_{j=1}^{N} w_{ji} S_j(x_j)], \quad i = 1, \ldots, N,$$

where S_j is the output function for the unit i.

2.2. Error Propagation Dynamics

Neural learning algorithms may be divided into two classes, namely those which expect an error signal $\boldsymbol{y}$ and others which are based upon the principle of self-organization and do not need an error signal. We assume in this subsection that an error signal $\boldsymbol{y}$ exists at the net output which evaluates the current state $\boldsymbol{x}$ of the network. Like the activations $\boldsymbol{x}$, the errors $\boldsymbol{y}$ propagate through the network in a recurrent dynamics. The propagation rule takes a form similar to that of the activation dynamics, namely,

$$y_i(t+1) - y_i(t) = G_i(\boldsymbol{x}(t), \boldsymbol{y}(t), \boldsymbol{w}(t), \boldsymbol{K}), \quad i = 1, \ldots, N,$$

in discrete time, or

$$\dot{y}_i\,(t) = G_i(\boldsymbol{x}(t), \boldsymbol{y}(t), \boldsymbol{w}(t), \boldsymbol{K}), \quad i = 1, \ldots, N,$$

in continuous time. An example of the propagation dynamics is of the form

$$\dot{y}_i = -d_i y_i + f_i' \cdot (h_i \cdot \delta_i + \sum_{j=1}^{N} w_{ji} y_i), \quad i = 1, \ldots, N,$$

where δ_i is defined to be 1 if the ith neuron is an output neuron, and 0 otherwise.

2.3. Learning Dynamics

Neural networks have learning ability. The rule for learning is given by its learning dynamics which describes the modification of the patterns of connectivity, i.e., the changes of weights with respect to the time t. The learning dynamics takes the form

$$\Delta w_{ij} = w_{ij}(t+1) - w_{ij}(t) = H_{ij}(\boldsymbol{x}(t), \boldsymbol{y}(t), \boldsymbol{w}(t), \boldsymbol{K}), \quad i, j = 1, \ldots, N,$$

in discrete time, or

$$\dot{w}_{ij} = H_{ij}(\boldsymbol{x}(t), \boldsymbol{y}(t), \boldsymbol{w}(t), \boldsymbol{K}), \quad i, j = 1, \ldots, N,$$

in continuous time as well. There are many different learning rules which correspond to different dynamics. For example, the dynamics for Widrow-Hoff's delta rule has the form

$$\dot{w}_{ij} = \alpha y_i S_j(x_j), \quad i, j = 1, \ldots, N,$$

and for the Kosko learning rule [22] has the form

$$\dot{w}_{ij} = -\alpha[\lambda_{ij} w_{ij} - S_i(x_i) S_j(x_j)], \quad i, j = 1, \cdots, N,$$

where α is a positive constant.

2.4. Characterization of Neurodynamics via Objective Functions

The activation dynamics, learning dynamics and error propagation dynamics all together form the dynamics of neural networks. Of course, such dynamics cannot be arbitrarily chosen. The proposal of an adaptive neurodynamics depends on a deep study of several issues: it should be based on the research and understanding of the brain by neural scientists, it needs the support of extensive experiments from biophysics, and theoretically, it is subject to possess certain mathematical properties. One important property the neurodynamics should have is its stability which is characterized by the existence of a specific function L, called *objective function* or *Liapunov function*. Under reasonable mathematical assumptions the stability of the neurodynamics is guaranteed if the differential of L along trajectories of the system is negative, i.e.,

$$\dot{L} \equiv \frac{dL}{dt} = \sum_{i=1}^{N} \frac{\partial L}{\partial x_i} \cdot \dot{x}_i + \sum_{i=1}^{N} \frac{\partial L}{\partial y_i} \cdot \dot{y}_i + \sum_{i,j=1}^{N} \frac{\partial L}{\partial w_{ij}} \cdot \dot{w}_{ij} < 0.$$

Therefore, the stability analysis for a given dynamics amounts to exactly construct the Liapunov function, which requires specific mathematical knowledge and necessary technical derivations. A great amount of work on the stability analysis of neural networks has existed [12, 13, 16, 17, 18, 19, 21, 22, 27, 36].

2.5. Derivation of Adaptive Neurodynamics

One may wish to derive adaptive neurodynamics so as to introduce new models. This may be done by examining possible objective functions. Often the activation dynamics for the $\boldsymbol{x}$ is specified first. The objective function need then be chosen in order to derive an adaptive dynamics for the weights, namely, the learning rule. This is frequently achieved by minimizing the function L in weight space. It is proposed in [33] to use objective functions of the form

$$E = \sum_{i;p} E_x(x_{ip}) + \sum_{i,j} E_w(w_{ij}) + \sum_{i,j,k;p} E_{xw}(x_{ip}, w_{jk})$$

$$+ \sum_{i,j;p,q} E_{xx}(x_{ip}, x_{jq}) + \sum_{i,j,k,l} E_{ww}(w_{ij}, w_{kl}) + \text{higher order terms},$$

where p and q are pattern indices. Then by minimizing E and requiring $\dot{w}_{ij} = -\alpha \partial E / \partial w_{ij}$ for positive α we would get the learning dynamics through the calculation of the implicit and explicit derivatives of E with respect to w_{ij}. This approach allows to derive hybrid learning rules [33, 37].

3. Computer Algebra Applied to Neurodynamics

Computer algebra is applicable whenever mathematical computations are involved. From the review in the preceding section we see that neurodynamics is mathematically characterized by differential or difference equations. The investigation of the dynamic behaviour consists then in manipulating these equations and the related mathematical objects, and in studying their mathematical properties and practical performance. This nature permits to apply computer algebra to the study of neurodynamics. In this section we point out some of the possible applications.

3.1. Manipulation of Neurodynamics-Related Objects

Neurodynamics and its related objects such as objective functions are symbolic mathematical expressions. CA systems can be used to represent these objects and to perform various basic operations and manipulations on them such as arithmetic operations, selection of parts, substitutions, partial evaluation, pattern searching and specification of unspecified parameters and functions. CA systems also provide the feasibility of creating databases collecting popular neural dynamics represented in symbolic form for a systematic investigation on their theoretical behaviour and practical performance. The neurodynamics stored in the databases can be directly accessed by the CA system for basic operations, manipulations and advanced computations. Such databases can be made to have interface with textprocessing utilities and neural simulators (see Subsections 3.4, 4.1 and 4.5). In addition to our work to be mentioned in later sections, we may consider the algebraic transformations of objective functions exhibited in [30] as a further example. In order to reduce network cost and increase the set of neurally implementable objective functions, it is desirable to introduce transformations which transform objective functions into other forms but preserve their original meaning. Such transformations include the simplification of squares and products of expressions, functions of one or two expressions, sparse matrix products, and the minimum and maximum of a set of expressions. It is answered in [30] whether and how the algebraic manipulation can be done, by providing a collection of algebraic transformations. It seems likely that the involved algebraic manipulation and transformations can be done more efficiently and/or in an automatic manner by using CA systems (with some extension).

3.2. Analysis and Derivation of Neurodynamics

A theoretical analysis of neurodynamics requires mathematical methods from nonlinear differential equations such as Liapunov's stability method, perturbation method and bifurcation method. As indicated in the preceding section, the stability analysis of neurodynamics consists in constructing specific Liapunov functions from the dynamics, for which no general algorithmic method exists, and the derivation of adaptive neurodynamics can proceed by specifying possible objective functions and minimizing them in weight space. In this approach, it is necessary to incorporate intelligent ideas, mathematical knowledge with computational techniques form computer algebra, and CA systems with proper extension can be of great help. The study may be of two cases: one considers dynamic systems of a given dimension, 2 or 3 for instance, and the other considers systems of an indefinite, unspecified dimension n. In the first case, current CA systems have been quite suitable for applications. Actually, the investigation of such applications in differential equations has already started (see [34, 43]), but the content was not specifically direct to neurodynamics. We have noticed that many calculations occurring in the neural networks literature can be done more efficiently by using CA systems. For example, we may mention Cohen's disproof of a conjecture about content addressable memory which involves extensive calculations [12] and the derivations and calculations made in [31]. It might be desirable to achieve the same results by using systematic computer algebraic methods as those described in [34, 43]. Some investigation seems to be worthwhile. In the case of dealing with neurodynamics of an indefinite dimension such as the verification of a function to be a Liapunov function and the minimization of an objective function in weight space, computations such as differentiation and integration of symbolic expressions involving indefinite sums and products and basic operations on matrices of an indefinite dimension are often required, but they are not provided for in general CA systems. To make this possible, it turns out that further research on the extension of methods and systems of computer algebra is necessary (see Section 4).

3.3. Detection of Critical Points

While studying the local behaviour of neurodynamics (cf. [3, 24, 27, 31]), it is necessary to locate the critical (equilibrium) points of the dynamics. In general such points are given as roots of systems of nonlinear algebraic equations. For example, the so-called *interior critical points* of a neurodynamics given in [31] are positive solutions $(x_1, \ldots, x_n)$ of the algebraic equations

$$1 - cx_i + \sum_{j=1}^{N} \delta_{ij} x_i x_j^2 = 0, \quad i = 1, \ldots, N,$$

where c is a constant and (δ_{ij}) is a connection matrix with δ_{ij} taking value 0, 1, or -1. Therefore, to locate the interior critical points, we need to solve the above system of nonlinear algebraic equations. In [31] Noonburg solved such a system analytically for a restricted form of the connection matrix. In computer algebra, solving systems of algebraic equations is a long-standing research subject and there are general and efficient methods implemented in CA systems which can be used to solve a variety of algebraic systems from diverse application areas. However, the general implementation of methods is still not sufficiently efficient to solve some large systems of algebraic equations of high degree. But, while dealing with a certain special class of systems, research efforts may make their resolution feasible. For example, solving the above mentioned system for $N = 3$ by the CA built-in solvers may run far beyond a reasonable time limit, but some improvement to the method may result in solving the system in one minute (see [15]).

3.4. Generation of Simulation Codes from Neurodynamics

The observation of the behavior of neural models has been largely based on extensive computer simulations, for which a great amount of tedious, error-prone program writing is involved. The researchers and programmers often have to go through the whole process of editing source code, correcting syntax errors and producing executable image for a slight modification of a neural network. We may note that a neural network is fixed when its topology and dynamics are fixed. For a fixed network, the writing of simulation codes is no more than a straightforward transformation of the network into the concrete programming codes according to the language's syntax. Therefore, we may develop software tools by using CA systems to do this transformation automatically. In this case, we need a general framework of neural networks and a high level specification of the network topology and dynamics. The characterization of major aspects of neural models in [35] and the unified description of neural algorithms in [33] appear appropriate for the first need. While having in mind the ideas expressed in Subsections 3.1 and 3.2, the languages of CA systems can be considered suitable as specification languages for neural network code generation. For example, we may specify the Kosko learning dynamics by

```
df(w[i,j],t)=-alpha*(lambda[i,j]*w[i,j]-S[i](x[i])*S[j](x[j]))
```

with some declarations on the meaning of the occurring variables by using the MAPLE language. In such a setting, the code generation tools can be developed to be interactive with the CA systems and the databases and to allow the generation of numerical computation codes from the specification of the network architecture and neurodynamics in full automation. In [23] a declarative language is introduced to describe the neural network architecture for code generation. However, the neurodynamics has not been taken into account there. We are pursuing a project on developing tools in connection with CA systems for generating C^{++} codes from the the network architecture and dynamics.

3.5. Other Possible Applications

Many mathematical methods have been employed and developed for artificial neural networks. We may refer to [2, 4, 5, 20, 28, 40], in addition to the above cited references, for further information. As computer algebra can be an assistant for research and applications of both elementary and advanced mathematics, there are surely many possible applications of computer algebra to the investigation of neurodynamics which the authors have not yet been aware of and which are not mentioned in the presented paper. However, we would like to mention a few more such applications which seem to be possible.

- Application of CA systems with extension to the problem of higher-order connections, optimization of network architecture/topology and trajectory learning in connection with neurodynamics.

- Application of differential-equations-solvers in CA systems to solve neurodynamic equations for some special cases in order to observe their analytic behaviour.

- Combination of computer algebraic methods and neural networks to analyze robot inverse kinematics of redundant manipulators (cf. [6]).

- Application of methods and systems of computer algebra to support the statistical and geometric study of neural networks (cf. [2, 28] and the relevant work).
- Application of methods and systems of computer algebra to support conjecturing and the proof and disproof of conjectures in neural networks.
- Application of routines on mathematical analysis available in CA systems to support the study of approximation of neural networks to real functions (cf. [7, 20]).

4. Symbolic Manipulation for Neurodynamics

We have started investigating some of the applications discussed in the preceding section in a joint project between Siemens AG Munich and RISC-LINZ since early 1989. A software package has been developed in the CA systems MACSYMA and MAPLE for dealing with neurodynamics and its related objects, in particular, the differentiation of objective functions with respect to indexed variables, based on some new rules we designed. In this section, we summarize part of our work in this direction.

4.1. Symbolic Representation of Neurodynamics-Related Objects

We have examined the capabilities of CA systems for representing symbolically the neurodynamics and its related objects such as Liapunov functions with consideration of the possibility and feasibility of their extension for performing symbolic manipulation and advanced computations on the involved objects. It was seen that the CA systems MACSYMA [25] and MAPLE [11] are proper candidates, for the former has implemented a good number of functions for indefinite sums and products which are essential for dealing with neurodynamics, and the latter which has less functions for indefinite sums and products, has a wide availability and portability. We have then created a database which collects the dynamics of a number of popular neural models including the Cohen and Grossberg, the Hopfield, the standard backpropagation, the recurrent backpropagation models together with (possibly Liapunov functions and) literature entry in MACSYMA and MAPLE. The neurodynamics and Liapunov functions can be easily read to MACSYMA or MAPLE session for processing. We note that most neurodynamics can be represented in MACSYMA and MAPLE, but some of them related to indefinite sums with non-interval domain, infinitely many indefinite sums and products, piecewise defined functions cannot be represented by the systems. We have made some extension which is not quite nice yet. Further work in ongoing.

4.2. Design of Rules for Manipulating Indefinite Sums and Products

To deal with neurodynamics and its related objects of an indefinite dimension, we need rules for manipulating expressions involving indefinite sums and products. For example, in the stability analysis and derivation of neurodynamics, a fundamental computation is the differentiation and integration of such expressions with respect to indexed variables, which have not been taken into account of computer algebra and are thus neither possible in MACSYMA nor in MAPLE, nor in others. The indefinite sums we have to handle are of the form

$$S = \sum_{i_1 \in \Delta_1} \sum_{i_2 \in \Delta_2} \cdots \sum_{i_r \in \Delta_r} A_{i_1 i_2 \cdots i_r},$$

where the summand $A_{i_1 i_2 \cdots i_r}$ is a function involving indexed variables of the form $x_{j_1 j_2 \cdots j_q}$ $(q \geq 1)$. The question is to calculate the partial derivative and the integral of the above S with respect to an indexed variable $x_{j_1 j_2 \cdots j_q}$, where the indices $j_1, j_2, \ldots, j_q$ may or may not be specified. In [41] the rules for the differentiation and integration were given for the case that the number r of sums is definitely given. The rule for the differentiation has been recently extended in [39] to the case that r is indefinite while all involved indexed variables have definitely many indices. The rule for the differentiation of indefinite products was also proposed in [39]. Using these rules we are able to calculate the derivatives and integrals of expressions involving indefinite sums and products for a wide variety of summands and product expressions. We have also considered several other basic rules needed for the manipulation of sums and products (cf. [29, 42]).

4.3. A Package for the Manipulation of Neurodynamics

Based on the rules for differentiation and integration and some other rules for sum and product manipulation, we have developed a toolkit in MACSYMA [42] and extended it to a package MANISUMP in MAPLE [39] for dealing with neurodynamics and its related objects. They have extended the capability of MACSYMA and MAPLE for manipulating indefinite sums and products. The package MANISUMP is linked to the database and to TEX previewing utility. Currently it contains 23 functions

call	df	dfintosum	factorp	help	sumint	multisum	nicedisplay
niceindex	rewritedf	sumcomb	sumcont	sumconv	sumdist	sumdistp	sumfactor
sump	sumsimp	sumsimpd	sumsimpm	sumsplit	where	who	

(which are newly implemented or modified from the base system) with on-line help facility. This package allows to perform basic manipulations and advanced computations necessary for the analysis and derivation of neurodynamics (see the next two subsections).

4.4. Computer Aided Derivation of Neurodynamics

The toolkit and package mentioned above have been a useful tool to assist our derivation of neurodynamics. Let us consider as an example the objective function

$$L = \frac{1}{4}\sum_{s,k=1}^{n} \lambda_{sk} w_{sk}^2 - \frac{1}{2}\sum_{s,k=1}^{n} w_{sk} S_s(z_s^\infty) S_k(z_k^\infty) + \sum_{k=1}^{n} \int_0^{z_k^\infty} g_k(\xi_k, K_k) \frac{dS_k(\xi_k)}{d\xi_k} d\xi_k.$$

In order to derive a learning rule from it, we need to calculate its partial derivative with respect to z_r^∞ and with respect to w_{ij}. This can be done within our package by simply typing the commands `SUMSIMP(DF(L, z[r]));` and `SUMSIMP(DF(L, w[i,j]));` The former results in

$$\frac{\partial L}{\partial z_r^\infty} = -\frac{1}{2} S_r'[\sum_{s=1}^{n} S_s(z_s^\infty)(w_{rs} + w_{sr}) - 2g_r(z_r^\infty, K_r)],$$

and the latter results in

$$\frac{\partial L}{\partial w_{ij}} = \frac{1}{2}[\lambda_{ij} w_{ij} - S_i(z_i^\infty) S_j(z_j^\infty)]$$

provided that $1 \leq r, i, j \leq n$.

In general, for an objective function L and a fixed activation dynamics we can derive the error propagation dynamics and learning dynamics given as the second and third of the following three equations

$$z_k^\infty = \frac{\sum_{s=1}^{n} w_{sk} S_s(z_s^\infty) + K_k \phi_{k\ 1\ n}}{c_k},$$

$$y_r^\infty = \frac{1}{c_r} S_r'(\sum_{j=1}^{n} w_{rj} y_j^\infty + J_r \phi_{r\ 1\ n}),$$

with $J_r = \frac{\partial L}{\partial x_r^\infty}$, $x_r^\infty = S_r(z_r^\infty)$ and $\phi_{k\ 1\ n}$ being defined to be 1 if $1 \leq k \leq n$, 0 otherwise, and

$$\dot{w}_{ij} = -\alpha \frac{dL}{dw_{ij}} = -\alpha[y_i^\infty S_j(z_j^\infty) + (\frac{\partial L}{\partial w_{ij}})_{expl}],$$

where the last term denotes the explicit derivative of L with respect to w_{ij} (see [44] for more details). Taking, for example, the above L with symmetric w_{ij} and λ_{ij}, we have

$$\dot{w}_{ij} = -\alpha[\lambda_{ij} w_{ij} - S_i(z_i^\infty) S_j(z_j^\infty)],$$

which is Kosko's learning rule [22].

4.5. Some Examples

In this subsection we supply a few examples to further illustrate the manipulation of neurodynamics-related objects. The xwindow dump of a DECstation below shows a screen during a MANISUMP session in Maple V. One can see two windows in which MANISUMP is running, one contains the message appearing right after the start of the program, the other shows the calculation of the partial derivative of an objective function L (which is somewhat more complicated than the one given in the last subsection). The TEX preview window which pops up after the call of the function **nicedisplay** displays the computed derivative dL of the objective function.

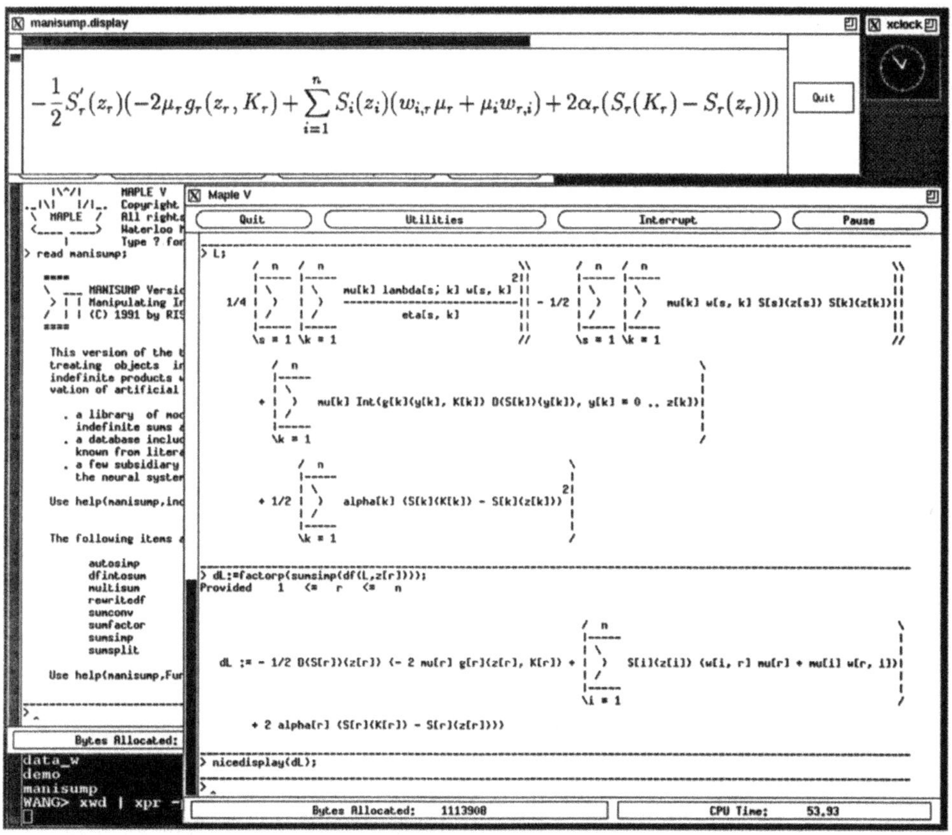

The examples below are chosen to be simple so as not to expend much page space. The output is printed in the **nicedisplay** form.

```
> E1:=b[i](x[i])-sum(c[i,k]*d[k](x[k]),k=1..n);
```

$$\mathrm{E1} := b_i(x_i) - \sum_{k=1}^{n} c_{i,k} d_k(x_k)$$

```
> sumint(E1,x[i]);
```

$$\int b_i(x_i)dx_i - x_i \sum_{k=1}^{n} c_{i,k} d_k(x_k) - c_{i,i} \int d_i(x_i)dx_i + x_i c_{i,i} d_i(x_i)$$

provided $1 \le i \le n$.

```
> E2:=sum(sum(Df(z[l],w[i,j])*L[k,l]*L[r,k]^`-1`,k=1..n),l=1..n)
    =S[i](z[i])*sum(Del[k,j]*L[r,k]^`-1`,k=1..n);
```

$$\mathrm{E2} := \sum_{l=1}^{n} \sum_{k=1}^{n} \frac{dz_l}{dw_{i,j}} L_{k,l} L_{r,k}^{-1} = S_i(z_i) \sum_{k=1}^{n} \delta_{k,j} L_{r,k}^{-1}$$

```
> sumsimpd(sumsimpm(E2));
```

$$\frac{dz_r}{dw_{i,j}} = S_i(z_i) L_{r,j}^{-1}$$

provided $1 \le r, j \le n$.

```
> E3:=-1/K!*multisum([i,1,N],1..N,w[i[1], .., i[K]]*product(V[i[j]],j=1..K))
    ..-1/2*sum(sum(w[i[1],i[2]]*V[i[1]]*V[i[2]],i[2]=1..N),i[1]=1..N)
    -sum(I[i[1]]*V[i[1]],i[1]=1..N);
```

$$\mathrm{E3} := -\frac{1}{K!} \sum_{i_1=1}^{N} \cdots \sum_{i_k=1}^{N} w_{i_1,\dots,i_K} \prod_{j=1}^{K} V_{i_j} - \cdots - \frac{1}{2} \sum_{i_1=1}^{N} \sum_{i_2=1}^{N} w_{i_1,i_2} V_{i_1} V_{i_2} - \sum_{i_1=1}^{N} I_{i_1} V_{i_1}.$$

5. Outlook

5.1. Future Challenges

Recent developments in neural network research like, amongst others, the optimization of network topologies (reducing complexity while maintaining generalization ability) or the treatment of time-dependent learning and recognition (important, e.g., for speech processing and robot trajectory learning) as indicated in Subsection 3.5 present further challenges to CA systems. For example, to derive a learning rule in the case of time-dependent pattern presentation, instead of Liapunov function L, a Liapunov *functional* $J = \int_{t_0}^{t_1} L dt$ has to be minimized with respect to the weights [32]. Here, t_0 and t_1 are the initial and final times of the temporal sequence to be learned. It appears that our package in MACSYMA and MAPLE is a suitable tool for the involved manipulations and derivations. We are further extending our package to attack various challenging problems of this kind in neurodynamics and exploiting potentialities of computer algebra for its application to problems discussed in Section 3. It is our belief that the application of computer algebra as well as symbolic computation in a broader sense to neural network research is promising and will contribute much to the development of neural theory and simulation.

5.2. Neural Networks Applied to Computer Algebra

Having discussed the applications of computer algebra to neural networks, the reverse question of applications of neurodynamics, or more generally neural networks, to computer algebra naturally arises. Research in this direction is still lacking or just at the very beginning. Now, let us mention two of such applications. The first is to use neural networks for operating CA systems with handprinted input as reported in [26]. The current CA systems have a discrepancy on the input which uses one-dimensional representation by using artificial linearized notations. To develop advanced I/O-device that allows a natural operation by hand in a traditional, two-dimensional fashion for CA systems, we can employ neural networks for pattern recognition. The second application is to use neural networks to predicate the practical complexity of algebraic algorithms implemented in CA systems. The estimation of practical complexity of algorithms is of importance in various situations, for example, real-time programming in robotics and job partition in parallel processing, whereas the theoretical complexity analysis does not result in much help. The basic idea of using a neural network for complexity predication is to train the network by sample inputs analogous to human beings who get a feeling about the complexity of algorithms from experience. Assume that an algorithm has been implemented on computer but we do not need any specific knowledge about the algorithm itself. We choose a set of sample inputs which may be from application areas or randomly generated, record the computing times of using the implemented algorithm for these inputs and use these data to train the neural network. A well trained network may be then used to predicate the computing times of using the algorithm for unseen inputs. This work is undergoing at RISC-LINZ.

Acknowledgement. The work of the first author was supported by a grant from Siemens AG, Munich (Dept. ZFE IS INF 2).

References

[1] Akritas, A. G. (1989), *Elments of Computer Algebra with Applications*, John Wiley & Sons, New York.

[2] Amari, S.-I. (1990), *Mathematical Foundations of Neurocomputing*, Proc. IEEE, **78**/9, 1443-1463.

[3] Amari, S.-I. and Arbib, M. A. (1982, eds.), *Competition and Cooperation in Neural Networks*, Springer-Verlag, New York.

[4] Anderson, J. A., Pellionisz, A. and Rosenfeld, E. (1990, eds.), *Neurocomputing 2: Directions for Research*, The MIT Press, Cambridge.

[5] Anderson, J. A. and Rosenfeld, E. (1988, eds.), *Neurocomputing: Foundations of Research*, The MIT Press, Cambridge.

[6] Barhen, J., Gulati, S. and Zak, M. (1989), *Neural Learning of Constrained Nonlinear Transformations*, IEEE Computer, **22**/6, 67-76.

[7] Blum, E. K. and Li, L. K. (1991), *Approximation Theory and Feedforward Networks*, Neural Networks, **4**, 511-515.

[8] Buchberger, B. (1991), *Symbolic Computation: Pure Computer Mathematics*, Trans. 8th Army Conf. Applied Math. Comput., 795-816.

[9] Buchberger, B., Collins, G. E. and Loos, R. (1982, eds.), *Computer Algebra: Symbolic and Algebraic Computation*, Springer-Verlag, Wien-New York.

[10] Caviness, B. F. (1986), *Computer Algebra: Past and Future*, J. Symbolic Computation, **2**, 217-236.

[11] Char, B. W. et al. (1988), *Maple: Reference Manual*, 5th Ed., WATCOM Publications Limited, Waterloo.

[12] Cohen, M. A. (1988), *Sustained Oscillations in a Symmetric Cooperative-Competitive Neural Network: Disproof of a Conjecture About Content Addressable Memory*, Neural Networks, **1**, 217-221.

[13] Cohen, M. A. and Grossberg, S. (1983), *Absolute Stability of Global Pattern Formation and Parallel Memory Storage by Competitive Neural Networks*, IEEE Trans. Systems, Man and Cybernetics, **SMC-13**, 815-826.

[14] Davenport, J. H., Siret, Y. and Tournier, E. (1988), *Computer Algebra: Systems and Algorithms for Algebraic Computation*, Academic Press, London.

[15] Gatermann, K. (1990), *Symbolic Solution of Polynomial Equation Systems with Symmetry*, Proc. ISSAC (Tokyo, August 20-24, 1990), 112-119.

[16] Goles, E. (1987), *Lyapunov Functions Associated to Automata Networks*, Automata Networks in Computer Science (F. Fogelman, Y. Robert and M. Tchuente eds.), Manchester Univ. Press, Manchester, 58-81.

[17] Grossberg, S. (1988), *Nonlinear Neural Networks: Principles, Mechanisms, and Architectures*, Neural Networks, **1**, 17-61.

[18] Hirsch, M. W. (1989), *Convergent Activation Dynamics in Continuous Time Networks*, Neural Networks, **2**, 331-349.

[19] Hopfield, J. J. (1984), *Neurons with Graded Response Have Collective Computational Properties Like Those of Two-State Neurons*, Proc. Nat. Acad. Sci. USA, **81**, 3088-3092.

[20] Hornik, K., Stinchcombe, M. and White, H. (1989), *Multilayer Feedforward Networks are Universal Approximators*, Neural Networks, **2**, 359-366.

[21] Kobuchi, Y. (1991), *State Evaluation Functions and Lyapunov Functions for Neural Networks*, Neural Networks, **4**, 505-510.

[22] Kosko, B. (1987), *Adaptive Bidirectional Associative Memories*, Applied Optics, **26**, 4947-4960.

[23] Leighton, R. and Wieland, A. (1991), *The Aspirin/MIGRAINES Software Tools: User's Manual*, Release V4.0, The MITRE Corporation, McLean, USA.

[24] Levine, D. S. (1983), *Neural Population Modeling and Psychology: A Review*, Math. Biosciences, **66**, 1-86.

[25] MAC (1985), *VAX UNIX MACSYMA^{TM} Reference Manual*, Version 11, Symbolic, Inc.

[26] Marzinkewitsch, M. (1991), *Operating Computer Algebra Systems by Handprinted Input*, Proc. ISSAC'91 (Bonn, July 15-17, 1991), 411-413.

[27] Michel, A. N. Farrel, J. A. and Porod, W. (1989), *Qualitative Analysis of Neural Networks*, IEEE Trans. Circuits Systems, **CAS-36**, 229-243.

[28] Minsky, M. L. and Papert, S. A. (1988), *Percetrons: An Introduction to Computational Geometry*, Expanded Ed., The MIT Press, Cambridge.

[29] Miola, A. (1984), *An Interactive System for Manipulation of Sums and Series*, Proc. Int. Conference on Artificial Intelligence (Varna, September 1984).

[30] Mjolsness, E. and Garrett, C. (1990), *Algebraic Transformations of Objective Functions*, Neural Networks, **3**, 651-669.

[31] Noonburg, V. W. (1989), *A Neural Network Modeled by an Adaptive Lotka-Volterra System*, SIAM J. Appl. Math., **49**/6, 1779-1792.

[32] Pearlmutter, B. A. (1989), *Learning State Space Trajectories in Recurrent Neural Networks*, Neural Computation, **1**, 263-269.

[33] Ramacher, U. and Schürmann, B. (1990), *Unified Description of Neural Algorithms for Time Independent Pattern Recognition*, VLSI Design of Neural Networks (U. Ramacher and U. Rückert, eds.), Kluwer Academic Publishers, 255-270.

[34] Rand, R. H. and Armbruster, D. (1987), *Perturbation Methods, Bifurcation Theory and Computer Algebra*, Springer-Verlag, New York.

[35] Rumelhart, D. E. and McClelland, J. L. (1986), *Parallel Distributed Processing: Explorations in the Microstructure of Cognition*, Vol. 1: Foundations; Vol. 2: Psychological and Biological Models, The MIT Press, Cambridge.

[36] Schürmann, B. (1989), *Stability and Adaptation in Artificial Neural Systems*, Physical Review A, **40**/5, 2681-2688.

[37] Schürmann, B., Hollatz, J. and Ramacher, U. (1990), *Adaptive Recurrent Neural Networks and Dynamic Stability*, Statistical Mechanics of Neural Networks (L. Garrido, ed.), Lecture Notes in Physics, **368**, 49-63.

[38] Schürmann, B. and Wang, D. M. (1990), *Stable Neurodynamics and Symbolic Computation*, Proc. INNC-90 (Paris, July 9-13, 1990), 1013; RISC-Linz Series no. 90-38.0, Johannes Kepler University, Austria.

[39] Shen, W., Wall, B. and Wang, D. M. (1991), *Manipulating Uncertain Mathematical Objects: The Case of Indefinite Sums and Products*, Preprint, RISC-LINZ, Johannes Kepler University, Austria.

[40] Tanaka, S (1990), *Theory of Self-Organization of Cortical Maps: Mathematical Framework*, Neural Networks, **3**, 625-640.

[41] Wang, D. M. (1990), *Differentiation and Integration of Indefinite Summations with Respect to Indexed Variables*, RISC-Linz Series no. 90-37.0, Johannes Kepler University, Austria.

[42] Wang, D. M. (1991), *A Toolkit for Manipulating Indefinite Summations with Application to Neural Networks*, Proc. ISSAC'91 (Bonn, July 15-17, 1991), 462-463; Extended version in ACM SIGSAM Bulletin, **25**/3, 18-27.

[43] Wang, D. M. (1991), *Mechanical Manipulation for a Class of Differential Systems*, J. Symbolic Computation, **12**, 233-254.

[44] Wang, D. M. and Schürmann, B. (1991), *Computer Aided Investigations of Artificial Neural Systems*, Proc. IJCNN'91 (Singapore, November 18-21, 1991), to appear.

Formale Theorie partieller Differentialgleichungen

Werner M. Seiler

Institut für Algorithmen und Kognitive Systeme, Universität Karlsruhe, 7500 Karlsruhe 1
Bitnet: kg04 at dkauni2

1 Einleitung

Ein klassisches Problem der angewandten Mathematik von großer praktischer Bedeutung liegt in der Berechnung der allgemeinen Lösung einer Differentialgleichung. Mit dem Aufkommen der Computeralgebra erfuhr diese Aufgabe noch eine Verschärfung: Die Konstruktion der Lösung soll *algorithmisch* erfolgen. Diese Forderung erweist sich jedoch als zu stark, so daß man sich mit geringeren Zielen zufrieden geben muß.

In der moderneren mathematischen Literatur findet sich nicht viel zu dieser Thematik. Hier liegt der Schwerpunkt auf Existenz- und Eindeutigkeitssätzen. Solche Sätze in größerer Allgemeinheit zu formulieren, erfordert aber eine Abschwächung des Begriffs einer Lösung und führt so z.B. zur Distributionentheorie oder zur Theorie der Sobolev-Räume. Das erlaubt jedoch kein konstruktives Arbeiten mehr.

In der älteren Literatur (damit ist gemeint etwa Mitte bis Ende des letzten Jahrhunderts) erfolgte eine viel stärkere Betonung konstruktiver Verfahren. So stammen viele der Methoden und Tricks zum Lösen von Differentialgleichungen aus jener Zeit. Es fehlte jedoch eine allgemeine Theorie, die die Behandlung größerer Klassen von Gleichungen ermöglicht. Die Fundamente der meisten, heute in der Computeralgebra verwendeten Algorithmen wurden aber bereits damals gelegt. Als Beispiel möchte ich nur die Entwicklung der Theorie der Lie-Gruppen nennen, die im Zuge der Suche nach einer Galois-Theorie für Differentialgleichungen erfolgte.

Die vorliegende Übersichtsarbeit soll kurz die Probleme von Lösungsalgorithmen darstellen und einige in der Computeralgebra verwendete mathematische Verfahren zur Behandlung von Differentialgleichungen vorstellen. Der Schwerpunkt wird dabei auf formalen Methoden liegen, d.h. auf Methoden, die keinen Gebrauch von expliziten Lösungen machen. Die formale Theorie im engeren Sinne behandelt der letzte Abschnitt. Einen ausführlichen Überblick mit zahlreichen Referenzen vermittelt der Artikel von Singer [21], der auch auf viele offene Fragen hinweist.

2 Aufbau und Probleme von Lösungsalgorithmen

Üblicherweise versteht man unter einem Algorithmus ein Verfahren, das aus zulässigen Eingaben nach einer festen Vorschrift in endlich vielen Schritten Ausgabewerte berechnet. Um ein solches Verfahren zum Lösen von Differentialgleichungen zu entwickeln, reicht jedoch die in der Einleitung gegebene Formulierung des Problems nicht aus. Vor allem die Forderung des Terminierens bewirkt starke Einschränkungen.

Als erstes muß daher eine *genaue Spezifikation* des Problems erfolgen. Das erfordert insbesondere die Beantwortung folgender zwei Fragen: Welche Klasse von Differentialgleichungen soll der Algorithmus behandeln und zu welcher Funktionenklasse sollen die Lösungen gehören? Eine Antwort könnte z.B. sein: homogene lineare gewöhnliche Differentialgleichungen mit Koeffizienten aus einem Differentialkörper und liouvillesche Funktionen. Dies sind – vereinfacht ausgedrückt – alle Funktionen, die sich mit Hilfe von $+, -, *, /, \sqrt{}, \exp, \log$ sowie $\exp\int$ schreiben lassen. Im Prinzip gehören alle Ausdrücke dazu, die keine speziellen Funktionen wie z.B. Bessel-Funktionen enthalten.

Wie schon die angegebenen Beispiele zeigen, ist es hierbei wichtig, Klassen zu wählen, die auch effektiv behandelt werden können. Dies erklärt die große Bedeutung der Differentialalgebra für Lösungsalgorithmen. Algebraische Funktionenkörper eignen sich zum algorithmischen Rechnen wesentlich besser als die in der Funktionalanalysis üblichen Banach-Räume.

Das Terminieren des Algorithmus sichert in der Regel ein *Strukturtheorem*. Es liefert zum einen eine Entscheidungsprozedur für die Existenz von Lösungen, zum anderen gibt es eine endliche, konstruierbare Lösungsdarstellung an. Ein aus Anfängervorlesungen bekanntes Beispiel bildet die Integration rationaler Funktionen. Grob gesprochen, sagt hier das Strukturtheorem aus, daß sich die Stammfunktion einer rationalen Funktion als die Summe einer rationalen Funktion, mehrerer Logarithmen und mehrerer Arcustangens-Terme schreiben läßt.

Der eigentliche Lösungsalgorithmus nimmt dann die Form eines *konstruktiven Beweises* des Strukturtheorems an, denn dieser liefert gerade die gesuchte Vorschrift zur Konstruktion von Lösungen in der gewünschten Form. Die Mathematik, die dabei auftritt, hat meistens wenig mit Differential- und Integralrechnung zu tun. Die bekannten Algorithmen zur Integration algebraischer Funktionen beruhen z.B. wesentlich auf der algebraischen Geometrie; zur Behandlung gewöhnlicher Differentialgleichungen benötigt man die Darstellungstheorie endlicher Gruppen.

Sucht man nach Lösungsalgorithmen in diesem strengen Sinne, so erlebt man eine Enttäuschung. Für größere Klassen von Differentialgleichungen existiert bisher *ein einziger*: der Singer-Algorithmus [20,23] (mittlerweile auch häufig als Kovačic-Singer-Ulmer-Algorithmus bezeichnet) zur Berechnung liouvillescher Lösungen homogener linearer gewöhnlicher Differentialgleichungen n-ter Ordnung mit Koeffizienten aus $K(x)$, wobei K eine endlich erzeugte Erweiterung von $\mathbb{Q}$ ist. Dieser Algorithmus entzog sich bisher einer Implementierung und seine Komplexität ist wohl auch zu groß, als daß er für Gleichungen höherer als dritter Ordnung vernünftig angewendet werden könnte.

Als Spezialfall des Singer-Algorithmus kann der unabhängig entwickelte Kovačic-Algorithmus [11] angesehen werden. Er behandelt Gleichungen zweiter Ordnung und wurde von Saunders in MACSYMA implementiert [18] (später auch von Bronstein in MAPLE). Die Tatsache, daß erst vor kurzem gezeigt werden konnte, daß er tatsächlich dem Singer-Algorithmus für den Fall $n = 2$ entspricht, demonstriert die Komplexität dieser Algorithmen. Wenn man andererseits bedenkt, daß es hier immer noch um einzelne *lineare gewöhnliche* Differentialgleichungen geht, kann man sich leicht ausmalen, was für eine Komplexität bei Algorithmen für Systeme nichtlinearer partieller Differentialgleichungen zu erwarten ist.

Die Komplexität bildet aber nicht das einzige Problem. Elementare Überlegungen zeigen bereits, daß für allgemeinere Klassen von Differentialgleichungen in der Regel keine endlichen Lösungsdarstellungen mehr existieren. So besitzen z.B. lineare partielle Differentialgleichungen meistens unendlich dimensionale Lösungsräume. Chaotische Lösungen nichtlinearer Gleichungen passen sicherlich nicht in einen differentialalgebraischen Formalismus.

Eine große Rolle für den Singer-Algorithmus spielt auch die Tatsache, daß der Lösungsraum einer linearen gewöhnlichen Differentialgleichung die Struktur eines Vektorraums mit a priori bekannter endlicher Dimension besitzt. Bei nichtlinearen Gleichungen gilt dies nicht mehr. Es gibt lediglich die Klasse der Gleichungen mit einem nichtlinearen Superpositionsprinzip, d.h. Gleichungen, deren allgemeine Lösung sich (nichtlinear) durch endlich viele partikuläre Lösungen ausdrücken läßt. Ein Theorem von Lie bestimmt ihre allgemeine Form [12]. Für sie könnte man sich am ehesten eine Weiterentwicklung des Singer-Algorithmus vorstellen.

Denef und Lipshitz gelangten für Potenzreihenlösungen sogar zu wesentlich stärkeren Aussagen [6]. So konnten sie beweisen, daß das Problem, ob ein System polynomialer gewöhnlicher Differentialgleichungen eine *konvergente* Potenzreihenlösung besitzt, nicht entscheidbar ist. Dasselbe gilt ereits für die Frage, ob eine nichtverschwindende Lösung existiert. Ferner zeigten sie, daß es Systeme partieller Differentialgleichungen gibt mit unendlich vielen Potenzreihenlösungen, von denen keine berechenbar ist.

Selbst wenn für diese theoretischen Fragen befriedigende Antworten gefunden werden könnten, blieben noch eine Reihe von praktischen Problemen bei der Anwendung. Das beginnt bereits mit dem Speicherplatz. Schon bei den Integrationsalgorithmen zeigte sich, daß die Behandlung größerer Klassen von Funktionen zu extrem umfangreichen Programmen führt, die auf kleineren Anlagen nicht mehr installiert werden können.

Aus der Sicht eines Physikers liegt eine große Schwäche von Lösungsalgorithmen in dem Zwang, die Klasse der möglichen Lösungsfunktionen nach Kriterien der algebraischen Konstruierbarkeit auszuwählen. Es gibt keinen physikalischen Grund, von der Wellenfunktion eines Elektrons zu verlangen, daß sie liouvillesch sein soll. Ganz im Gegenteil ist es sogar zu erwarten, daß in Lösungen physikalisch interessanter Differentialgleichungen spezielle Funktionen der mathematischen Physik wie eben z.B. Bessel-Funktionen auftreten. Folgerichtig schreiben auch Davenport, Siret, Tournier [4]: *Solutions in finite form to a differential equation are very useful, if there are any. But there are no such solutions to most differential equations in physics.*

Als letztes möchte ich noch das Problem der Zweckmäßigkeit allgemeiner Lösungen von Differentialgleichungen erwähnen. Die Hauptmotivation für die Berechnung einer symbolischen Lösung an Stelle einer numerischen liegt in dem Gewinn an Einsicht und Verständnis. Es stellt sich nun die Frage, wieviel Einsicht eine Lösung vermitteln kann, deren Ausdruck mehrere Seiten Papier füllt? Die Lösung einer Differentialgleichung, die realistische Effekte beschreiben soll, wird in der Regel keine einfache Funktion sein, da sie sonst nicht die natürliche Komplexität der Phänomene wiedergeben kann.

Aus diesen Überlegungen lassen sich mehrere Konsequenzen für das praktische Arbeiten mit Differentialgleichungen ziehen:

- Lösungsalgorithmen sind zwar mathematisch sehr interessant aber für Anwender zu ineffizient und zu aufwendig.
- Man sollte versuchen, soweit wie möglich ohne explizite Lösungen auszukommen.
- Falls man wirklich eine Differentialgleichung lösen muß, ist es in der Regel günstiger, entweder numerische Methoden oder heuristische Algorithmen zu benutzen.

Die formale Theorie knüpft an den zweiten Punkt an. Bei vielen Anwendungen benötigt man eigentlich keine Lösung in geschlossener Form sondern daraus abgeleitete Größen wie z.B. Erhaltungssätze. Im weitesten Sinne beschäftigt sich die formale Theorie damit, solche Größen ohne die Kenntnis von Lösungen direkt durch Manipulationen der Gleichungen zu gewinnen.

3 Mathematische Methoden für Differentialgleichungen

Obwohl zur Zeit also nur ein echter Lösungsalgorithmus für Differentialgleichungen existiert, kennt die Computeralgebra zahlreiche Verfahren zu ihrer Behandlung. Die wenigsten davon sind algorithmisch; viele hängen von Heuristiken ab. Die angestrebten Ziele können sehr unterschiedlich seien. Zahlreiche Ansätze streben eine Vereinfachung des Ausgangsproblems an: nichtlineare Gleichungen sollen in lineare, partielle in gewöhnliche überführt werden; gekoppelte Gleichungen sollen entkoppelt werden; die Ordnung der Gleichungen soll reduziert werden.

Ein anderes Ziel liegt in der Entwicklung von Normalformen, die Klassifikationen ermöglichen. Eng damit verwandt ist das Äquivalenzproblem. In der mathematischen Physik interessiert man sich häufig für eine Überprüfung der (vollständigen) Integrabilität oder für die Konstruktion von Erhaltungsgrößen.

Die zur Zeit in der Computeralgebra für Differentialgleichungen benutzte Methoden lassen sich grob in drei Gruppen einteilen: differentialalgebraische, differentialgeometrische und sonstige. Viele Verfahren kombinieren natürlich Elemente aus mehreren Richtungen.

Innerhalb der Differentialalgebra [9,10] können wiederum zwei Strömungen unterschieden werden. Die differentielle Galois-Theorie versucht analog zur rein algebraischen Galois-Theorie, aus der differentiellen Galois-Gruppe Informationen über die Lösbarkeit der Gleichung zu gewinnen und diese zur Konstruktion der Lösung einzusetzen. Der Singer-Algorithmus stellt das prominenteste Ergebnis dieser Theorie dar.

Daneben gibt es die Theorie der Differentialpolynome, in der versucht wird, die kommutative Algebra und hier insbesondere die konstruktive Idealtheorie zu kopieren. Z.B. sucht man nach Gröbner-Basen für Differentialideale [2]. Das bereitet aber deutlich mehr Probleme als im algebraischen Fall, da der Ring der Differentialpolynome nicht mehr noethersch ist.

Trotzdem konnte diese Theorie bereits mit einigem Erfolg zum automatischen Schliessen und zum automatischen Beweisen eingesetzt werden. Die Grundlage bildet dabei die sogenannte Wu-Ritt-Methode [27]. Wu gelang es mit ihr z.B., das Newtonsche Gravitationsgesetz aus den Kepler-Gesetzen herzuleiten [28].

Charakteristisch für differentialalgebraische Ansätze ist, daß nur die algebraischen Eigenschaften der Differentiation (d.h. Linearität und Leibnitz-Regel) eingehen. Differentialpolynome sind für beliebige Derivationen definiert. Im allgemeinen treten daher in der Differentialalgebra keine unabhängigen Variablen (nach denen abgeleitet wird) auf. In konkreten Anwendungen besitzt der zugrunde liegende Differentialkörper allerdings praktisch immer die Form $k(x)$ und die Ableitung nach x definiert die Derivation.

Die differentialgeometrischen Methoden orientieren sich an dem Ansatz der algebraischen Geometrie. Jeder Differentialgleichung[1] wird ein geometrisches Objekt (eine Untermannigfaltigkeit) im sogenannten Jetbündel [19] zugeordnet. Das Jetbündel kann man sich als einen Raum vorstellen, der neben den abhängigen und unabhängigen Variablen der Differentialgleichung auch die Ableitungen der abhängigen Variablen nach den unabhängigen als zusätzliche Koordinaten[2] besitzt. Die Differentialgleichung kann in diesem Raum als eine *algebraische* Gleichung angesehen werden, zu der natürlich eine Nullstellenvarietät gehört.

Dieser geometrische Ansatz erlaubt sofort in natürlicher Weise Symmetrien von Differentialgleichungen zu definieren: Symmetrien sind Transformationen, die die Untermannigfaltigkeit invariant lassen. Die Symmetrieanalyse von Differentialgleichungen ist rechentechnisch sehr aufwendig und fand daher erst mit dem Aufkommen der Computeralgebra starke Beachtung. Mittlerweile existieren bereits zahllose Programme hierfür. Ich möchte an dieser Stelle nur ein neueres Packet erwähnen, CRACKSTAR von Wolf [26]. Es stellt wohl das zur Zeit leistungsstärkste auf diesem Gebiet dar.

Im Rahmen dieser kurzen Übersicht kann nicht näher auf die Konstruktion und die Anwendung von Symmetrien eingegangen werden. Eine kurze Aufzählung von Stichpunkten sollte aber genügen, um die Vielzahl der Möglichkeiten zu demonstrieren; für genauere Informationen sei auf die Literatur [1,14] verwiesen:

- Symmetrien transformieren Lösungen wieder in Lösungen. Eine partikuläre Lösung erzeugt so eine ganze Familie von Lösungen.
- Durch Symmetriereduktion kann bei partiellen Differentialgleichungen die Anzahl der unabhängigen Variablen, bei gewöhnlichen die Ordnung reduziert werden.
- Das Noether-Theorem erlaubt die Konstruktion von Erhaltungsgrößen aus Symmetrien.
- Es gibt einen Zusammenhang zwischen Symmetrien und Separationsansätzen [13].
- Symmetrien besitzen eine große Bedeutung für das Äquivalenzproblem von Differentialgleichungen.
- Es gibt Zusammenhänge zwischen der Existenz von Symmetrien und der vollständigen Integrabilität [7].
- ...

Man kann ohne Übertreibung feststellen, daß für nichtlineare partielle Differentialgleichungen praktisch alle systematischen Ansätze und Techniken auf Symmetriemethoden beruhen.

---

[1] Wenn im folgenden von einer Differentialgleichung gesprochen wird, ist damit immer auch ein System von Differentialgleichungen gemeint.

[2] Diese neuen Koordinaten sind von den alten *algebraisch* unabhängig. *Differentiell* sind sie natürlich abhängig, was den großen Unterschied zur algebraischen Geometrie bewirkt.

Zu den sonstigen Methoden zählen in erster Linie Potenzreihenansätze und die Analyse von Singularitäten. Reihenlösungen in der Umgebung von Singularitäten berechnet z.B. das Programm DESIR der Gruppe um Della Dora [4,5]. Eine ganz andere Anwendung von Singularitäten besteht in der Painlevé-Analyse [8], die in letzter Zeit vor allem im Zusammenhang mit der vollständigen Integrabilität [25] wieder viel Aufmerksamkeit gefunden hat. Hierbei wird untersucht, ob die Lösung bewegliche Singularitäten enthält, d.h. Singularitäten, deren Lage von den Anfangsbedingungen abhängt. Daneben gibt es z.B. noch Versuche mit Integraltransformationen und vieles mehr (siehe [21]).

4 Formale Theorie

Im Sinne der obigen Definition von formalen Methoden als Verfahren, die ohne explizite Lösungen auskommen, gehören praktisch alle im letzten Abschnitt erwähnten Methoden zur formalen Theorie. Meistens versteht man jedoch unter der formalen Theorie ein viel engeres Gebiet. Den größten Teil davon enthalten die Bücher von Pommaret [15,16], die auch Anwendungen in der theoretischen Physik bringen. Etwas andere Schwerpunkte setzt die Schule von Vinogradov [22,24], auf die hier jedoch nicht näher eingegangen werden kann.

Die formale Theorie benutzt die differentialgeometrische Definition einer Differentialgleichung, verwendet aber auch viele abstrakte algebraische Methoden wie z.B. die Spencer-Cohomologie. Während ihr theoretischer Rahmen schon weit ausgebaut ist, fehlt es noch an praktischen Anwendungen. Ansätze hierzu existieren bisher in erster Linie in der Kontrolltheorie. Durch ihre starke Algebraisierung erscheint die formale Theorie jedoch sehr gut für die Computeralgebra geeignet. Der größte Teil der Theorie ist auch konstruktiv.

Im Zentrum der Theorie steht der Begriff der *formalen Integrabilität*[3]. Eine anschauliche Interpretation liefert die Aussage, daß es nur bei formal integrablen Systemen möglich ist, für das Cauchy-Problem eine formale Potenzreihenlösung Ordnung für Ordnung auszurechnen. Die Betonung liegt dabei auf *Ordnung für Ordnung* und nicht auf Potenzreihe. Die formale Theorie hat nichts mit den oben erwähnten Reihenansätzen zu tun.

Aus Platzgründen kann ich wieder nur eine Reihe von Stichpunkten erwähnen und verweise für alle Details auf die oben angegebene Literatur. Obige, anschauliche Deutung der formalen Integrabilität legt nahe, daß für analytische, formal integrable Systeme ein sehr allgemeiner Existenzsatz gilt (Cartan-Kähler-Theorem), da dann die Konvergenz der formalen Potenzreihenlösung gezeigt werden kann. Alleine deshalb ist es schon interessant, mit formal integrablen Systemen zu arbeiten.

Alles andere als naheliegend ist dagegen der Satz, daß jede Differentialgleichung durch endlich viele Differentationen und Projektionen (Eliminationen) in ein lösungsäquivalentes, formal integrables System überführt werden kann (Cartan-Kuranishi-Theorem). Bei den Projektionen entstehen neue, von den Ausgangsgleichungen unabhängige Differentialgleichungen. Man bezeichnet sie häufig als *Integrabilitätsbedingungen*, da sie durch Überkreuzableiten entstehen. Jede Lösung des ursprünglichen Systems erfüllt sie automatisch. Ein formal integrables System enthält alle seine Integrabilitätsbedingungen.

---

[3] Nicht zu verwechseln mit der oben erwähnten *vollständigen* Integrabilität!

In dem klassischen Beispiel von Janet [17]

$$u_{zz} + y u_{xx} = 0\,, \tag{1}$$

$$u_{yy} = 0 \tag{2}$$

ergeben sich die Bedingungen

$$u_{xxy} = 0\,, \tag{3}$$

$$u_{xxxx} = 0\,. \tag{4}$$

Das zugehörige, formal integrable System besteht daher aus allen Gleichungen vierter Ordnung, die aus (1)–(4) durch Differentation erzeugt werden können. Tatsächlich nimmt man hier jedoch sogar die Gleichungen fünfter Ordnung, um ein *involutives* System zu erhalten. Ich kann hier allerdings nicht auf den recht technischen Unterschied zwischen formal integrabel und involutiv[4] eingehen.

Nicht zu verwechseln mit den Integrabilitätsbedingungen sind die *Kompatibilitätsbedingungen* für die rechte Seite bei inhomogenen Systemen. So besitzt das System

$$\partial_i u(x) = f_i(x) \tag{5}$$

trivialerweise nur Lösungen, wenn für die Funktionen f_i

$$\partial_i f_j(x) = \partial_j f_i(x) \tag{6}$$

gilt. Ohne die Voraussetzung, daß das System involutiv ist, kann keine Aussage über die Anzahl und die Ordnung der auftretenden Kompatibilitätsbedingungen getroffen werden. Bei involutiven System bereitet weder das noch die explizite Konstruktion der Bedingungen Schwierigkeiten.

Durch die Überführung in ein formal integrables System kann die Dimension des Lösungsraums eines linearen Systems partieller Differentialgleichungen bestimmt werden. Insbesondere kann sofort entschieden werden, ob er endlich oder unendlich dimensional ist. Eine Information, die für viele Anwendungen eine große Bedeutung besitzt.

Eine hervorstechende Eigenschaft der formalen Theorie ist, daß für die konkrete Durchführung all dieser Rechnungen (Konstruktion eines äquivalenten, involutiven Systems, Bestimmung der Integrabilitäts- sowie der Kompatibilitätsbedingungen) lineare Algebra ausreicht. Die Verwendung formaler Methoden führt sie auf die Behandlung linearer Gleichungssysteme zurück. Ihre Anzahl und ihre Größe sowie das Auftreten symbolischer Einträge legen allerdings den Einsatz eines Computeralgebrasystems nahe.

In [17] werden all diese Fragen für obiges Beispiel von Janet abgehandelt. Eine der laufenden Arbeiten unserer Gruppe besteht in der Umsetzung des Cartan-Kuranishi-Theorems in einen Algorithmus und dessen Implementation in dem Computeralgebrasystem AXIOM. Dabei sollen auch sofort alle Kompatibilitätsbedingungen konstruiert werden.

---

[4] Der hier verwendete Begriff der Involutivität ist schärfer als der der Cartanschen Theorie [3]. Ein nach Cartan involutives System ist formal integrabel, da es alle seine Integrabilitätsbedingungen enthält. Es kann durch eine endliche Anzahl von Differentationen in ein involutives System im Sinne der formalen Theorie überführt werden.

Koordinatentransformationen $\bar{y} = \bar{y}(y)$, die durch partielle Differentialgleichungen beschrieben werden, bilden sogenannte Lie-Pseudogruppen. Ein prominentes Beispiel liefern die volumenerhaltenden Transformationen, die der Differentialgleichung $|\partial\bar{y}/\partial y| = 1$ ($|\partial\bar{y}/\partial y|$ bezeichne die Jacobi-Determinante) gehorchen. Als weiteres Beispiel seien die symplektischen Abbildungen der Mechanik genannt.

Diese und weitere für die theoretische Physik interessante Transformationsgruppen wurden bisher kaum für die Symmetrieanalyse von Differentialgleichungen herangezogen. Die formale Theorie bietet hierzu den nötigen Formalismus. Man muß dabei allerdings beachten, daß sie nur Transformationen der abhängigen Variablen betrachtet, während die herkömmliche Symmetrietheorie abhängige und unabhängige Variablen gleich behandelt. Dies rührt von dem Wunsch her, auch differentialalgebraische Methoden einsetzen zu können. Dort gibt es ja, wie bereits oben erwähnt, im allgemeinen keine unabhängigen Variablen.

Lie-Gruppen gehören als Spezialfall zu den Lie-Pseudogruppen. Bei ihnen sind die zugehörigen Differentialgleichungen vom endlichen Typ. Vernünftiges Arbeiten mit Lie-Pseudogruppen erfordert die formale Integrabilität der definierenden Gleichungen. Bei einer Symmetrieanalyse muß auch das zu untersuchende System formal integrabel sein. Die Verallgemeinerung der für Lie-Symmetrien entwickelten Methoden auf Lie-Pseudogruppen und die Entwicklung von Algorithmen zu ihrer praktischen Handhabung werden ein Schwerpunkt der künftigen Arbeit unserer Gruppe bilden.

Literatur

1. G.W. Bluman, S. Kumei: Symmetries and Differential Equations. Applied Mathematical Sciences 81, Springer, New York 1989
2. G. Carrà Ferro: Gröbner Bases and Differential Algebra. *In:* Proc. AAECC-5, L. Huguet, A. Poli (Eds.), Lecture Notes in Computer Science 350, Springer, Berlin 1987, S. 129–140
3. E. Cartan: Sur la Theorié des Systémes en Involution et ses Applications à la Relativité. *In:* Oeuvres Complètes d'Elie Cartan, Partie II, Vol. 2, Gauthier-Villais, Paris 1953, S. 1199–1230 (Orginal: Bull. Soc. Math. France 59 (1931) 88–118)
4. J.H. Davenport, Y. Siret, E. Tournier: Computer Algebra. Academic Press, London 1988
5. J. Della Dora, E. Tournier: Formal Solutions of Differential Equations in the Neighbourhood of Singular Points. *In:* Proc. SYMSAC '81, P.S. Wang (Ed.), ACM, New York 1981, S. 25–29
6. J. Denef, L. Lipshitz: Power Series Solutions of Algebraic Differential Equations. Math. Ann. 267 (1984)
7. A.S. Fokas: Symmetries and Integrability. Stud. Appl. Math. 77 (1987) 253–299
8. E.L. Ince: Ordinary Differential Equations. Dover, New York 1956
9. I. Kaplansky: An Introduction to Differential Algebra. Hermann, Paris 1957
10. E.R. Kolchin: Differential Algebra and Algebraic Groups. Academic Press, New York 1973
11. J.J. Kovačic: An Algorithm for Solving Second Order Linear Homogeneous Differential Equations. J. Symb. Comp. 2 (1986) 3–43
12. S. Lie, G. Sheffers: Vorlesungen über Continuierliche Gruppen. Teubner, Leipzig 1893
13. W. Miller: Mechanisms for Variable Separation in Partial Differential Equations and their Relationship to Group Theory. *In:* Symmetries and Nonlinear Phenomena, D. Levi, P. Winternitz (Eds.), World Scientific, Singapur 1988, S. 188–221
14. P.J. Olver: Applications of Lie Groups to Differential Equations. Graduate Texts in Mathematics 107, Springer, New York 1986

15. J.F. Pommaret: Systems of Partial Differential Equations and Lie Pseudogroups. Gordon&Breach, London 1978
16. J.F. Pommaret: Differential Galois Theory. Gordon&Breach, London 1983
17. J.F. Pommaret, A. Haddak: Effective Methods for Systems of Algebraic Partial Differential Equations. *In:* Proc. MEGA '90, T. Mora, C. Traverso (Eds.), Birkhäuser, Boston 1991, S. 411–426
18. B.D. Saunders: An Implementation of Kovačic's Algorithm for Solving Second Order Linear Homogeneous Differential Equations. *In:* Proc. SYMSAC '81, P.S. Wang (Ed.), ACM, New York 1981, S. 105–108
19. D.J. Saunders: The Geometry of Jet Bundles. London Mathematical Society Lecture Notes Series 142, Cambridge University Press, Cambridge 1989
20. M.F. Singer: Liouvillian Solutions of n-th Order Homogeneous Linear Differential Equations. Am. J. Math. 103 (1981) 661–682
21. M.F. Singer: Formal Solutions of Differential Equations. J. Symb. Comp. 10 (1990) 59–94
22. T. Tsujishita: Formal Geometry of Systems of Differential Equations. Sugaku Expositions 3 (1990) 25–73
23. F. Ulmer: On Liouvillian Solutions of Linear Differential Equations. *Erscheint in:* Appl. Alg. Eng. Comm. Comp.
24. A.M. Vinogradov: The $\mathcal{C}$-Spectral Sequence, Langrangian Formalism, and Conservation Laws. I. The Linear Theory. II. The Nonlinear Theory. J. Math. Anal. Appl. 100 (1984) 1–40, 41–129
25. J. Weiss, M. Tabor, G. Carnevale: The Painlevé Property for Partial Differential Equations. J. Math. Phys. 24 (1983) 522–526
26. T. Wolf: A Package for the Analytic Investigation and Exact Solution of Differential Equations. *In:* Proc. EUROCAL '87, J.H. Davenport (Ed.), Lecture Notes in Computer Science 378, Springer, Berlin 1987, S. 479–490
27. Wu W.-T.: A Constructive Theory of Differential Algebraic Geometry based on works of J.F. Ritt with Particular Applications to Mechanical Theorem Proving of Differential Geometries. *In:* Proc. Differential Geometry and Differential Equations, Shanghai 1985, G. Chaohao, M. Berger, R.L. Bryant (Eds.), Lecture Notes in Mathematics 1255, Springer, Berlin 1987, S. 173–189
28. Wu W.-T.: Automatic Derivation of Newton's Gravitational Laws from Kepler's Laws. *In:* Academica Sinica Mathematics-Mechanization Research Preprints 1 (1987) 53

Informatik – Fachberichte

Band 206: P. Horster, H. Isselhorst, Approximative Public-Key-Kryptosysteme. VII, 174 Seiten. 1989.

Band 207: J. Knop (Hrsg.), Organisation der Datenverarbeitung an der Schwelle der 90er Jahre. 8. GI-Fachgespräch, Düsseldorf, März 1989. Proceedings. IX, 276 Seiten. 1989.

Band 208: J. Retti, K. Leidlmair (Hrsg.), 5. Österreichische Artificial-Intelligence-Tagung, Igls/Tirol, März 1989. Proceedings. XI, 452 Seiten. 1989.

Band 209: U.W. Lipeck, Dynamische Integrität von Datenbanken. VIII, 140 Seiten. 1989.

Band 210: K. Drosten, Termersetzungssysteme. IX, 152 Seiten. 1989.

Band 211: H.W. Meuer (Hrsg.), SUPERCOMPUTER '89. Mannheim, Juni 1989. Proceedings, 1989. VIII, 171 Seiten. 1989.

Band 212: W.-M. Lippe (Hrsg.), Software-Entwicklung. Fachtagung, Marburg, Juni 1989. Proceedings. IX, 290 Seiten. 1989.

Band 213: I. Walter, Datenbankgestützte Repräsentation und Extraktion von Episodenbeschreibungen aus Bildfolgen. VIII, 243 Seiten. 1989.

Band 214: W. Görke, H. Sörensen (Hrsg.), Fehlertolerierende Rechensysteme / Fault-Tolerant Computing Systems. 4. Internationale GI/ITG/GMA-Fachtagung, Baden-Baden, September 1989. Proceedings. XI, 390 Seiten. 1989.

Band 215: M. Bidjan-Irani, Qualität und Testbarkeit hochintegrierter Schaltungen. IX, 169 Seiten. 1989.

Band 216: D. Metzing (Hrsg.), GWAI-89. 13th German Workshop on Artificial Intelligence. Eringerfeld, September 1989. Proceedings. XII, 485 Seiten. 1989.

Band 217: M. Zieher, Kopplung von Rechnernetzen. XII, 218 Seiten. 1989.

Band 218: G. Stiege, J. S. Lie (Hrsg.), Messung, Modellierung und Bewertung von Rechensystemen und Netzen. 5. GI/ITG-Fachtagung, Braunschweig, September 1989. Proceedings. IX, 342 Seiten. 1989.

Band 219: H. Burkhardt, K. H. Höhne, B. Neumann (Hrsg.), Mustererkennung 1989. 11. DAGM-Symposium, Hamburg, Oktober 1989. Proceedings. XIX, 575 Seiten. 1989

Band 220: F. Stetter, W. Brauer (Hrsg.), Informatik und Schule 1989: Zukunftsperspektiven der Informatik für Schule und Ausbildung. GI-Fachtagung, München, November 1989. Proceedings. XI, 359 Seiten. 1989.

Band 221: H. Schelhowe (Hrsg.), Frauenwelt – Computerräume. GI-Fachtagung, Bremen, September 1989. Proceedings. XV, 284 Seiten. 1989.

Band 222: M. Paul (Hrsg.), GI – 19. Jahrestagung I. München, Oktober 1989. Proceedings. XVI, 717 Seiten. 1989.

Band 223: M. Paul (Hrsg.), GI – 19. Jahrestagung II. München, Oktober 1989. Proceedings. XVI, 719 Seiten. 1989.

Band 224: U. Voges, Software-Diversität und ihre Modellierung. VIII, 211 Seiten. 1989

Band 225: W. Stoll, Test von OSI-Protokollen. IX, 205 Seiten. 1989.

Band 226: F. Mattern, Verteilte Basisalgorithmen. IX, 285 Seiten. 1989.

Band 227: W. Brauer, C. Freksa (Hrsg.), Wissensbasierte Systeme. 3. Internationaler GI-Kongreß, München, Oktober 1989. Proceedings. X, 544 Seiten. 1989.

Band 228: A. Jaeschke, W. Geiger, B. Page (Hrsg.), Informatik im Umweltschutz. 4. Symposium, Karlsruhe, November 1989. Proceedings. XII, 452 Seiten. 1989.

Band 229: W. Coy, L. Bonsiepen, Erfahrung und Berechnung. Kritik der Expertensystemtechnik. VII, 209 Seiten. 1989.

Band 230: A. Bode, R. Dierstein, M. Göbel, A. Jaeschke (Hrsg.), Visualisierung von Umweltdaten in Supercomputersystemen. Karlsruhe, November 1989. Proceedings. XII, 116 Seiten. 1990.

Band 231: R. Henn, K. Stieger (Hrsg.), PEARL 89 – Workshop über Realzeitsysteme. 10. Fachtagung, Boppard, Dezember 1989. Proceedings. X, 243 Seiten. 1989.

Band 232: R. Loogen, Parallele Implementierung funktionaler Programmiersprachen. IX, 385 Seiten. 1990.

Band 233: S. Jablonski, Datenverwaltung in verteilten Systemen. XIII, 336 Seiten. 1990.

Band 234: A. Pfitzmann, Diensteintegrierende Kommunikationsnetze mit teilnehmerüberprüfbarem Datenschutz. XII, 343 Seiten. 1990.

Band 235: C. Feder, Ausnahmebehandlung in objektorientierten Programmiersprachen. IX, 250 Seiten. 1990.

Band 236: J. Stoll, Fehlertoleranz in verteilten Realzeitsystemen. IX, 200 Seiten. 1990.

Band 237: R. Grebe (Hrsg.), Parallele Datenverarbeitung mit dem Transputer. Aachen, September 1989. Proceedings. VIII, 241 Seiten. 1990.

Band 238: B. Endres-Niggemeyer, T. Hermann, A. Kobsa, D. Rösner (Hrsg.), Interaktion und Kommunikation mit dem Computer. Ulm, März 1989. Proceedings. VIII, 175 Seiten. 1990.

Band 239: K. Kansy, P. Wißkirchen (Hrsg.), Graphik und KI. Königswinter, April 1990. Proceedings. VII, 125 Seiten. 1990.

Band 240: D. Tavangarian, Flagorientierte Assoziativspeicher und -prozessoren. XII. 193 Seiten. 1990.

Band 241: A. Schill, Migrationssteuerung und Konfigurationsverwaltung für verteilte objektorientierte Anwendungen. IX, 174 Seiten. 1990.

Band 242: D. Wybranietz, Multicast-Kommunikation in verteilten Systemen. VIII, 191 Seiten. 1990.

Band 243: U. Hahn, Lexikalisch verteiltes Text-Parsing. X, 263 Seiten. 1990.

Band 244: B. R. Kämmerer, Sprecherunabhängigkeit und Sprecheradaption. VIII, 110 Seiten. 1990.

Band 245: C. Freksa, C. Habel (Hrsg.), Repräsentation und Verarbeitung räumlichen Wissens. VIII, 353 Seiten. 1990.

Band 246: Th. Bräunl, Massiv parallele Programmierung mit dem Parallaxis-Modell. XII, 168 Seiten. 1990

Band 247: H. Krumm, Funktionelle Analyse von Kommunikationsprotokollen. IX, 122 Seiten. 1990.

Band 248: G. Moerkotte, Inkonsistenzen in deduktiven Datenbanken. VIII, 141 Seiten. 1990.

Band 249: P. A. Gloor, N. A. Streitz (Hrsg.), Hypertext und Hypermedia. IX, 302 Seiten. 1990.

Band 250: H.W. Meuer (Hrsg.), SUPERCOMPUTER '90. Mannheim, Juni 1990. Proceedings. VIII, 209 Seiten. 1990.

Band 251: H. Marburger (Hrsg.), GWAI-90. 14th German Workshop on Artificial Intelligence. Eringerfeld, September 1990. Proceedings. X, 333 Seiten. 1990.

Band 252: G. Dorffner (Hrsg.), Konnektionismus in Artificial Intelligence und Kognitionsforschung. 6. Österreichische Artificial-Intelligence-Tagung (KONNAI), Salzburg, September 1990. Proceedings. VIII, 246 Seiten. 1990.

Band 253: W. Ameling (Hrsg.), ASST'90. 7. Aachener Symposium für Signaltheorie. Aachen, September 1990. Proceedings. XI, 332 Seiten. 1990.

Band 254: R. E. Großkopf (Hrsg.), Mustererkennung 1990. 12. DAGM-Symposium, Oberkochen-Aalen, September 1990. Proceedings. XXI, 686 Seiten. 1990.

Band 255: B. Reusch, (Hrsg.), Rechnergestützter Entwurf und Architektur mikroelektronischer Systeme. GME/GI/ITG-Fachtagung, Dortmund, Oktober 1990. Proceedings. X, 298 Seiten. 1990.

Band 256: W. Pillmann, A. Jaeschke (Hrsg.), Informatik für den Umweltschutz. 5. Symposium, Wien, September 1990. Proceedings. XV, 864 Seiten. 1990.

Band 257: A. Reuter (Hrsg.), GI – 20. Jahrestagung I. Stuttgart, Oktober 1990. Proceedings. XVIII, 602 Seiten. 1990.

Band 258: A. Reuter (Hrsg.), GI – 20. Jahrestagung II. Stuttgart, Oktober 1990. Proceedings. XVIII, 602 Seiten. 1990.

Band 259: H.-J. Friemel, G. Müller-Schönberger, A. Schütt (Hrsg.), Forum '90 Wissenschaft und Technik. Trier, Oktober 1990. Proceedings. XI, 532 Seiten. 1990.

Band 260: B. J. Frommherz, Ein Roboteraktionsplanungssystem. XI, 134 Seiten. 1990.

Band 261: W. Zimmermann, Automatische Komplexitätsanalyse funktionaler Programme. VII, 194 Seiten. 1990.

Band 262: W. Gerth, P. Baacke (Hrsg.), PEARL 90 - Workshop über Realzeitsysteme. 11. Fachtagung, Boppard, November 1990. Proceedings. X, 187 Seiten. 1990.

Band 263: H. Eckhardt, Entwurfstransaktionen für modulare Objektsysteme. VIII, 144 Seiten. 1990.

Band 264: T. Härder, H. Wedekind, G. Zimmermann (Hrsg.), Entwurf und Betrieb verteilter Systeme. Fachtagung, Dagstuhl, September 1990. Proceedings. XII, 283 Seiten. 1990.

Band 265: U. Herrmann, Mehrbenutzerkontrolle in Nicht-Standard-Datenbanksystemen. VIII, 183 Seiten. 1991.

Band 266: R. Cunis, A. Günter, H. Strecker (Hrsg.), Das PLAKON-Buch. VIII, 279 Seiten. 1991

Band 267: W. Effelsberg, H. W. Meuer, G. Müller (Hrsg.), Kommunikation in verteilten Systemen. GI/ITG-Fachtagung, Mannheim, Februar 1991. Proceedings. X, 589 Seiten. 1991.

Band 268: J. Raczkowsky, Multisensordatenverarbeitung in der Robotik. X, 168 Seiten. 1991.

Band 269: G. Hommel (Hrsg.), Prozeßrechensysteme '91. Berlin, Februar 1991. Proceedings. XIV, 449 Seiten. 1991.

Band 270: H.-J. Appelrath (Hrsg.), Datenbanksysteme in Büro, Technik und Wissenschaft. GI-Fachtagung, Kaiserslautern, März 1991. Proceedings. XIII, 507 Seiten. 1991.

Band 271: A. Pfitzmann, E. Raubold (Hrsg.), VIS '91, Verläßliche Informationssysteme. GI-Fachtagung, Darmstadt, März 1991. Proceedings. VIII, 355 Seiten. 1991.

Band 272: R. Grebe, C. Ziemann, Parallele Datenverarbeitung mit dem Transputer. Aachen, September 1990. Proceedings. X, 300 Seiten 1991.

Band 273: M. Timm (Hrsg.), Requirements Engineering '91. VIII, 208 Seiten. 1991.

Band 274: R. Denzer, H. Hagen, K.-H. Kutschke (Hrsg.), Visualisierung von Umweltdaten. Workshop, Rostock, November 1990. Proceedings. VII, 97 Seiten. 1991.

Band 275: D. P. F. Möller, O. Richter (Hrsg.), Analyse dynamischer Systeme in Medizin, Biologie und Ökologie. Bad Münster, April 1990. Proceedings. XI, 258 Seiten. 1991.

Band 276: H. Maurer (Hrsg.), Hypertext/Hypermedia '91. Tagung der GI, SI und OCG, Graz, Mai 1991. Proceedings. VIII, 299 Seiten. 1991.

Band 277: U. Borgolte, Flexible, realzeitfähige Kollisionsvermeidung in Mehrroboter-Systemen. XIII, 105 Seiten. 1991.

Band 278: H. W. Meuer (Hrsg.), SUPERCOMPUTER '91. Proceedings. VIII, 266 Seiten. 1991.

Band 279: G. Schwichtenberg (Hrsg.), Organisation und Betrieb von Informationssystemen. 9. GI – Fachgespräch über Rechenzentren, Dortmund, März 1991. Proceedings. IX, 337 Seiten. 1991.

Band 280: B. Westfechtel, Revisions- und Konsistenzkontrolle in einer integrierten Softwareentwicklungsumgebung. X, 321 Seiten. 1991.

Band 281: W. Emde, Modellbildung, Wissensrevision und Wissensrepräsentation im Maschinellen Lernen. XI, 204 Seiten. 1991.

Band 282: P. Buchholz, Die strukturierte Analyse Markovscher Modelle. VII, 192 Seiten 1991.

Band 283: M. Dal Cin, W. Hohl (Hrsg.), Fault-Tolerant Computing Systems. 5th International GI/ITG/GMA Conference, Nürnberg, September 1991. Proceedings. XII, 425 Seiten. 1991.

Band 284: R. Stadler, Ausführbare Spezifikation von Directory-Systemen in einer logischen Sprache. X, 142 Seiten. 1991.

Band 285: T. Christaller (Hrsg.), GWAI-91. 15. Fachtagung für Künstliche Intelligenz, Bonn, September 1991. IX, 273 Seiten. 1991.

Band 286: A. Lehmann, F. Lehmann (Hrsg.), Messung, Modellierung und Bewertung von Rechensystemen. 6. GI/ITG-Fachtagung, Neubiberg, September 1991. Proceedings. VIII, 338 Seiten. 1991.

Band 287: H. Kaindl (Hrsg.), 7. Österreichische Artificial-Intelligence-Tagung, Wien, September 1991. Proceedings. VIII, 180 Seiten. 1991.

Band 288: G. Helm, Symbolische und konnektionistische Modelle der menschlichen Informationsverarbeitung. X, 161 Seiten. 1991.

Band 289: N. Fuhr (Hrsg.), Information Retrieval. GI/GMD-Workshop, Darmstadt, Juni 1991. Proceedings. VII, 162 Seiten. 1991.

Band 290: B. Radig (Hrsg.), Mustererkennung 1991. 13. DAGM-Symposium, München, Oktober 1991. Proceedings. XVIII, 584 Seiten. 1991.

Band 291: W. Brauer, D. Hernández (Hrsg.) Verteilte künstliche Intelligenz und kooperatives Arbeiten. 4. Internationaler GI-Kongreß, München, Oktober 1991. Proceedings. IX, 546 Seiten. 1991.

Band 292: P. Gorny (Hrsg.), Informatik und Schule 1991. GI-Fachtagung, Oldenburg, Oktober 1991. Proceedings. IX, 335 Seiten. 1991.

Band 293: J. Encarnação (Hrsg.) Telekommunikation und multimediale Anwendungen der Informatik. GI-21. Jahrestagung, Darmstadt, Oktober 1991. Proceedings. XII, 710 Seiten. 1991.

Band 294: R. Möller (Hrsg.), 2. Workshop Sichtsysteme. Bremen, November 1991. Proceedings, 1991. VII, 118 Seiten. 1991.

Band 295: W. A. Halang (Hrsg.), PEARL 91 – Workshop über Realzeitsysteme. 12. Fachtagung, Boppard, November 1991. Proceedings, 1991. X, 197 Seiten. 1991.

Band 296: M. Hälker, A. Jaeschke (Hrsg.), Informatik für den Umweltschutz – Computer Science for Environmental Protection. 6. Symposium, München, Dezember 1991. Proceedings. XX, 657 Seiten. 1991.

Band 297: Th. Ruf, Featurebasierte Integration von CAD/CAM – Systemen. XVII, 314 Seiten. 1991.

Band 298: A. Kemper, Zuverlässigkeit und Leistungsfähigkeit objekt-orientierter Datenbanksysteme. VIII, 207 Seiten. 1992.

Band 299: P. Baumann, Software-Bewertung. VIII, 106 Seiten. 1992.

Band 300: M. Mohnhaupt, Prinzipien piktorieller Repräsentationssysteme. XVI, 186 Seiten. 1992.

Band 301: O. Günther, H. Kuhn, R. Mayer-Föll, F. J. Radermacher (Hrsg.), Konzeption und Einsatz von Umweltinformationssystemen. Proceedings. X, 494 Seiten. 1992.

Band 303: R. Studer (Hrsg.), Informationssysteme und Künstliche Intelligenz: Modellierung. 2. Workshop, Ulm, Februar 1992. Proceedings. VII, 168 Seiten. 1992.

Band 306: D. Krönig, M. Lang (Hrsg.), Physik und Informatik – Informatik und Physik. Arbeitsgespräch, München, November 1991. Proceedings. XIII, 326 Seiten. 1992.

Band 307: G. Klose, E. Lang, Th. Pirlein (Hrsg.), Ontologie und Axiomatik der Wissensbasis von LILOG. X, 255 Seiten. 1992.

GPSR Compliance

The European Union's (EU) General Product Safety Regulation (GPSR) is a set of rules that requires consumer products to be safe and our obligations to ensure this.

If you have any concerns about our products, you can contact us on ProductSafety@springernature.com

In case Publisher is established outside the EU, the EU authorized representative is:

Springer Nature Customer Service Center GmbH
Europaplatz 3
69115 Heidelberg, Germany

Batch number: 10683531

Printed by Printforce, the Netherlands